Graphene-Polymer Composites

Graphene-Polymer Composites

Special Issue Editor
Fernão D. Magalhães

MDPI • Basel • Beijing • Wuhan • Barcelona • Belgrade

MDPI

Special Issue Editor
Fernão D. Magalhães
University of Porto
Portugal

Editorial Office
MDPI
St. Alban-Anlage 66
Basel, Switzerland

This is a reprint of articles from the Special Issue published online in the open access journal *Polymers* (ISSN 2073-4360) from 2017 to 2018 (available at: http://www.mdpi.com/journal/polymers/special_issues/graphene_polymer_composites)

For citation purposes, cite each article independently as indicated on the article page online and as indicated below:

LastName, A.A.; LastName, B.B.; LastName, C.C. Article Title. *Journal Name* **Year**, *Article Number*, Page Range.

ISBN 978-3-03897-041-5 (Pbk)
ISBN 978-3-03897-042-2 (PDF)

Cover image courtesy of Fernão D. Magalhães.

Contents

About the Special Issue Editor

Fernão D. Magalhães obtained his PhD in Chemical Engineering from the University of Massachusetts at Amherst, USA, in 1997. He is currently a Chemical Engineering Assistant Professor at the Faculty of Engineering of University of Porto, and a member of the LEPABE research group. His research interests include graphene-based materials for biomedical applications, high-performance industrial coatings, adhesive systems for lignocellulosic composites, and adhesives based on natural products. He teaches classes on Process Control, Polymer Materials and Industrial Informatics

Preface to "Graphene-Polymer Composites"

The mechanical, electrical, thermal, magnetic, optical and biological properties of graphene have attracted a significant amount of attention from the research community since the isolation of single-atom-thick graphene layers by Geim and co-workers in 2004. Presenting very high surface-to-volume ratio, relatively simple processability and being low cost, graphene and graphene-related materials were soon identified as promising nanofillers for polymer matrixes. Reports have shown notable property enhancements for graphene–polymer composites (GPC) at very low filler loadings. Uses of GPC in varied fields, such as energy, electronics, catalysis, separation and purification, biomedicine, aerospace, tribology, etc., have been demonstrated and, in some cases, put into industrial practice; however, challenges still exist. Platelet agglomeration within the polymer matrix often hinders performance. Poor interfacial adhesion between filler and matrix is also a limiting factor in many systems, necessitating tuning the surface chemistry to enhance physical or chemical interactions with the polymer chains. The various routes for fabrication of graphene-related materials, leading to different morphologies, oxidation states, and degrees of platelet exfoliation, have an impact on the final properties of the composites and this has not yet been fully researched. Some argue that the potential of graphene and its advantages in relation to other nanofillers, has not yet been clearly demonstrated for polymer composites. On the other hand, the relevance of graphene-related materials in polymer science is not restricted to physical–mechanical modifications of the matrix properties, but extends to other aspects, such as polymer synthesis kinetics and catalysis.

In summary, this Special Issue provides a state of the art view of the different facets of graphene–polymer composite materials, showing that this area of research is highly topical and major advances can still be expected.

Fernão D. Magalhães
Special Issue Editor

polymers

Review

Poly(lactic acid) Composites Containing Carbon-Based Nanomaterials: A Review

Carolina Gonçalves [1], Inês C. Gonçalves [2,3], Fernão D. Magalhães [1] and Artur M. Pinto [1,2,3,*]

[1] LEPABE—Faculdade de Engenharia, Universidade do Porto, rua Dr. Roberto Frias, 4200-465 Porto, Portugal; carol.goncalves8827@gmail.com (C.G.); fdmagalh@fe.up.pt (F.D.M.)
[2] INEB—National Institute of Biomedical Engineering, University of Porto, Rua do Campo Alegre, 823, 4150-180 Porto, Portugal; icastro@ineb.up.pt
[3] i3S—Institute for Innovation and Health Research, University of Porto, Rua Alfredo Allen, 208, 4200-135 Porto, Portugal
* Correspondence: arturp@fe.up.pt; Tel.: +351-22-508-1400

Academic Editor: Alexander Böker
Received: 15 June 2017; Accepted: 4 July 2017; Published: 6 July 2017

Abstract: Poly(lactic acid) (PLA) is a green alternative to petrochemical commodity plastics, used in packaging, agricultural products, disposable materials, textiles, and automotive composites. It is also approved by regulatory authorities for several biomedical applications. However, for some uses it is required that some of its properties be improved, namely in terms of thermo-mechanical and electrical performance. The incorporation of nanofillers is a common approach to attain this goal. The outstanding properties of carbon-based nanomaterials (CBN) have caused a surge in research works dealing with PLA/CBN composites. The available information is compiled and reviewed, focusing on PLA/CNT (carbon nanotubes) and PLA/GBM (graphene-based materials) composites. The production methods, and the effects of CBN loading on PLA properties, namely mechanical, thermal, electrical, and biological, are discussed.

Keywords: PLA; graphene-based materials; carbon nanotubes; composites; mechanical properties; thermal properties; electrical properties; biological properties

1. Introduction

The growing environmental awareness and new rules and regulations are forcing the industries to seek more ecologically friendly materials for their products [1]. In the last two decades, industrial and academic research on polymer composites was pursued to provide added value properties to the neat polymer without sacrificing its processability or adding excessive weight [2].

Poly(lactic acid) (PLA), which is derived from natural sources, biodegradable, and bioabsorbable, has had significant demand due to presenting versatile applications in packaging, pharmaceutical, textiles, engineering, chemical industries, automotive composites, biomedical and tissue engineering fields [3]. Its biodegradation time can be tuned, depending on the molecular weight, crystallinity, and material geometry [4]. However, the relatively low glass transition temperature, low thermal dimensional stability, and mechanical ductility limit the number of its applications. A significant body of research has dealt with the use of fillers for improving the properties of PLA [5–7]. In this context, carbon based nanomaterials (CBN), offer the potential to combine PLA properties with several of their unique features, such as high mechanical strength, electrical conductivity, thermal stability and bioactivity [8–16]. Carbon nanotubes (CNT) and graphene-based materials (GBM) are state of the art and very promising representatives of these materials. CNT have exceptional mechanical properties, aspect ratio, electrical and thermal conductivities, and chemical stability. However, their production methods are usually more complex and expensive, often leaving toxic metal residues [17–20]. Hence,

GBM provide an alternative option to produce functional composites due to their excellent properties and the natural abundance of their precursor, graphite. Moreover, GBM can be produced by simple and inexpensive physico-chemical methods [21–24].

In the last years there has been a surge of research works on PLA/CNT and PLA/GBM composites. Due to the large amount of information available, there is the need to congregate, compare and withdraw conclusions.

Several recent reviews have addressed PLA [3,25–30] and CBN [30–46] production, applications and properties, however, none of these focus on PLA/CBN composites. This work presents a comprehensive review on the current knowledge regarding the production of PLA/CBN composites and the resulting properties, namely mechanical, electrical, thermal and biological.

2. Poly(lactic acid) (PLA)

PLA is a thermoplastic aliphatic polyester commonly produced by direct condensation polymerization of lactic acid or by ring-opening polymerization of lactide. As lactic acid is a chiral molecule, existing in L and D isomers, the term "poly(lactic acid)" refers to a family of polymers: poly-L-lactic acid (PLLA), poly-D-lactic acid (PDLA), and poly-D,L-lactic acid (PDLLA). The 2 optically active configurations of lactic acid, the L (+) and D (−) stereoisomers are produced, respectively by bacterial homo- or hetero-fermentation of carbohydrates. A great variety of carbohydrate sources can be used to produce lactic acid, like molasses, corn syrup, whey, dextrose, and cane or beet sugar. Nowadays, industry only uses the fermentation process, because the synthetic routes have major limitations, as the inability of selective production of the L-lactic acid stereoisomer, and high manufacturing costs [47,48].

PLA can be polymerized by diverse methods, like polycondensation, ring opening polymerization, azeotropic dehydration condensation, and enzymatic polymerization. Direct polymerization and ring opening polymerization are the most used. Controlling polymerization parameters is important, since PLA properties vary with isomer composition, temperature, and reaction time used [3,25,28,29,48–51].

Increasing interest in PLA is related to some characteristics that are lacking in other polymers, namely regarding renewability, biocompatibility, processability, and energy saving [29]. PLA is derived from renewable and biodegradable resources, and its degradation products are non-pollutant and non-toxic. Thus, PLA is a green alternative to petrochemical commodity plastics, used in packaging, agricultural products, disposable materials, textiles, and automotive [25]. Furthermore, PLA has several bioapplications, such as biodegradable matrix for surgical implants, and in drug delivery systems [3].

The use of PLA has some shortcomings, related to poor chemical modifiability (absence of readily reactive side-chain groups), mechanical ductility [50], and relatively high price [28]. To overcome some of these issues, some approaches are commonly used, like blending with other polymers [52–59], functionalization [60–64], and addition of nanofillers [6,7,48,65–70]. The last is an interesting approach, since with small filler amounts it is possible to enhance desired features, keeping PLA's key properties intact. The most used nanofillers are nanoclays [5,71–80], nanosilicas [6,68,69,73,81,82], and carbon nanomaterials [7,77,83–88].

3. Carbon-Based Nanomaterials (CBN)

There are several types of carbon-based nanomaterials (carbon nanotubes, graphene-based materials, fullerenes, nanodiamonds) and most have been tested to improve PLA properties. This review is focused on the most widely tested and available: CNT and GBM. The high specific area of these materials allows for low loadings to be sufficient to tune key properties concerning mechanical, thermal, electrical, and biological performance.

CBN Production Methods and Modifications

Graphene is the elementary structure of graphite, being a one carbon atom thick sheet, composed of sp^2 carbon atoms arranged in a flat honeycomb structure composed of two equivalent sub-lattices of carbon atoms bonded together with σ bonds (in plane) and a π bond (out-of-plane), which contributes to a delocalized network of electrons [39,46,89]. These unique characteristics explain its unmatched electronic, mechanical, optical and thermal properties. For that reason, this material has been studied to be applied in many fields, such as electronics [90–95], energy [96–99], membrane [100–103], composite [21,22,24,104], and biomedical technology [11,105–107].

The intrinsic properties of graphene, and GBM in general, are affected by the production or modification methods. For example, structural integrity of graphene sheets is disrupted by oxidation and some other chemical modifications. The dimensions (diameter and thickness) of the final GBM also depend on the raw materials and methods employed [11,34,35,46,90]. Thus, those should be chosen according to desired applications.

GBM can be obtained by top-down and bottom-up approaches [104]. The first involves exfoliating graphite to obtain few or single layer graphene sheets [38,108]. The second, consists in assembling graphene from deposition of carbon atoms from other sources [109,110]. The main difficulty in top-down methods is to overcome the van der Waals forces that hold the graphene layers together in graphite, preventing reagglomeration and avoiding damages in the honeycomb carbon structure [111,112]. Some examples of such methods are micromechanical exfoliation, direct sonication, electrochemical exfoliation, and superacid dissolution. Bottom-up methods include chemical vapor deposition (CVD), arc discharge, and epitaxial growth on silicon carbide [104].

The structure of CNT can be conceptualized by wrapping graphene into a cylinder. Typically, CNT are classified as either single-walled carbon nanotubes (SWCNT) or multi-walled carbon nanotubes (MWCNT). SWCNT exhibit better electrical properties, while MWCNT display better chemical resistance [113].

CNT can be produced using different methods, which mainly involve gas phase processes [114,115], like CVD, arc discharge, and laser ablation [116]. The most commonly used and efficient methods are the ones involving CVD, in which a carbon containing source (e.g., methane, acetylene, ethylene) reacts with a metal catalyst particle (e.g., iron, cobalt, nickel) which act as growth nuclei for CNT, at temperatures above 600 °C. There are several substrate materials for catalyst particles, as graphite, quartz, silicon, silicon carbide, amongst others. It is pertinent to mention that for graphene production by this technique, no catalyst particles are used, being the substrate itself a catalytic metal, often copper for monolayer or nickel for few layer graphene. Generally, CVD has the advantages of allowing mild and controllable synthesis in large scale [117–120].

CNT are strong, flexible, electrically conductive, and can be functionalized [121]. Potential applications of CNT have been reported such as in composite materials [122], electrochemical devices [123], hydrogen storage [124], field emission devices [125], nanometer-sized electronic devices, sensors and probes [126]. Determining the toxicity of CNT has been one of the most pressing questions in nanotechnology [127]. There is still some controversy on this subject, thus continued research is needed to assure that these materials are safe for biomedical applications [128,129]. Parameters such as structure, size distribution, surface area, surface chemistry, surface charge and agglomeration state, as well as the sample purity, have considerable impact on CNT properties [121].

In the research works reported in this review, CBN are both commercial products or lab-made by the authors. Most commercial CNT are produced by CVD, with suppliers often making available information about material dimensions and sometimes type of CVD used. On the other hand, researchers usually produce GBM from graphitic precursors, using top-down methods involving chemical oxidation and exfoliation, namely the Staudenmaier and modified Hummers methods (Figure 1). Commercial GBM are also used, with suppliers giving information about dimensions, and sometimes production methods. These involve direct exfoliation in a liquid, with or without the use of a surfactant, or in the solid state by edge functionalization, or by first inserting a chemical species between the graphene layers in graphite to weaken their interaction, followed by expansion/exfoliation [130]. Commercial products offer insured reproducibility and widespread availability. Moreover, with the optimization of the production processes, the costs of GBM are coming closer to its precursor, graphite [11].

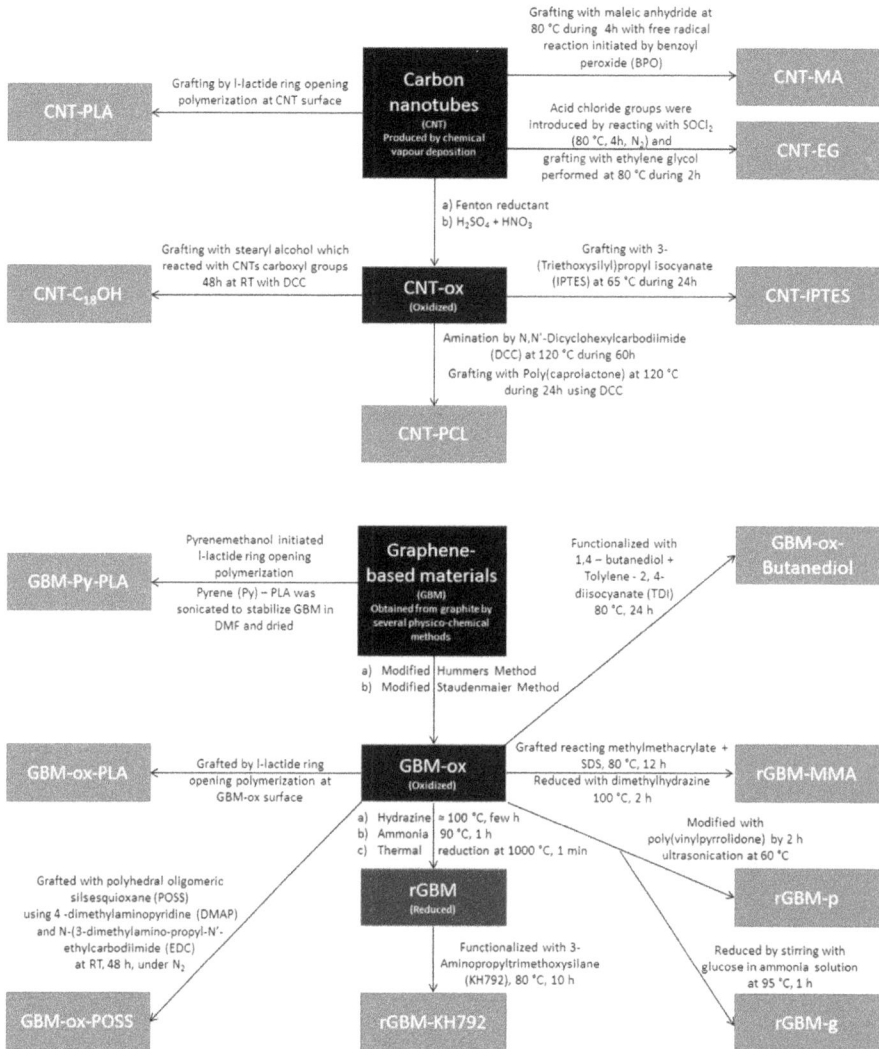

Figure 1. Scheme showing the different types of modifications performed on carbon-based nanomaterials (CBN) prior to incorporation in poly(lactic acid) (PLA).

CBN have been extensively used in polymer composites. In order to take advantage of their large surface area maximizing its effectiveness as filler, dispersion must be efficient, so as to maximize the amount of deagglomerated primary units. Functionalization is often used to improve compatibility with the polymer matrix. However, this can disrupt the sp^2 hybridization of CBN carbon structure and subsequently hinder their properties [131]. Some examples of CBN modifications used on the research works reported in this review are compiled in Figure 1. Some of these involve simple chemical oxidation, prior to surface modification with isocyanates, polymers (ethylene glycol, poly(caprolactone), methyl methacrylate, poly(vinyl pyrrolidone), and PLA), polyols or silanes. The impact of these on the composite properties is discussed in Section 5.

4. Production of PLA/CBN Composites

Three methods are most frequently used to obtain a dispersion of CBN into a polymer matrix: solution mixing, melt blending, and in situ polymerization [22,104]. Mechanical milling, also called ball milling, has been gaining recognition as an alternative technique with specific advantages, but it has not yet been reported for PLA/CBM composites. High impact milling is performed at room temperature on dry powders, prior to melt processing. Its effectiveness and benefits in relation to other methods have been shown for different polymer/filler systems [132].

4.1. Solution Mixing

Solution mixing is a simple procedure, requiring no special equipment, and allowing for straightforward scale-up. This method typically consists of three steps: (i) dispersion of the nanomaterial in a suitable solvent using sonication or mechanical stirring; (ii) dissolution of the polymer in the previous dispersion, under appropriate stirring; and (iii) removal of the solvent by distillation or lyophilization. Often the dispersion is cast into a flat mold, and then the solvent is evaporated. Flat composite slabs are therefore obtained. For this reason, the procedure is often called "solvent casting". As an alternative, the dispersion may be cast onto a low surface energy material (e.g., PTFE coated surface) using a blade applicator (doctor blading). After solvent evaporation, thin composite films are obtained. The viscosity of the dispersion needs to be adjusted for this procedure, which can be done by changing the concentration of polymer [133]. If production of fibers is desired, the third step can be replaced by electrospinning. This technique allows obtaining fibers that are much smaller in diameter (ranging from micrometers to nanometers) than those produced by conventional techniques. The basis of electrospinning is to charge the polymer solution in the spinneret tip with a high voltage, so that the electrostatic repulsion overcomes the surface tension of the solution, causing its ejection. The solvent vaporizes while the jet is in the air, producing a continuous fiber which deposits on the ground collector [27].

Complete solvent removal is a critical issue when using solution mixing to prepare composites, since toxicity concerns may arise when organic solvents are used. In addition, presence of residual solvent induces plasticization of the polymer matrix, which may alter significantly its mechanical properties [134–136].

PLA is soluble in organic solvents such as chlorinated solvents, benzene, tetrahydrofuran (THF), dimethyl formamide (DMF) and dioxane, but insoluble in ethanol, methanol, and aliphatic hydrocarbons. CBN are hydrophobic, therefore cannot be easily dispersed in polar solvents. However, they can be oxidized or modified with hydrophilic groups in order to allow dispersion in such solvents. Solubility limitations can also be overcome to a certain point by using ultrasonication to produce short-time metastable dispersions of CBN in organic solvents, which can then be mixed with polymer solutions [137].

Chloroform is the most used solvent to prepare PLA/CNT composites [138–143]. Despite, some authors obtain good results with THF [88,144], and dichloromethane [145,146]. McCullen et al. [147] conclude that a combination of chloroform and DMF is beneficial. Sometimes the introduction of new functional groups may originate incompatibility with the polymer matrix. To elude this problem, improvement of CNT dispersion by surfactant addition (e.g., polyoxyethylene 8 lauryl, dodecyl octaethylene) may be used, which allows preserving the chemical structure of the nanofiller [148]. GBM have been often incorporated in PLA by solution mixing using chloroform [135,149–151] or DMF [152–157] as solvents. Agglomeration of CBN may take place during solvent evaporation. Composite formation by electrospinning allows minimizing this problem, but leads to formation of fibers and not films [27,147,158].

4.2. Melt Blending

Melt blending is an economically attractive, environmentally friendly and highly scalable method for preparing nanocomposites. This strategy involves direct addition of the nanomaterial into the molten polymer, allowing optimization of the state of dispersion by adjusting operating parameters such as mixing speed, time and temperature. Due to the absence of solvent, the only compatibility issue is placed in terms of the nanofiller towards the polymer matrix [27,48]. The drawbacks of this procedure are the low bulk density of CBN, that makes the feeding of the melt-mixer a troublesome task and the lower degree of dispersion that is usually attained when compared to solvent mixing [137,159].

Most published research works use a lab-scale melt mixer to melt PLA and mix it with the nanofillers. Typical processing conditions correspond to temperatures between 160 °C and 180 °C [160–166], mixing times of 5 to 10 min [160–162,164,165,167], and rotation speeds between 50 and 100 rpm [160–164,166–169]. After mixing, the composite materials are almost always molded into flat sheets with controlled thickness in a hot press, however, other methods are also used (e.g., injection molding and piston spinning). Typically, the pressing is performed between 160 °C and 190 °C for 2 to 5 min, under 110 to 150 Kgfcm^{-2} pressure [160,165–170].

In addition to melt blending not being as effective as the solution mixing method or in situ polymerization in terms of the ability to achieve good filler dispersion, damage to the nanofillers or polymer may occur under severe conditions. Some studies have shown that processing conditions can have an impact on the molecular weight of PLA [171]. This can be mainly attributed to the presence of impurities such as acidic species, peroxide groups, metallic ions or other residual products that can increase the degradation of PLA during melt mixing [172].

4.3. In Situ Polymerization

In situ polymerization for production of polymer composites generally involves mixing the filler in neat monomer, or a solution of monomer, in the presence of catalysts and under proper reaction conditions [173]. The polymer chains grow on the filler surface, being covalently bonded. In situ polymerization generally results in more homogeneous particle dispersion than melt blending [174]. Use of this approach for polymerizing lactide in the presence of CNT has been reviewed by Brzeziński and Biela [175]. Contrary to CNT, that usually are post-treated, GBM already present some chemical groups that can be used in further functionalization, such as grafting polymer chains via atom transfer radical polymerization. Examples of in situ polymerization on GBM include polymers such as polyaniline (PANI), polyurethane (PU), polystyrene (PS), poly(methyl methacrylate) (PMMA) and polydimethylsiloxane (PDMS) [24].

Concerning the particular case of PLA/CBN, only a few examples of in situ polymerization can be found in the literature. Ring opening polymerization of L-lactide in presence of GBM has been reported by Yang et al. [176] and Promoda et al. [177]. Carboxyl-functionalized CNT have been grafted with PLA by Li and co-workers [178].

The above-mentioned composite production methods can be used both with GBM and CNT, and are congregated in Figure 2.

Solution Mixing

Melt blending

In-situ polymerization

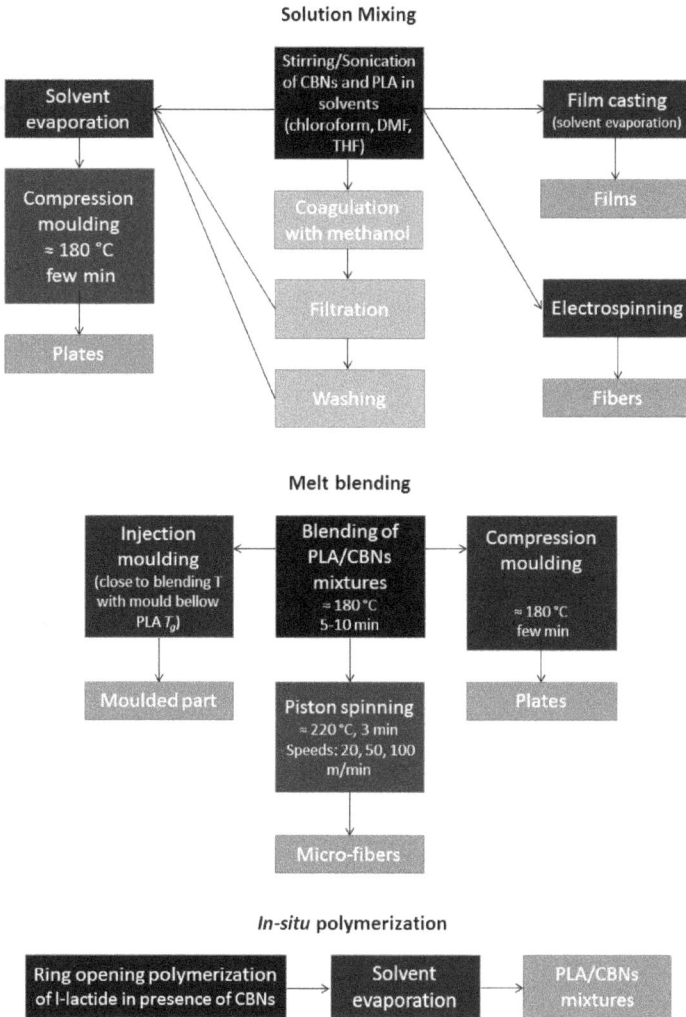

Figure 2. Scheme showing the different production methods of PLA/CBN composites.

5. Properties of PLA/CBN Composites

Numerous researchers have studied the properties of PLA combined with other materials, in order to tune key properties regarding specific applications [48]. The current review is focused on the effect of incorporating two carbon-based nanomaterials, CNT and GBM, in PLA. CNT are known for two decades and have well established large-scale production methods. GBM, which have been raising a growing interest from the scientific community, are cheaper and, in principle, comparable in properties to CNT [177].

5.1. Mechanical Properties

Physico-chemical interactions between fillers and polymer phase contribute to load transfer and distribution along the CBN network. Table 1 shows that solution mixing is the most commonly reported method for incorporation of CBN in PLA. The most frequently used solvents are chloroform, DMF and

THF. The filler concentrations most often tested are between 0.1–2 wt %. Maximum improvements in Young's modulus (E), storage modulus (E'), and tensile strength (σ_{max}) are found for concentrations between 0.25–5 wt % for CNT, and between 0.1–1 wt % for GBM. The larger improvement in E, relative to unfilled PLA, is of 372%, for 0.25 wt % MWCNT sonicated in a PLA/chloroform dispersion, followed by compression molding of the dried mixture [138]. For GBM, the best performance is an increase of 156% with incorporation of 0.4 wt % GNP-M, also by sonication, but followed by film casting using doctor blading. In this study, comparison is made with GO, which yields a maximum E increase at 0.3 wt % loading. Figure 3 presents microscopy images demonstrating good dispersion of the fillers in the PLA matrix [135].

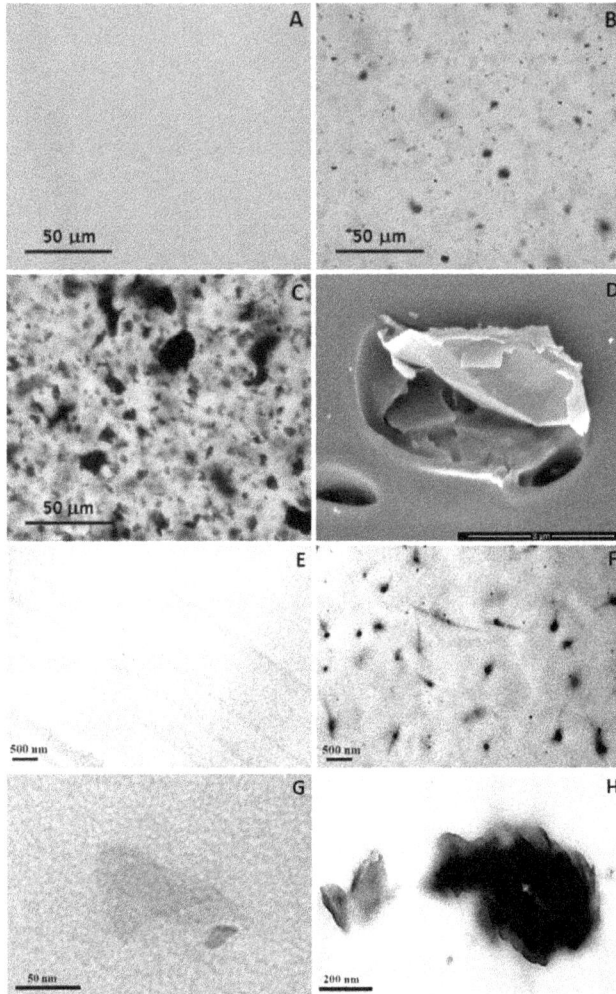

Figure 3. Microscopy images of PLA, PLA/GNP and GO 0.4 wt % films produced by solution mixing followed by film casting using doctor blading, displaying good filler dispersion and interaction with polymer matrix. Optical microscopy images of PLA (**A**); PLA/GO (**B**); and PLA/GNP (**C**); Scanning electron microscopy image of PLA/GNP (**D**); Transmission electron microscopy images of PLA (**E**) and PLA/GO (**F–H**) [179].

The maximum increase on E' is of 1500%, achieved with incorporation of 0.5 wt % rGO-KH792 in PLLA, by simple stirring, casting on PTFE mold, and vacuum drying the resultant films at 120 °C for 48 h [157]. However, this increase only occurs around PLA transition temperature (60–65 °C). At ambient temperature, the best result is an increase of 67% with incorporation of 3 wt % A-SWCNT-Si (acid treated and grafted with 3-isocyanatoporpyl triethoxysilane) in PLA by sonication, followed by drying and compression molding at 190 °C [144]. The maximum increase in σ_{max} is of 129 wt %, obtained with incorporation of 0.4 wt % GNP-M in PLA by sonication and film casting by doctor blading [135]. For CNT the best result is an increase of 47% obtained with MWCNT grafted with PLA, and then incorporated at a loading of 1 wt % in PLA by sonication in chloroform, separation, drying and compression molding at 180 °C [141]. When considering CNT without modification, the best result reported is an increase of 9% for 1.2 wt % MWCNT incorporated in PLA by solution mixing, followed by drying and compression molding at 180 °C with a pressure of 1000 Kg [142].

Melt-blending is less frequently reported than solution mixing for production of PLA/CBN composites, probably due to the lower availability of the necessary equipment. Results show that it tends to be not as effective in improvement of mechanical properties, as solution mixing. The best performance in terms of E (↑88%) and E' (↑76%) is reported by Lin et al. [160] for an incorporation of 3 wt % MWCNT grafted with stearyl alcohol (MWCNT-C_{18}OH) in PLA by melt blending (180 °C, 5 min, 50 rpm), using $Ti(OBu)_4$ for transesterification, followed by compression molding at the same temperature. When PLA is not transesterified, E and E' increases were of 74% and 44%, respectively. The maximum increase in σ_{max} (40%) is obtained incorporating 0.08 wt % rGO using a twin-screw mixer (175 °C, 8 min, 60 rpm), followed by compression molding at 180 °C [168]. The incorporation by melt blending (180 °C, 20 min, 50 rpm) of 0.25 wt % GNP-M5 and C in PLA followed by compression molding at 190 °C, prevented its mechanical properties decay after 6 months degradation in phosphate-buffered saline at 37 °C [180].

In situ polymerization is the least used technique. It has been reported by Pramoda et al. [177], who performed PLA ring-opening polymerization in presence of 1 wt % of GO functionalized with butanediol and GO modified with POSS silsesquioxane. In the first case, improvements of 1% and 14% in E and hardness are obtained, respectively. In the second, the performance is increased by 33% and 45%, in the same order.

Comparing the results for CNT and GBM, we can conclude that both can effectively improve PLA mechanical properties, whether by solution mixing and melt blending. However, use of GBM usually implies lower amounts of GBM than of CNT. Several chemical modifications have been tried to improve compatibility with the polymer matrix, with ineffective results is some cases. Functionalization with carboxyl groups is the most common and effective procedure to improve CNT compatibility with PLA matrix [146]. On the other hand, no relation has been observed between CBN morphological properties (size, length, and diameter) and the mechanical performance of the composites.

Table 1. Mechanical properties of PLA/CBN composites in comparison with non-modified PLA. Production methods and CBN characteristics.

Method	Procedure	CNT Characteristics	CNT Content (wt %)	Mechanical Properties Relative to Neat Polymer ΔE: maximum Young's modulus improvement $\Delta E'$: maximum storage modulus improvement $\Delta\sigma_{max}$: maximum tensile strength improvement	References
	Sonication in chloroform and DMF, electrospinning	MWCNT Diameter (d) 15 ± 5 nm Length (l) 5–20 μm 95% purity Produced by plasma enhanced CVD	MWCNT: 0.25, 0.5, 1	$\Delta E\uparrow$372% (0.25 wt %)	[147]
	Sonication in chloroform, drying and compression molding (200 °C, 150 Kgf cm^{-2}, 15 min)	MWCNT d not given l ± 2000 μm	MWCNT: 0.5, 3, 5, 10	$\Delta E\uparrow$150% (5 wt %)	[138]
	Sonication in chloroform, film casting	Unzipped CNT (uCNT) Diameter 30 nm l = 10 μm 95% purity	uCNT: 1, 2, 3, 4, 5	$\Delta E'\uparrow$14% (3 wt %)	[139]
Solution mixing	PLA was modified with benzoyl chloride and pyridine (PLAm), then acid chloride groups were added by reaction with thionyl chloride and triethylamine, then fMWCNT were added and the mixture centrifuged and filtered to remove excess filler and salts. Finally, sonication in chloroform and film casting was performed	MWCNT functionalized with COOH using Fenton reactant and then reacted with SOCl$_2$ and ethylene glycol (fMWCNT) d = 9.5 nm l = 1.5 μm 95% purity	Not clear	$\Delta E\uparrow$17%, $\Delta\sigma_{max}\uparrow$8% (comparing to PLAm)	[140]
	Sonication in chloroform, coagulation with methanol, filtration, vacuum drying, and compression molding (180 °C)	MWCNT (thermal CVD, d = 10–15 nm, l = 10–20 μm, 95% purity) MWCNT carboxyl-functionalized (MWCNT-COOH) by H$_2$SO$_4$ 1:3 HNO$_3$, 3 h, 120 °C MWCNT grafted with PLA (MWCNT-g-PLA): MWCNT-COOH + L-lactide, 12 h, 150 °C, + tin(II) chloride, 20 h, 180 °C, under vacuum, filtration, vacuum drying	MWCNT: 1 MWCNT-COOH: 1 MWCNT-g-PLA: 0.1, 0.2, 0.5, 1, 5	PLA/MWCNT-g-PLA: $\Delta E\uparrow$32%, $\Delta\sigma_{max}\uparrow$47% (1 wt %)	[141]
	Solution mixing in chloroform, drying and compression molding (180 °C)	MWCNT grafted with PLLA after reaction with SOCl$_2$ and ethylene glycol (MWCNT-PLLA) Dimensions not given 95% purity	MWCNT and MWCNT-g-PLLA: 0.1, 0.2, 0.4, 0.6, 0.8, 1.2	PLA/MWCNT: $\Delta E\uparrow$46%, $\Delta\sigma_{max}\uparrow$9% (1.2 wt %) PLA/MWCNT-g-PLLA: $\Delta E\uparrow$86%, $\Delta\sigma_{max}\uparrow$13% (1.2 wt %)	[142]

Table 1. *Cont.*

Method	Procedure	CNT Characteristics	CNT Content (wt %)	Mechanical Properties Relative to Neat Polymer ΔE: maximum Young's modulus improvement ΔE': maximum storage modulus improvement Δσ_max: maximum tensile strength improvement	References
Solution mixing	Solution mixing in chloroform, filtered, washed, dried under vacuum, and compression molded (180 °C, 500 psi)	MWCNT, MWCNT-COOH (both as in [101]), and MWCNT grafted with PLA chains of 122–530 g mol^{-1} by ring open polymerization (MWCNT-g-PLA530). d = 10–15 nm l = 10–20 μm 95% purity	MWCNT-COOH: 1 MWCNT-g-PLA530: 1	PLA/MWCNT-COOH: $\Delta E\uparrow 4\%$, $\Delta \sigma_{max} = 9\%$ PLA/MWCNT-g-PLA530: $\Delta E\uparrow 44\%$, $\Delta \sigma_{max} = 44\%$	[143]
	Solution mixing in THF, vacuum drying, thermal compression	SWCNT (d < 2 nm, l = 5–15 μm, 95% purity) treated with 3:1 H$_2$SO$_4$/HNO$_3$ (A-SWCNT), and functionalized (1:2 v/v) with 3-isocyanatopropyl triethoxysilane (IPTES)—A-SWCNT-Si	SWCNT, A-SWCNT and A-SWCNT-Si: 0.1, 0.3, 0.5, 1, 3	PLA/SWCNT: $\Delta E'\uparrow 20\%$ PLA/A-SWCNT: $\Delta E'\uparrow 33\%$ PLA/A-SWCNT-Si: $\Delta E'\uparrow 67\%$ (3 wt % for all conditions)	[144]
	Sonication in dichloromethane and THF, vacuum drying, and compression molding (190 °C)	MWCNT (d = 9–20 nm, l = 5 μm) functionalized with 3:1 H$_2$SO$_4$/HNO$_3$ (MWCNT-COOH)	MWCNT-COOH 0.5, 1, 2.5 MWCNT: 2.5	PLA/MWCNT-COOH: $\Delta E'\uparrow 80\%$, $\Delta E'\uparrow 35\%$, $\Delta \sigma_{max}\uparrow 28\%$ (2.5 wt %) PLA/MWCNT: $\Delta E'\uparrow 25\%$, $\Delta E'\downarrow 6\%$, $\Delta \sigma_{max}$ (not reported) (2.5 wt %)	[146]
	Internal mixer (180 °C, 50 rpm, 5 min) with and without transesterification with Ti(OBu)$_4$, compression molding (180 °C)	MWCNT (l = 1–10 μm) functionalized with HNO$_3$ (120 °C, 40 min)—MWCNT-COOH, and modified with DCC and stearyl alcohol (MWCNT-C$_{18}$OH)	PC: MWCNT/PLA CNT-C$_{18}$OH/PLA PC-18T: MWCNT-C$_{18}$OH/PLA transesterified 0.5, 1.5, 3	(3 wt %) PLA/PC: $\Delta E\uparrow 73\%$, $\Delta E'\uparrow 34\%$ PLA/PC-18: $\Delta E\uparrow 74\%$, $\Delta E'\uparrow 44\%$ PLA/PC-18T: $\Delta E\uparrow 88\%$, $\Delta E'\uparrow 76\%$	[160]
Melt blending	Twin-screw extrusion (150–190 °C, 100 rpm), injection molding (160–190 °C) High-crystalline PLA (HC-PLA) and low-crystalline PLA (LC-PLA) were tested	MWCNT (l = 5–20 μm, d = 40–60 nm) functionalized with maleic anhydride (MWCNT-g-MA) at 80 °C, 4h, +benzoyl peroxide	LC-PLA/MWCNT, HC-PLA/MWCNT and MWCNT-g-MA: 0.25, 0.5, 0.75, 1, 2, 4	PLA/LC-PLA/MWCNT: $\Delta \sigma_{max}\uparrow 23\%$ PLA/HC-PLA/MWCNT: $\Delta \sigma_{max}\uparrow 13\%$ PLA/MWCNT-g-MA: $\Delta \sigma_{max}\uparrow 27\%$ (4 wt % for all conditions)	[163]
	Twin-screw extrusion (180 °C, 150 rpm, 5 min), compression molding at 180 °C	MWCNT (d = 6–13 nm, l = 2.5–20 μm, specific surface area = 220 m^{2}g^{-1}) produced by CVD	MWCNT: 1.5, 3, 5	PLA/MWCNT: $\Delta E'\uparrow 28\%$, $\Delta \sigma_{max}\uparrow 27\%$ (5 wt %)	[165]
	Twin-screw extrusion (160–190 °C)	Carboxyl-functionalized (MWCNT-COOH) d = 10–11 nm, l = 12–15 μm	MWCNT-COOH: 1	ΔE and $\Delta \sigma_{max}\uparrow 8\%$ (1 wt %)	[181]

Table 1. *Cont.*

Method	Procedure	GBM Characteristics	GBM Content (wt %)	Mechanical Properties Relative to Neat Polymer ΔE: maximum Young's modulus improvement $\Delta E'$: maximum storage modulus improvement $\Delta\sigma_{max}$: maximum tensile strength improvement	References
	Sonication in chloroform, casting and doctor blading. GO was pre-dispersed in acetone while GNP was directly dispersed in chloroform	GNP grade M (commercial product) t = 6–8 nm, d ≈ 5 µm. GO (MHM) d ≈ 100 nm	GO and GNP: 0.2, 0.4, 0.6	PLA/GO: $\Delta E\uparrow$115%, $\Delta\sigma_{max}\uparrow$95% (0.3 wt %) PLA/GNP: $\Delta E\uparrow$156%, $\Delta\sigma_{max}\uparrow$129% (0.4 wt %)	[135]
	Sonication in chloroform, filtration, vacuum drying, compression molding (170 °C, 10 min)	GO (from natural graphite, MHM + lyophilization) d ≈ 300 nm GO-g-PLLA (GO + L-lactide (Sn(oct)$_2$), filtration, vacuum drying)	GO and GO-g-PLLA: 0.5	PLA/GO: $\Delta\sigma_{max}\uparrow$51% PLA/GO-g-PLLA: $\Delta\sigma_{max}\uparrow$106%	[150]
Solution mixing	Stirring and sonication in DMF; coagulation with methanol, filtration, and vacuum drying	GO (MHM) from expandable graphite, chemically reduced with hydrazine, and lyophilized (GNSs—solvent free graphene nanosheets) t < 1 nm, d < 50 nm	GNSs: 0.2	$\Delta E'\uparrow$18%, $\Delta\sigma_{max}\uparrow$26%	[152]
	Sonication in DMF, coagulation with methanol, drying, compression molding (185 °C)	TRG (commercial product, t = few layer, d = hundreds of nm) TRG/PLA/Py-PLA: Py-PLA-OH (1-Pyrenemethanol + L-lactide, Sn(oct)$_2$) + TRG (10:1)—sonication + PLA—coagulation and drying	TRG and TRG/PLA/Py-PLA: 0.25, 1	PLA/TRG: $\Delta E'\uparrow$1%–3%, $\Delta\sigma_{max}\uparrow$8% PLA/TRG/PLA/Py-PLA: $\Delta E'\uparrow$10%–15%, $\Delta\sigma_{max}\uparrow$19%	[154]
	Solution mixing in DMF, film casting	GO prepared according to MHM, reduced to rGO and functionalized with N-(aminoethyl)-aminopropyltrimethoxysilane (KH792)	rGO-KH792: 0.1, 0.2, 0.5	$\Delta E'\uparrow$1500% around the T_g (0.5 wt %)	[157]

Table 1. Cont.

Method	Procedure	GBM Characteristics	GBM Content (wt %)	Mechanical Properties Relative to Neat Polymer ΔE: maximum Young's modulus improvement $\Delta E'$: maximum storage modulus improvement $\Delta\sigma_{max}$: maximum tensile strength improvement	References
	Twin-screw mixer (175 °C, 60 rpm, 8 min), compression molding at 180 °C	GO prepared by MHM and reduced with hydrazine and ammonia (rGO)t = 0.4–0.6 nm, d = 0.1–0.5 μm	rGO: 0.02, 0.04, 0.08, 0.2, 0.5, 1, 2	$\Delta E\uparrow27\%$, $\Delta\sigma_{max}\uparrow40\%$ (0.08 wt %) $\Delta E'\uparrow54\%$, $\Delta\sigma_{max}\downarrow40\%$ (2 wt %)	[168]
	Internal mixer (160 °C, 25 rpm, 10 min), compression molding (160 °C, 10 min) (*Polymer was PLA/PEG 9:1 blend*)	GNP grade M15 (commercial product) t = 6–8 nm, d ≈ 15 μm	GNP-M15: 0.1, 0.3, 0.5, 0.7, 1	$\Delta E'\uparrow84$ and 70%, $\Delta\sigma_{max}\uparrow20$ and 33% (0.1 and 0.3 wt %) (*relative to pristine PLA/PEG blend*)	[167]
Melt blending	Internal mixer (180 °C, 80 rpm, 10 min) Compression molding (180 °C)	GO (MHM) + SDS; ultrasounds, stirring 12 h, 25 °C Methylmethacrylate (MMA), stirring 12 h + ammonium persulfate (APS) 12 h, 80 °C + reduction with dimethyl hydrazine, 100 °C, 2 h (PFG—polymer-functionalized graphene nanoparticles) t = 2.4 nm	PFG: 1, 2, 3, 4, 5	$\Delta E'\uparrow80\%$, $\Delta\sigma_{max}\uparrow10\%$ (5 wt %)	[164]
	Internal mixer (180 °C, 50 rpm, 20 min) Compression molding (190 °C, 2 min, 150 Kg cm^{-2})	GNP grade M5 (t = 6–8 nm, d ≈ 5 μm) and C (t = up to 2 single layers, d < 2 μm) (commercial products)	GNP-M5 and C: 0.1, 0.25, 0.5	PLA/GNP-M5: $\Delta E'\uparrow14\%$, $\Delta\sigma_{max}\uparrow6\%$ (0.25 wt %) PLA/GNP-C: $\Delta E'\uparrow14\%$, $\Delta\sigma_{max}\uparrow20\%$ (0.25 wt %) The incorporation of both fillers prevented mechanical properties decay after 6 months degradation	[180,182]
In situ polymerization	Sonication of L-lactide + filler in toluene, addition of Tin(II)-2-ethylhexanoate under N$_2$, stirring at 110 °C, 3 days	Expanded graphite (MHM) to GO GO-functionalized: GO + TDI + 1,4-butanediol, 80 °C, 24 h GO-g-POSS: GO + POSS—polyhedral oligomeric silsesquioxane + DMAP—4-(dimethylaminopyridine) + EDC—N-(3-dimethylamino-propyl-N'-ethylcarbodiimide), 2 days, room temperature. N$_2$ (dimensions not given)	GO-functionalized, GO-g-POSS, GO+POSS (physical mixture): 1	PLA/GO-functionalized: $\Delta E'\uparrow1\%$, *Hardness*↑14% PLA/GO-g-POSS: $\Delta E'\uparrow33\%$, *Hardness*↑45% PLA/GO + POSS: $\Delta E'\uparrow29\%$, Hardness↑36%	[177]

13

5.2. Electrical Properties

Neat PLA is electrically insulating with a low electrical conductivity ($\sigma \approx 1 \times 10^{-16}$ S m^{-1}), and high sheet resistance ($\rho_\square \approx 5 \times 10^{12}$ Ω sq^{-1}) [144,160]. Since CNT and reduced forms of GBM present high electrical conductivity, they can be incorporated in PLA to improve its conductivity. This sort of composites have potential to be used as electrical stimulating implants, since PLA is used as a biodegradable matrix in orthopedic material. Other advantages of increasing PLA conductivity are the possibility of using it as antistatic coating/material or for electromagnetic shielding [104]. The minimum amount of filler required to form a conductive network within the polymer is called percolation threshold, and should be as low as possible in order to keep processing simple (relatively low viscosity of the melt) and low costs. Table 2 shows that, once again, the most used method to incorporate CBN on PLA for electrical properties evaluation is solution mixing. The amount of fillers ranges from 0.01 to 10 wt %. The best result, considering electrical conductivity (σ) with CNT is 3.5×10^{-3} S m^{-1}, obtained incorporating 10 wt % MWCNT in PLA by sonication in chloroform, followed by drying and compression molding at 200 °C during 15 min [138]. Results are also often presented in terms of sheet resistance ($\rho_\square$), being the lowest value reported by Shao et al. [183], of 1×10^2 Ω sq^{-1} achieved incorporating 5 wt % MWCNT previously oxidized (treated with HCl and HNO$_3$) in PLA by solution mixing, followed by electrospinning of aligned nanofibers (d $\approx$ 250 nm). The alignment of the fibers slightly improved sheet resistance, comparing with random meshes. Interestingly, Yoon et al. [143] observe a considerable sheet resistance of 1×10^5 Ω sq^{-1}, with incorporation of 1 wt % MWCNT-COOH, also oxidized by treatment with strong acids (H$_2$SO$_4$ and HNO$_3$). For GBM, the maximum conductivity reported is 2.2 S m^{-1}, higher than for CNT, obtained incorporating 1.25 wt % rGO-g (reduced with ammonia) in PLA by sonication in DMF. Interestingly, the solvent used for dispersion of CNT in PLA is always chloroform and for GBM is always DMF.

Melt-blending is the second most used approach to disperse CBN in PLA in order to improve its electrical properties, being most often performed by twin-screw extrusion, followed by compression molding. The highest σ considering CNT is 50 S m^{-1}, which is reported by Pötschke et al. [184]. These authors prepare MWCNT mixtures by twin-screw extrusion, followed by piston spinning at different speeds. They conclude that non-spun mixtures with 5 wt % MWCNT in PLA present the same conductivity as 3 wt % mixtures after piston spinning at a speed of 20 m min^{-1}. Microscopy images in Figure 4 allow to observe good MWCNT dispersion and orientation due to spinning process.

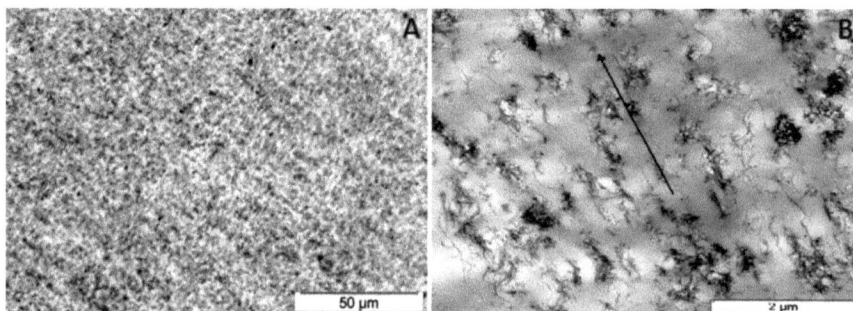

Figure 4. Optical microscopy image of a PLA/MWCNT 3 wt % mixture produced by twin-screw extrusion (**A**)—illustrating the high degree of macroscopic filler dispersion. Transmission electron microscopy image of a PLA/MWCNT 3 wt % mixture produced by twin-screw extrusion, followed by piston spinning; (**B**)—arrow indicates that fillers are strongly oriented in fiber direction due to the spinning process [185].

Table 2. Electrical properties of PLA/CBN composites in comparison with non-modified PLA. Production methods and CBN characteristics.

Method	Procedure	CNT Characteristics	CNT Content (wt %)	Electrical Properties σ: electrical conductivity $\rho_\square$: sheet resistance (PLA $\sigma \approx 1 \times 10^{-16}$ S m^{-1}, $\rho_\square \approx 5 \times 10^{12}$ Ω sq^{-1}) [106,122]	References
	Sonication in chloroform, drying and compression molding (200 °C, 150 Kgf cm^{-2}, 15 min)	MWCNT Diameter (d) not given Length (l) = ±2000 μm	MWCNT: 0.5, 3, 5, 10	$\sigma = 1.8 \times 10^{-3}$ and 3.5×10^{-3} S m^{-1} (3 and 10 wt %)	[138]
	Sonication in chloroform, coagulation with methanol, filtration, vacuum drying, and compression molding (180 °C)	MWCNT (thermal CVD, d = 10–15 nm, l = 10–20 μm, 95% purity) MWCNT carboxyl-functionalized (MWCNT-COOH) by H$_2$SO$_4$ 1:3 HNO$_3$, 3 h, 120 °C MWCNT grafted with PLA (MWCNT-g-PLA): MWCNT-COOH + L-lactide, 12 h, 150 °C, + tin(II) chloride, 20 h, 180 °C, under vacuum, filtration, vacuum drying	MWCNT: 1 MWCNT-COOH: 1 MWCNT-g-PLA:0.1, 0.2, 0.5, 1, 5	PLA/MWCNT: $\rho_\square = 1 \times 10^{12}$ Ω sq^{-1} (for 0.1 and 0.2 wt % is similar to PLA), 1×10^5 and 1×10^4 Ω sq^{-1} (0.5 wt %, and 1–5 wt %) PLA/MWCNT-g-PLA: $\rho_\square = 1 \times 10^{12}$ Ω sq^{-1} (0.1–5 wt %—always similar to PLA)	[141]
Solution mixing	Solution mixing in chloroform, drying and compression molding (180 °C)	MWCNT, MWCNT grafted with PLLA after reaction with SOCl$_2$ and ethylene glycol (MWCNT-g-PLLA) Dimensions not given 95% purity	MWCNT and MWCNT-g-PLLA: 0.1, 0.2, 0.4, 0.6, 0.8, 1.2	PLA/MWCNT: $\sigma = 2 \times 10^{-13}$ S m^{-1} (0.1–0.4 wt %), 3×10^{-9} S m^{-1} (0.6 wt %), and 2×10^{-5} S m^{-1} (1.2 wt %) PLA/MWCNT-g-PLLA: $\sigma = 2 \times 10^{-13}$ S m^{-1} (0.1–0.4 wt %), 5×10^{-13} S m^{-1} (0.6 wt %), and 3×10^{-8} S m^{-1} (1.2 wt %) Increases with filler amount	[142]
	Solution mixing in chloroform, filtered, washed, dried under vacuum, and compression molded (180 °C, 500 psi)	MWCNT, MWCNT-COOH (both as in [101]), and MWCNT grafted with PLA chains of 122–530 g mol^{-1} by ring open polymerization (MWCNT-g-PLA122–530). d = 10–15 nm l = 10–20 μm 95% purity	MWCNT-COOH: 1 MWCNT-g-PLA122-530: 1	PLA/MWCNT-COOH: $\rho_\square = 1 \times 10^5$ Ω sq^{-1} PLA/MWCNT-g-PLA112-530: $\rho_\square = 2 \times 10^6$, 2×10^{12}, and 1×10^{12} Ω/sq (122, 250, 530 g mol^{-1})	[143]
	Sonication in THF, vacuum drying, thermal compression	MWCNT (d = 8–15 nm, l = 50 μm) purified by sonication with H$_2$SO$_4$ and HNO$_3$ at 50 °C, filtration, and washing	MWCNT purified/non-purified: 1, 3, 5, 7	PLA/MWCNT purified: $\sigma = 4 \times 10^{-9}$, 1×10^{-9}, and 2×10^{-6} S m^{-1} (1, 5, and 7 wt %), PLA/MWCNT non-purified: $\sigma = 7 \times 10^{-11}$, 2×10^{-8}, and 5×10^{-8} S m^{-1} (1, 5, and 7 wt %) Increases with filler amount	[88]
	Solution mixing in THF, vacuum drying, thermal compression	SWCNT (d < 2 nm, l = 5–15 μm, 95% purity) treated with 3:1 H$_2$SO$_4$/HNO$_3$ (A-SWCNT), and functionalized (1:2 v/v) with 3-isocyanatoporpyl triethoxysilane (IPTES)—A-SWCNT-Si	SWCNT, A-SWCNT and A-SWCNT-Si: 0.3, 0.5, 1, 3	PLA/SWCNT: $\sigma = 2 \times 10^{-16}$, 3×10^{-9}, and 5×10^{-8} S m^{-1} (0.3, 1, 3 wt %) PLA/A-SWCNT-Si: $\sigma = 5 \times 10^{-15}$, 5×10^{-8}, and 2×10^{-6} S m^{-1} (0.3, 1, 3 wt %) Increases with filler amount	[144]

Table 2. *Cont.*

Method	Procedure	CNT Characteristics	CNT Content (wt %)	Electrical Properties σ: electrical conductivity $\rho_\square$: sheet resistance (PLA $\sigma \approx 1 \times 10^{-16}$ S m^{-1}, $\rho_\square \approx 5 \times 10^{12}$ Ω sq^{-1}) [106,122]	References
Solution mixing	MWCNT-ox (HCl, 2 h at 25 °C + HNO$_3$, 4 h at 110 °C) Nanofibers (MWCNT-ox sonicated in DMF 2 h + SDS, adding to PLA in dicloromethane, 1 h sonication before electrospinning)	MWCNT (l = 10–20 μm, d = 10–20 nm) Nanofibers (PLA ≈ 400 nm, PLA/MWCNT-ox ≈ 250 nm)	PLA/MWCNT-ox (3 wt %) random (R) and aligned (A) nanofibers: 1, 2, 3, 4, 5 wt %	PLA/MWCNT-ox-R: $\rho_\square = 1 \times 10^4$, 5×10^2 Ω sq^{-1} (3 and 5 wt %) PLA/MWCNT-ox-A: $\rho_\square = 5 \times 10^3$, 1×10^2 Ω sq^{-1} (3 and 5 wt %) Increases with both fillers amount	[183]
	Internal mixer (180 °C, 50 rpm, 5 min) with and without transesterification with Ti(OBu)$_4$, compression molding (180 °C)	MWCNT (l = 1–10 μm) functionalized with HNO$_3$ (120 °C, 40 min)—MWCNT-COOH, and modified with DCC and stearyl alcohol (MWCNT-C$_{18}$OH)	PC: MWCNT/PLA PC-18: MWCNT-C18OH/PLA PC-18T: MWCNT-C18OH/PLA transesterified 0.5, 1.5, 3	PLA/PC: $\rho_\square = 2 \times 10^7$, 3×10^6, and 3×10^5 Ω sq^{-1} (0.5, 1.5, 3 wt %) PLA/PC-18: $\rho_\square = 8 \times 10^5$, 9×10^4, and 1×10^{-1} Ω sq^{-1} (0.5, 1.5, 3 wt %) PLA/PC-18T: $\rho_\square = 5 \times 10^{12}$, 9×10^5, and 9×10^{-2} Ω sq^{-1} (0.5, 1.5, 3 wt %)	[160]
Melt blending	Twin-screw extruder (180, 215 and 250 °C; 100, 200 and 500 rpm; 5 min) 1st—masterbatch production 2nd—dilution of masterbatches and composites production	MWCNT d = 9.5 nm l = 1.5 μm 90% purity	MWCNT: 0.5, 0.75, 1, 2	σ is below 2.5×10^{-1} S m^{-1} (0.5–2 wt %) slightly decreasing with filler wt % increase	[162]
	Twin-screw extrusion (150–190 °C, 100 rpm), injection molding (160–190 °C) High-crystalline PLA (HC-PLA) and low-crystalline PLA (LC-PLA) were tested	MWCNT (l = 5–20 μm, d = 40–60 nm) functionalized with maleic anhydride (MWCNT-g-MA) at 80 °C, 4 h, + benzoyl peroxide	LC-PLA/MWCNT, HC-PLA/MWCNT and MWCNT-g-MA: 0.25, 0.5, 0.75, 1, 2, 4	LC-PLA/MWCNT: $\rho_\square = 2 \times 10^{13}$, 5×10^3, and 5×10^2 Ω sq^{-1} (0.5, 2, 4 wt %) HC-PLA/MWCNT: $\rho_\square = 1 \times 10^{14}$, 9×10^{10}, and 8×10^{10} Ω sq^{-1} (0.5, 2, 4 wt %) LC-PLA/MWCNT-g-MA: $\rho_\square = 3 \times 10^2$, 2×10^2, and 7×10^1 Ω sq^{-1} (0.5, 2, 4 wt %)	[163]
	Twin-screw extrusion (180 °C, 150 rpm, 5 min), compression molding at 180 °C	MWCNT d = 6–13 nm, l = 2.5–20 μm, specific surface area = 220 m^2 g^{-1} produced by CVD	MWCNT: 1.5, 3, 5	$\sigma = 1 \times 10^{-9}$, 1×10^{-2}, and 1 S m^{-1} (1.5, 3, 5 wt %)	[165]
	Twin-screw extruder (180–220 °C, 500 rpm) Piston spinning (20, 50, 100 m min^{-1}) to produce micro-fibers (220 °C, 3 min)	MWCNT d = 9.5 nm l = 1.5 μm 90% purity	MWCNT: 0.5, 1, 2, 3, 5	Extruded composites: $\sigma = 4, 14$, and 50 S m^{-1} (2, 3,5 wt %) Fibers (3 wt %): $\sigma = 50, 40$, and 1 S m^{-1} (spinning speeds of 20, 50, and 100 m min^{-1})	[184]

Table 2. *Cont.*

Method	Procedure	GBM Characteristics	GBM Content (wt %)	Electrical Properties σ: electrical conductivity $\rho_\square$: sheet resistance (PLA $\sigma \approx 1 \times 10^{-16}$ S m^{-1}, $\rho_\square \approx 5 \times 10^{12}$ Ω sq^{-1}) [106,122]	References
Solution mixing	Sonication in DMF, coagulation with methanol, drying, compression molding (185 °C)	TRG (commercial product, t = few layer, d = hundreds of nm) TRG/PLA/Py-PLA: Py-PLA-OH (1-Pyrenemethanol + L-lactide, Sn(oct)$_2$) + TRG (10:1)—sonication + PLA—coagulation and drying	TRG and TRG/Py-PLA-OH: 0.25, 1	PLA/TRG: $\sigma = 1 \times 10^{-16}$ and 1×10^{-6} S m^{-1} (0.25 and 1 wt %) PLA/TRG/PLA/Py-PLA-OH: $\sigma = 1 \times 10^{-16}$ and 1×10^{-7} S m^{-1} (0.25 and 1 wt %)	[154]
	Sonication in DMF, coagulation with methanol, drying, and compression molding (210 °C)	GO: from graphite flakes (modified Staudenmaier method) rGO-p: GO + Polyvinylpyrrolidone (1:5), sonication at 60 °C rGO-g: reduced by stirring with glucose in ammonia solution at 95 °C, 60 min Dimension not given	GO rGO-p rGO-g (0.5–2.5 vol %)	PLA/GO: $\sigma = \uparrow 6.5 \times 10^{-13}$ S m^{-1} PLA/rGO-p: $\sigma = \uparrow 4.7 \times 10^{-8}$ S m^{-1} PLA/rGO-g: $\sigma = 2.2$ S m^{-1} (for 1.25 vol % for all) Increases with filler amount	[155]
Melt blending	Twin-screw mixer (175 °C, 60 rpm, 8 min), compression molding at 180 °C	GO prepared according to MHM and chemically reduced to rGO. Thickness 0.4–0.6 nm and lateral dimension 0.1–0.5 mm.	rGO: 0.02, 0.04, 0.06, 0.2, 0.5, 1, 2	$\sigma = 1 \times 10^{-13}$ and 1×10^{-9} S m^{-1} (0.2 and 2 wt %) Increases with filler amount	[168]
	Internal mixer (180 °C, 80 rpm, 10 min)	GO (MHM) + SDS, ultrasounds, stirring 12 h, 25 °C Methylmethacrylate (MMA), stirring 12 h + ammonium persulfate (APS) 12 h, 80 °C + reduction with dimethyl hydrazine, 100 °C, 2 h (PFG—polymer-functionalized graphene nanoparticles) t = 2.4 nm	PFG: 1, 2, 3, 4, 5	$\sigma = 5.6 \times 10^{-14}$ and 2.6×10^{-4} S m^{-1} (1 and 5 wt %) Increases with filler amount	[164]
In situ polymerization	Ring-opening melt polymerization of lactide in presence of trGO	GO prepared according to MHM and thermally reduced to trGO Dimensions not given	TrGO: 0.01, 0.1, 0.5, 1, 1.5, 2	$\sigma = 5 \times 10^{-6}$ and 1.6×10^{-2} S m^{-1}. (1.5 and 2 wt %) Increases with filler amount	[176]

Considering $\rho_\square$, the best performance is obtained incorporating 3 wt % MWCNT-C$_{18}$OH (MWCNT modified with DCC and stearyl alcohol) using and external mixer, followed by compression molding at 180 °C during 5 min, resulting in a $\rho_\square$ of 1×10^{-1} Ω sq^{-1} [160]. This is the most effective modification performed, considering the sheet resistance values obtained with incorporation of the same amount of non-modified MWCNT, which was 3×10^5 Ω sq^{-1}. For GBM, the higher σ is 2.6×10^{-4} S m^{-1}, resultant from dispersion using an internal mixer at 180 °C, of 5 wt % PFG (graphene nanoparticles functionalized with methylmethacrylate) [164]. For rGO, a non-functionalized GBM, the best conductivity value is obtained for 2 wt % incorporation in PLA using a twin-screw extruder and compression molding. The value obtained is of 1×10^{-9} S m^{-1}, being higher than for the other concentrations tested. It can be compared, for example, with a σ of 1×10^{-13} S m^{-1} for 0.2 wt % [168]. In most works evaluated, electrical properties improve with the increase of filler amount.

In situ polymerization is the least explored technique, despite interesting results being obtained by Yang et al. [176], which incorporate 0.01–2 wt % trGO (thermally reduced) in PLA by ring-opening melt polymerization of L-lactide in presence of the filler. As example, σ obtained is 5×10^{-6} and 1.6×10^{-2} S m^{-1} for 1.5 and 2 wt %, respectively.

An interesting study by Chiu et al. [88], shows that purification of MWCNT by sonication with strong acids improved fillers compatibility and dispersibility in PLA, resulting in better electrical conductivity. The values of σ for incorporations of 7 wt % are 5×10^{-8} and 2×10^{-6} S m^{-1}, respectively for non-purified and purified MWCNT. Purification introduced polar functional groups on the CNT surface, allowing better dispersion, which resulted in more deagglomerated particles that formed a wider conductive network on PLA matrix.

5.3. Thermal Properties

Several works studied thermal properties of PLA containing CBN. CNT incorporations range from 0.01 to 15 wt %, while for GBM lower amounts are needed 0.01–2 wt % (Table 3). However, for both CBN, slight or no changes are observed in the composites' thermal properties, especially when low fillers amounts are used [135,146,156,157,160–162,167]. The most frequently used techniques to evaluate thermal properties in polymer composites are thermogravimetric analysis (TGA), differential scanning calorimetry (DSC), and dynamic mechanical analysis (DMA). TGA allows determination of thermal degradation temperatures (T_d) and DSC and DMA phase transition temperatures (T_g—glass transition temperature, T_m—melting temperature, and T_c—cold crystallization temperature).

A positive deviation in T_d is expected when there is good compatibility between CBN and the polymer matrix, combined with good dispersion of the fillers. This leads to restriction of PLA's chains motions, delaying thermal decomposition. Also, CBN can induce the formation of a crystallization region on their surfaces, which absorbs some heat as temperature of the composite increases. However, the incorporation of too high amounts of CBN can lead to the formation of agglomerates, which represent structural defects in the matrix, decreasing thermal stability [145]. Some works also attribute improvements in thermal stability to the barrier effect caused by the CBN, which creates a "tortuous path" delaying permeation of oxygen and the escape of volatile degradation products, and also to char formation [146,150,167]. Increases in T_g are usually also associated with good interaction between CBN and polymer matrix, leading to constraint of PLA's molecular mobility by hydrogen bonding and electrostatic attraction [139,140,146,150]. T_m increases are usually attributed to a nucleation effect caused by the CBN, which increases the degree of crystallinity [146,150,176]. For the same reason, T_c usually decreases with CBN incorporation [141,146,153,162,170,176].

When using solution mixing, the highest variation in terms of T_g is an increase of 10 °C, obtained using 1 wt % MWCNT purified by treatment with strong acids. Comparing with non-purified filler at the same loading, the increase is 5 °C higher. This is explained by purified MWCNT having stronger interfacial interactions with PLA matrix, imposing increased restriction to the mobility of macromolecular chains, and therefore rising T_g. Also, T_d (decomposition temperature) presents an increase of 10 °C for purified materials [88]. For T_m, the higher increase is of 16 °C for 0.3 and

1 wt % MWCNT-PCL (functionalized with poly(caprolactone)) incorporated in PLA aligned fibers by sonication in dichloromethane and electrospinning. Also, T_c decreases more than 10 °C, due to MWCNT inducing heterogeneous crystallization [145]. However, the higher decrease in T_c (<20 °C), is obtained by Moon et al. [138], with the incorporation of 3–10 wt % MWCNT, with a length of about 2000 μm. In literature, the degradation temperatures of the polymeric materials determined by TGA are presented in different terms. For example, as T_{di} (beginning of thermal degradation), T_{d5} (decomposition temperature for 5 wt % loss), and T_{d50} (decomposition temperature for 50% weight loss). For T_{di}, the highest increase is of 20 °C, obtained incorporating 2.5 wt % MWCNT-COOH (carboxylated with strong acids) by sonication in PLA dispersed in dichloromethane and THF, followed by vacuum drying and compression molding [146]. Considering T_{d50}, the best result is an increase of 1–3 °C, in a work above described [145].

GBM incorporation also induces changes on thermal properties of PLA. For T_g, an increase of 7 °C was obtained sonicating 0.4 wt % GNP in PLA films prepared by solvent casting [135]. The highest increases in T_m have been of 5 °C, for samples obtained by compression molding of PLA with 0.5 wt % GO grafted with PLA, produced by vacuum drying a dispersion in chloroform [150]. Significant decrease in T_c, of 20 °C, is observed for PLA with 2 wt % GO, obtained by solvent mixing [153]. Thermal stability of PLA has been shown to improve with addition of GBM. 2 wt % GONSs (graphene oxide nanosheets) increases T_{di} by 16 °C in samples produced by solvent mixing [156]. Also, T_{d5} is increased by 11 °C sonication of 0.2 wt % GNSs (graphene nanosheets) in PLA dispersed in DMF, dried under vacuum to produce composites [152]. Finally, $T_{d\,max}$ (T of maximum degradation rate) increases 33 °C for PLA filled with TRG, produced by solution mixing [154]. Chemical modifications of MWCNT are reported to increase thermal properties of the composites. For example, directly comparing with PLA/MWCNT(non-modified), the incorporation of 1 wt % MWCNT grafted with PLA in the same PLA matrix, results in increases of about 3 °C in T_g and decreases of 9 °C in T_c [141]. Treatment with strong acids followed by silanization of SWCNT [144], which are incorporated in PLA at loading ranging from 0.1 and 3 wt %, results in increases of about 5 °C in T_g.

Concerning composites produced by melt-blending, the highest increases in T_g are of 5–6 °C, for PLA micro-fibers with 3 wt % MWCNT to PLA [184]. Also, T_c is observed to decrease at most 12 °C with incorporation of 0.5 and 2 wt % MWCNT [170]. Chieng et al. [167], study on the thermal properties of PLA/PEG (9:1) blends with addition of 0.1–1 wt % GNP, reveals no variations on T_g, T_m, and T_c. However, T_{di}, T_{max}, and T_{50}, increase by 56, 53, and 44 °C, respectively, for 0.5 wt % loadings.

In situ polymerization of L-lactide in presence of TRG in amounts from 0.01 to 2 wt % result in considerable increases on T_g, T_m, and T_{dmax}. For example, at 2 wt % loading, increases of 5, 14, and 18 °C are obtained, respectively [176]. In a different work reporting in situ polymerization of L-lactide, covalent functionalization of GO with both 1,4-butanediol, and polyhedral silsesquioxane results in increases in T_g (18, 20 °C), T_c (15, 8 °C), T_m (7, 5 °C), and T_{d5} (23, 11 °C) comparing with PLA/GO composites at 1 wt % loadings [177].

Table 3. Thermal properties of PLA/CBN composites in comparison with non-modified PLA. Production methods and CBN characteristics.

Method	Procedure	CNTs Characteristics	CNTs Content (wt %)	Thermal Properties Relative to Neat Polymer	References
	Sonication in chloroform, drying and compression molding (200 °C, 150 Kgf cm⁻², 15 min)	MWCNT Diameter (d) not given Length (l) ≈ 2000 μm	MWCNT: 0.5, 3, 5, 10	T_g (glass transition) ↓1–4 °C (3, 5 wt %) and = (10 wt %) T_c (crystallization) ↓>20 °C (3, 5, 10 wt %) T_m (melting) = (3, 5, 10 wt %) T_d (degradation) ↑10–20 °C (3, 5, 10 wt %)	[138]
	Sonication in chloroform, film casting	Unzipped CNT (uCNT) d = 30 nm l = 10 μm 95% purity	uCNT: 1, 2, 3, 4, 5	T_g ↑7, 8 °C (3, 5 wt %) T_m ↑5, 3 °C (3, 5 wt %)	[139]
	PLA was modified with benzoyl chloride and pyridine (PLAm), then acid chloride groups were added by reaction with thionyl chloride and triethylamine, then fMWCNT were added and the mixture centrifuged and filtered to remove excess filler and salts. Finally, sonication in chloroform and film casting was performed	MWCNT functionalized with COOH using Fenton reactant and then reacted with SOCl₂ and ethylene glycol (fMWCNT). d = 9.5 nm l = 1.5 μm 95% purity	Not clear	T_g (tanδ) ↑9 °C T_{di} (beginning of thermal degradation) ↑80 °C	[140]
Solution mixing	Sonication in chloroform, coagulation with methanol, filtration, vacuum drying, and compression molding (180 °C)	MWCNT (thermal CVD, d = 10-15 nm, l = 10-20 μm, 95% purity). MWCNT carboxyl-functionalized (MWCNT-COOH) by H₂SO₄ 1:3 HNO₃, 3 h, 120 °C MWCNT grafted with PLA (MWCNT-g-PLA): MWCNT-COOH + L-lactide, 12 h, 150 °C, + tin(II) chloride, 20 h, 180 °C, under vacuum, filtration, vacuum drying	MWCNT: 1 MWCNT-COOH: 1 MWCNT-g-PLA: 0.1, 0.2, 0.5, 1, 5	No significant changes in T_m for all materials PLA/MWCNT: T_g ↓3, T_c ↓3 °C (1 wt %) PLA/MWCNT-COOH: T_g ↑2, T_c ↓3 °C (1 wt %) PLA/MWCNT-g-PLA: T_g ↑5–6 T_c ↑1 ↓2, 6, 12, 19 °C (0.1, 0.2, 0.5, 1, 5 wt %)	[141]
	Sonication in dichloromethane, electrospinning	MWCNT (d = 8-15 nm, l—not given, 95% purity) were functionalized with -COOH by H₂SO₄ and HNO₃ (3:1). Then, MWCNT-NH₂ were produced reacting MWCNT-COOH with N,N'-dicyclohexylcarbodiimide (DCC). MWCNT-PCL were produced reacting 1 g MWCNT-NH₂, 10 g PCL, and 20 g DCC	MWCNT-PCL(0.3, 0.5, 1, 3)/PLA aligned composite fibers	T_{d50} (50% weight loss) ↑1–3 °C (0.3, 1 wt %) T_g = (0.3, 1 wt %) T_m ↑16 °C (0.3, 1 wt %) T_c ↓13 °C and 12 °C (0.3, 1 wt %)	[145]
	Sonication in THF, vacuum drying, thermal compression	MWCNT (d = 8-15 nm, l = 50 μm) purified by sonication with H₂SO₄ and HNO₃ at 50 °C, filtration, and washing	MWCNT purified/non-purified: 1, 3, 5, 7	PLA/MWCNT non-purified: T_g ↑5–6 °C (1, 3, 5, 7 wt %) PLA/MWCNT purified: T_g ↑10, 7, 5, 5 °C (1, 3, 5, 7 wt %) PLA/MWCNT non-purified vs. purified: T_d ↑10, 11, 7, 8 °C (1, 3, 5, 7 wt %)	[88]

Table 3. *Cont.*

Method	Procedure	CNTs Characteristics	CNTs Content (wt %)	Thermal Properties Relative to Neat Polymer	References
Solution mixing	Solution mixing in THF; vacuum drying, thermal compression	SWCNT (d < 2 nm, l = 5–15 μm, 95% purity) treated with 3:1 H_2SO_4/HNO_3 (A-SWCNT), and functionalized (1:2 v/v) with 3-isocyanatoporpyl triethoxysilane (IPTES)—A-SWCNT-Si	SWCNT, A-SWCNT, and A-SWCNT-Si: 0.1, 0.3, 0.5, 1, 3	T_{d5} (5 wt % loss) ↓ for PLA/SWCNT (poor interfacial interaction), = for PLA/A-SWCNT, and A-SWCNT-Si; T_g: (higher that pure PLA) PLA/SWCNT < PLA/A-SWCNT < PLA/A-SWCNT-Si (considering all loadings, increases are below 5 °C)	[144]
	Sonication in dichloromethane and THF; vacuum drying, and compression molding (190 °C)	MWCNT (d = 9–20 nm, l = 5 μm) functionalized with 3:1 H_2SO_4/HNO_3 (MWCNT-COOH)	MWCNT-COOH: 0.5, 1, 2.5	T_{di} ↑ 10–20 °C (0.5–2.5 wt %); T_g ↑ 0, 1, 2 °C (0.5, 1, 2.5 wt %); T_c ↑ 1, 2, 4 °C 0.5, 1, 2.5 wt %); T_m ↑ 3, 4, 5 °C 0.5, 1, 2.5 wt %)	[146]
	Internal mixer (180 °C, 50 rpm, 5 min) with and without transesterification with Ti(OBu)₄, compression molding (180 °C)	MWCNT (l = 1–10 μm) functionalized with HNO_3 (120 °C, 40 min)—MWCNT-COOH, and modified with DCC and stearyl alcohol (MWCNT-C₁₈OH)	PC: MWCNT/PLA PC-18: MWCNT-C18OH/PLA PC-18T: MWCNT-C18OH/PLA transesterified 0.5, 1.5, 3	PLA/PC, PLA/PC-18—No change in T_m PLA/PC-18T—2 melting peaks, 1 bellow T_m for pristine PLA (low M_w PLA from transesterification), other at the same T_m	[160]
	Sonication in THF; vacuum drying + Microextruder (180 °C, 50 rpm, 5 min)	MWCNT (d = 9.5 nm, l = 1.5 μm) produced by catalytic carbon vapor deposition (CCVD)	MWCNT: 0.1, 1	T_g ↑1 °C (0.1, 1 wt %)	[161]
Melt blending	Twin-screw extruder (180, 215 and 250 °C; 100, 200 and 500 rpm; 5 min) 1st—masterbatch production 2nd—dilution of masterbatches and composites production	MWCNT d = 9.5 nm l = 1.5 μm 90% purity	MWCNT: 0.5, 0.75, 1, 2, 7.5, 15	Similar T_g (7.5, 15 wt %)	[162]
	Twin-screw extruder (210 °C, 400 rpm), compression molding (210 °C)	MWCNT d = 5–20 nm l = 10 μm Specific surface area = 100–700 m² g⁻¹ CCVD	MWCNT: 0.5, 1, 2, 3, 5	T_g ↓1, 2 °C (0.5, 1–5 wt %); T_c ↓12, 10, 12, 7, 6 °C (0.5, 1, 2, 3, 5 wt %); T_m ↓1, 2 °C (0.5–3, 5 wt %)	[170]
	Twin-screw extruder (180-220 °C, 500 rpm) Piston spinning to produce micro-fibers (220 °C, 3 min)	MWCNT d = 9.5 nm l = 1.5 μm 90% purity	MWCNT: 0.5, 1, 2, 3, 5	T_g: pellet = (3 wt %) Fibers ↑ 5–6 °C (3 wt %)	[184]

Table 3. *Cont.*

Method	Procedure	GBM Characteristics	GBM Content (wt %)	Thermal Properties Relative to Neat Polymer	References
	Sonication in chloroform, casting and doctor blading; GO was pre-dispersed in acetone while GNP was directly dispersed in chloroform	GNP grade M (commercial product) t = 6–8 nm, d ≈ 5 μm. GO (MHM) d ≈ 100 nm	GO and GNP: 0.2, 0.4, 0.6	PLA/GO: T_g ↑3, 4, 3 °C (0.2, 0.4, 0.6 wt %); PLA/GNP: T_g ↑6, 7, 5 °C (0.2, 0.4, 0.6 wt %); Similar T_m for both GO and GNP	[135]
	Sonication in chloroform, filtration, vacuum drying, compression molding (170 °C, 10 min)	GO (from natural graphite, MHM + lyophilization) d ≈ 300 nm; GO-g-PLLA (GO + L-lactide (Sn(oct)$_2$), filtration, vacuum drying)	GO and GO-g-PLLA: 0.5	PLA/GO: T_g ↑6 °C; T_m ↑3 °C; PLA/GO-g-PLLA: T_g ↑6 °C; T_m ↑5 °C	[150]
	Stirring and sonication in DMF, coagulation with methanol, filtration, and vacuum drying	GO (MHM) from expandable graphite, chemically reduced with hydrazine, and lyophilized (GNSs—solvent free graphene nanosheets) t < 1 nm, d < 50 nm	GNSs: 0.2	T_{d5} ↑11 °C	[152]
Solution mixing	Sonication in DMF; film casting, vacuum drying	GO prepared according to Staudenmaier method (H_2SO_4 + HNO_3 + $KClO_3$) (dimensions not given)	GO: 0.5, 1, 2	(0.5, 1, 2 wt %); T_c ↓9, 15, 20 °C; T_g similar	[153]
	Sonication in DMF; coagulation with methanol, drying, compression molding (185 °C)	TRG (commercial product, t = few layer, d = hundreds of nm) TRG/PLA/Py-PLA: Py-PLA-OH (1-Pyrenemethanol + L-lactide, Sn(oct)$_2$) + TRG (10:1)—sonication + PLA—coagulation and drying	TRG and TRG/Py-PLA-OH: 0.25, 1	PLA/TRG: T_{d5} ↓32 °C; $T_{d\,max}$ (max. degradation) ↑33 °C; PLA/TRG/PLA/Py-PLA: T_{d5} ↓2 °C; $T_{d\,max}$ ↑25 °C; (loadings not clear)	[154]
	Sonication in DMF; coagulation with water, vacuum drying, compression molding (200 °C, 3 min)	Graphene oxide nanosheets—GONSs (MHM) from expandable graphite (t = few layer, d = 5–20 μm)	GONSs: 0.25, 0.5, 1, 2	(0.25, 0.5, 1, 2 wt %); T_{m1} ↓1, 4, 0, 1 °C; T_{m2} ↓0, 1, 1, 1 °C; T_c ↓3, 6, 2, 4 °C; T_{di} ↑2, 6, 11, 16 °C	[156]
	Sonication in DMF; film casting, vacuum drying	GNS (commercial product) t = 5–25 nm, d = 0.5–20 μm, specific surface area = 50 m^2 g^{-1}	GNS: 1	Similar T_g and $T_{m1\ and\ 2}$; T_c ↑3 °C	[157]

Table 3. *Cont.*

Method	Procedure	GBM Characteristics	GBM Content (wt %)	Thermal Properties Relative to Neat Polymer	References
Melt blending	Internal mixer (160 °C, 25 rpm, 10 min), compression molding (160 °C, 10 min) (*Polymer was PLA/PEG 9:1 blend*)	GNP grade M15 (commercial product) t = 6–8 nm, d ≈ 15 μm	GNP-M15: 0.1, 0.3, 0.5, 0.7, 1	*(relative to pristine PLA/PEG blend)* (0.1, 0.3, 0.5, 1 wt %) T_g ↓0, 0, 1, 1; T_m ↑2, 4 ↓1, 1; T_c ↑1, 2, 2, 1; T_{di}, $T_{d\,max}$, T_{50} ↑56, 53, 44 °C (0.5 wt %)	[167]
	Melt ring-opening polymerization of L-lactide in presence of TRG (Sn(oct)$_2$, 170 °C, 4 h), filtration, vacuum drying	Natural graphite (MHM + lyophilization)—GO GO thermal reduction (1000 °C, 1 min) to TRG t = few layers	TRG: 0.01, 0.1, 0.5, 1, 1.5, 2	(0.01, 0.1, 0.5, 1, 1.5, 2 wt %) T_g = ↑9, 6, 6, 7, 8, 5 °C; T_m = ↑11, 12, 13, 14, 14 °C; $T_{d\,max}$ = ↑4, 13, 10, 11, 16, 18 °C	[176]
In situ polymerization	Sonication of L-lactide + filler in toluene, addition of Tin(II)-2-ethylhexanoate under N$_2$, stirring at 110 °C, 3 days	Expanded graphite (MHM) to GOGO-functionalized: GO + TDI + 1,4-butanediol, 80 °C, 24 h. GO-*g*-POSS: GO + POSS—polyhedral oligomeric silsesquioxane + DMAP—4-(dimethylaminopyridine) + EDC—N-(3-dimethylamino-propyl-N'-ethylcarbodiimide), 2 days, room temperature. N$_2$ (dimensions not given)	GO-functionalized, GO-*g*-POSS, GO+POSS (physical mixture): 1	PLA/GO-functionalized: T_{d5} ↑8, T_g ↓8, T_c ↑14, T_m ↓2 °C PLA/GO-*g*-POSS: T_{d5} ↑31, T_g ↑10, T_c ↑29, T_m ↑5 °C PLA/GO+POSS: T_{d5} ↑19, T_g ↑12, T_c ↑22, T_m ↑3 °C	[177]

5.4. Biological Properties

Most nanomaterials may present toxicity at concentrations above a certain threshold when in isolated form, i.e., when not incorporated in a polymer matrix [40,186]. Biocompatibility of the composites must be tested when considering uses as biomaterials. Table 4 shows that PLA/CBN composites (films and nanofibers) do not tend to decrease in vitro metabolic activity of several cell types, or cause increases up to 40% until 72 h incubations. Also, the selection of production method used (melt blending or solvent mixing followed by casting, doctor blading, spin coating or electrospinning), does not seem to influence cell proliferation. For long term incubations, McCullen et al. [187] shows that scaffolds of PLA with 1 wt % MWNTs do not to influence metabolic activity of adipose-derived human mesenchymal stem cells (hMSCs) at 7 days. At 14 days, cells present increased metabolic activity and longitudinal alignment induced by the scaffolds. Sherrell et al. [188] reports PLGA (1:1) with a surface layer of graphene applied by CVD to increase PC-12 cells average length of neurites by 2.5 fold when electrical stimulated. Also, hemocompatibility improvements are reported with both incorporation of 0.4 wt % GNP by solvent mixing followed by doctor blading [149] and 4 wt % MWCNT by extrusion followed by injection molding [189] in PLA. In the last case, MWCNT alignment is associated with decreased platelet adhesion and activation. Thus, alignment seems to be generally benefit for biocompatibility. The bioeffectiveness of electrical stimulation together with nanofibers and its fillers alignment is confirmed by Shao et al. [183], which cultures osteoblasts at the surface of PLA/MWCNT-ox (3 wt %) produced by solution mixing followed by electrospinning. They observe improvements in cell elongation (190%) and metabolic activity (20%) for random nanofibers (d ≈ 250 nm) under DC 100 μA, comparing to unstimulated controls. For aligned fibers the previous values increase by 90 and 40%, respectively. The aspect ratio is higher for the latter, comparing with random stimulated fibers (Figure 5). Finally, An et al. [190] find that PLA composite films and nanofibers with 3 wt % PU and 5 wt % GO almost completely suppress *Escherichia coli* and *Staphylococcus aureus* growth after 24 h, not affecting MC3T3-E1 cells metabolic activity. This effect is attributed to GO potentially inducing oxidative stress or physical disruption on bacteria.

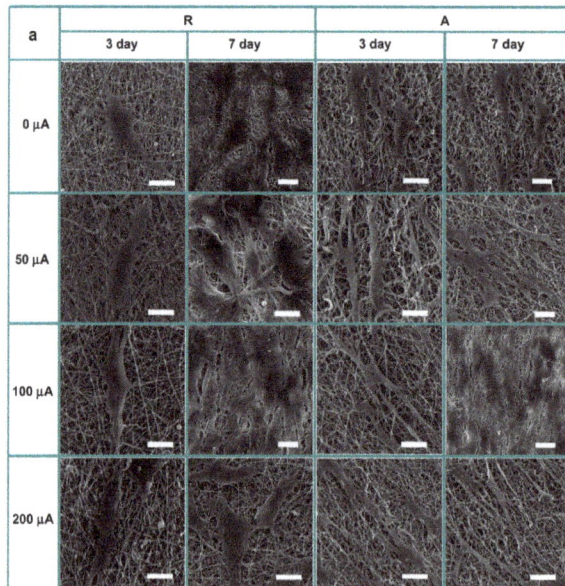

Figure 5. *Cont.*

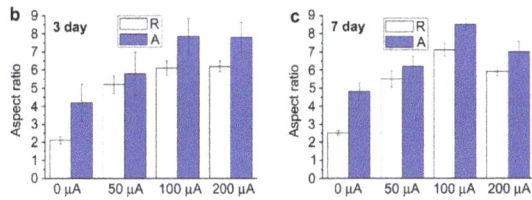

Figure 5. Scanning electron microscopy images of osteoblasts cultured on random (R) and aligned (A) nanofiber meshes of PLA/multi-walled carbon nanotubes (MWCNT)-ox 3 wt % produced by solution mixing followed by electrospinning, without or with electrical stimulation 0-200 μA (**a**); Osteoblast elongation is presented as the aspect ratio (**b,c**). Scale bars represent 30 μm [191].

In an in vivo study, Kanczler et al. [192] observe that PLA-CB 0.1 wt % scaffolds seeded or not with fetal femur-derived cells, when implanted in a murine critical-size femur segmental defect model aid the regeneration of bone defect. Pinto et al. [193] report both PLA/GNP-M5 (2 wt %) and CNT-COOH (0.3 and 0.7 wt %) to be biocompatible, both in vitro and in vivo (2 weeks subcutaneous implantation in C57Bl/6 mice). Also, PLA/GNP-M5 and C 0.25 wt % composites have not release toxic products after 6 months degradation in phosphate-buffered saline at 37 °C [180]. This is relevant considering that long-term biocompatibility must be assured for safe PLA/CBN composites implantation.

Table 4. Biological properties of PLA/CBN composites in comparison with non-modified PLA. Production methods and CBN characteristics.

Method	Procedure	CBN Characteristics	CBN Content (wt %)	Biocompatibility Properties	References
	GO—MHM Nanofibers (l = 11–14 μm) electrospinning	GO (thickness (t) = 1.5 nm, length (l) ≈ 1 μm)	PLGA (1:1)/GO 1 and 2 wt % nanofibers	Cell metabolic activity (MA): (PLGA = 100%, PLGA/GO 1 wt % ≈ 102%, PLGA/GO 2 wt % ≈ 108%, 48 h) (PC 12 cells)	[191]
	GO—MHM Films (t ≈ 5 μm)—spin coating	GO (not found)	PLGA (1:1)/GO films	Cell MA: Small increase (≈ 10%) comparing to PLGA for PLGA/GO 2 wt % (48 h) (Hela cells)	[179]
	GO—MHM Nanofibers (diameter (d) = 0.3–1.3 μm) electrospinning	GO (few layer)	PLA/HA(10 wt %)/GO nanofibers	Cell MA: 1, 2 and 5 wt % GO ↑, comparing to PLA/HA (24 h) Only nanofibers with 5 wt % GO presented higher MA than PLA/HA (48 h) (MC3T3-E1 cells)	[185]
	GO—MHM Films (t = 25–65 μm) solvent mixing + doctor blading	GO (d ≈ 500 nm)	PLA/GO films (0.4 wt %)	Cell MA: No variations until 48 h, except for PLA/GO after 24 h (more 13% than pristine PLA) (Mouse embryo fibroblasts 3T3)	[149]
	GNP—commercial product Films (t = 25–65 μm) solvent mixing + doctor blading	GNP-M5 (t ≈ 6–8 nm, l ≈ 5 μm)	PLA/GNP films (0.4 wt %)	Hemocompatibility: Less human platelets activated in PLA/GNP comparing with PLA in presence of plasma proteins	
Solution mixing	Graphene—CVD (chemical vapor deposition) Films (t = 25–65 μm) solvent casting over graphene	Graphene (t = 2 layers)	PLGA(1:1)/graphene surface layer	Cell MA: No significant changes until 4 days for PC-12 cells (rat adrenal gland pheochromocytoma) Cell differentiation: with electrical stimulation the average length of neurites increased 2.5-fold	[188]
	GO—MHM Films (dimensions not found)—solvent mixing + solvent casting Nanofibers (d ≈ 1 μm) electrospinning	GO (not found)	PLA/PU (3 wt %)/GO (5 wt %) films and nanofibers	Cell proliferation: not decreased (MC3T3-E1 cells) Antibacterial effect: E. coli and S. aureus growth 100% reduced at 24 h	[190]
	MWNTs—CVD Scaffolds (d = 0.7 μm, average porosity = 87%, void space = 89%)—electrospinning	MWNTs (l = 5–20 mm, d = 5–15 nm)	PLA/MWNTs (1 wt %) scaffolds	Cell MA: equal until day 7 and increased with MWNTs at day 14 (hMSCs) Cell morphology: MWNTs induced longitudinal alignment on cells at day 14	[187,189]
	MWCNT-ox (HCL 2 h at 25°C + HNO$_3$, 4 h at 110 °C) Nanofibers (MWCNT-ox sonicated in DMF 2h + SDS, adding to PLA in dicloromethane, 1h sonication before electrospinning) (PLA nanofibers, d ≈ 400 nm, PLA/MWCNT-ox nanofibers, d ≈ 250 nm)	MWCNT (l = 10–20 μm, d = 10–20 nm)	PLA/MWCNT-ox (3 wt %) random (R) and aligned (A) nanofibers	Cell MA: increased for osteoblasts at day 3 for PLA/MWCNT-ox (3 wt %) R—20% and A—40%, under DC = 100 μA Cell morphology: induced osteoblasts alignment at day 3 for PLA/MWCNT-ox (3 wt %) R—↑190% and A—↑90%, under DC = 100 μA	[183,187]

Table 4. *Cont.*

Method	Procedure	CBN Characteristics	CBN Content (wt %)	Biocompatibility Properties	References
Melt blending	MWNTs—CVD Composites (dimensions not found)—extrusion + injection moldingAligned composites—mechanical stretching at 90 °C	MWNTs (l = 10–30 mm, d = 20–40 nm)	PLA/MWNTs (5, 10, 15 wt %) composites	Hemolysis: bellow standard permissible (5%) in all cases, decreases with MWNTs incorporation and alignment Kinetic clothing time: increases with MWNTs incorporation and alignment (best was PLA/MWNTs 5 wt % which increased time by 480%) Platelet adhesion and activation: decreases with MWNTs incorporation and alignment	[183,189]
	GNP (commercial product) Composites (t ≈ 0.5 mm) Melt blending + compression molding	GNP-C (t = up to 2 single layers, l < 2 μm) GNP-M5 (t ≈ 6–8 nm, l ≈ 5 μm)	PLA/GNP-C and M5 (0.25 wt %) composites	Comparing with PLA: similar cell adhesion and growth at the surface No release of toxic products after 6 months degradation in phosphate-buffered saline at 37 °C	[180]
	GNP (commercial product) CNT-COOH—CVD, shortened, surface oxidized Composites (t ≈ 0.5 mm) Melt blending + compression molding	GNP-M5 (t ≈ 6–8 nm, l ≈ 5 μm) CNT-COOH (l < 1 μm, d = 9.5 nm, <8% COOH content)	PLA/GNP-M5 (2 wt %) PLA/CNT-COOH (0.3 and 0.7 wt %)	Biocompatible, both *in vitro* (human fibroblasts, HFF-1) and in vivo (2 weeks subcutaneous implantation in C57Bl/6 mice)	[193]
Laser sintering	CB (carbon black)—not found Scaffolds (several shapes)—surface selective laser sintering	(CB) Carbon black (d = 360 nm, surface area = 100 m² g⁻¹)	SSLS-PLA/CB 0.1 wt % scaffolds	SSLS-PLA/CB 0.1 wt % scaffolds seeded or not with fetal femur-derived cells aided regeneration of murine bone defect	[192]

6. Conclusions

Both CNT and GBM nanofillers are effective at improving PLA thermo-mechanical and electrical properties. However, lower amounts of GBM (0.1–1 wt %) are usually needed when comparing with CNT (0.25–5 wt %). Melt-blending is less reported than solution mixing for production of PLA/CBN composites, maybe because it implies use of specialized equipment. Moreover, results show that melt blending suffers from some drawbacks, since viscous shear is less effective than solvent sonication for promoting exfoliation/deagglomeration of CBN. In situ polymerization is the least reported technique, with further research being needed to demonstrate its advantages over the previous production methods.

Surface modifications of CBN can be used to improve compatibility with a polymer matrix. Functionalization with carboxyls is the most common and effective procedure to improve CNT dispersibility and compatibility with PLA. Some authors refer that purification with strong acids introduces polar groups in the carbon surface, which results in positive interaction with PLA. Besides straightforward chemical oxidation of CBN, other chemical modifications which lead to better performance after incorporation in PLA, comparing with non-modified CBN, include reaction with isocyanates, polyols, or silanes, and grafting with polymers (ethylene glycol, poly(caprolactone), poly(methyl methacrylate), poly(vinyl pyrrolidone), and PLA).

When comparing reduced and oxidized forms of GBM as PLA fillers, like rGO and GO, only in the case of increasing electrical conductivity the reduced forms show clearly better performance.

Based on the available data, no relation can be determined between CBN morphological properties (size, length, and diameter) and the composites performances.

The alignment of PLA/CNT fibers, has been shown to improve electrical conductivity. Electrical properties also improve with the increase of the amount of CBN incorporated.

Concerning biological properties, the composite production process does not influence cell metabolic activity, which does not decrease comparing to non-filled PLA. Furthermore, increases up to 40% in cell viability can be induced by GBM incorporation. Improvements in hemocompatibility are achieved with incorporation of both CNT and GBM. Also, both fiber/filler alignment and electrical stimulation, improve cell metabolic activity and elongation. Short term in vivo studies reveal PLA/CBN composites to be biocompatible, and no release of toxic degradation products is found up to 6 months in vitro degradation of PLA/GBM composites. Incorporation of GO has lead to suppression of *Escherichia coli* and *Staphylococcus aureus* growth, without compromising the composite biocompatibility. However, there is still no information on antimicrobial activity of these composites on other types of microorganisms or with other types of GBM. Also, long-term in vivo biocompatibility of PLA/CBN composites needs to be assured prior to their clinical use.

Some other relevant topics for future research include obtaining a better understanding of how the fillers physico-chemical properties, and their alignment inside the polymer matrix, affect the composites properties. In situ polymerization of PLA in presence of CBN is a not well developed topic, being worthwhile of further exploration due to the potential for optimization of the degree of interaction and dispersion of CBN in the polymer matrix. Mechanical milling is an increasingly interesting technique for mixing filler nanoparticles with a polymer matrix, but has not yet been reported for producing PLA/CBN composites. This is expected to change in the near future. Finally, emerging technologies, like 3D printing, will surely contribute to the conception of materials appropriate for the broad potential applications of PLA/CBN composites.

Acknowledgments: This work was financially supported by: Project POCI-01-0145-FEDER-006939 (Laboratory for Process Engineering, Environment, Biotechnology and Energy—LEPABE), Project POCI-01-0145-FEDER-007274 (Institute for Research and Innovation in Health Sciences), and Project PTDC/CTM-BIO/4033/2014 (NewCat), funded by FEDER funds through COMPETE2020—Programa Operacional Competitividade e Internacionalização (POCI)—and by national funds through FCT—Fundação para a Ciência e a Tecnologia; PhD grant SFRH/BD/86974/2012, funded by European Social Fund and Portuguese Ministry of Education and Science (MEC) through Programa Operacional Capital Humano (POCH).

Author Contributions: Artur M. Pinto and Carolina Gonçalves have compiled the literature and written the text. Inês C. Gonçalves and Fernão D. Magalhães have revised the text and made suggestions concerning its structure and contents.

Conflicts of Interest: The authors declare no conflict of interest.

References

1. Oksman, K.; Skrifvars, M.; Selin, J.F. Natural fibers as reinforcement in polylactic acid (PLA) composites. *Compos. Sci. Technol.* **2003**, *63*, 1317–1324. [CrossRef]
2. Vaia, R.A.; Wagner, H.D. Framework for nanocomposites. *Mater. Today* **2004**, *7*, 32–37. [CrossRef]
3. Lasprilla, A.J.R.; Martinez, G.A.R.; Lunelli, B.H.; Jardini, A.L.; Maciel, R. Poly-lactic acid synthesis for application in biomedical devices—A review. *Biotechnol. Adv.* **2012**, *30*, 321–328. [CrossRef] [PubMed]
4. Vieira, A.C.; Vieira, J.C.; Ferra, J.M.; Magalhaes, F.D.; Guedes, R.M.; Marques, A.T. Mechanical study of PLA-PCL fibers during in vitro degradation. *J. Mech. Behav. Biomed.* **2011**, *4*, 451–460. [CrossRef] [PubMed]
5. Chang, J.H.; An, Y.U.; Sur, G.S. Poly(lactic acid) nanocomposites with various organoclays. I. Thermomechanical properties, morphology, and gas permeability. *J. Polym. Sci.* **2003**, *41*, 94–103. [CrossRef]
6. Mittal, V. Polymer layered silicate nanocomposites: A review. *Materials* **2009**, *2*, 992–1057. [CrossRef]
7. Raquez, J.M.; Habibi, Y.; Murariu, M.; Dubois, P. Polylactide (PLA)-based nanocomposites. *Prog. Polym. Sci.* **2013**, *38*, 1504–1542. [CrossRef]
8. Bafekrpour, E.; Salehi, M.; Sonbolestan, E.; Fox, B. Effects of micro-structural parameters on mechanical properties of carbon nanotube polymer nanocomposites. *Sci. Iran.* **2014**, *21*, 403–413.
9. Coleman, J.N.; Khan, U.; Gun'ko, Y.K. Mechanical reinforcement of polymers using carbon nanotubes. *Adv. Mater.* **2006**, *18*, 689–706. [CrossRef]
10. Fiedler, B.; Gojny, F.H.; Wichmann, M.H.G.; Nolte, M.C.M.; Schulte, K. Fundamental aspects of nano-reinforced composites. *Compos. Sci. Technol.* **2006**, *66*, 3115–3125. [CrossRef]
11. Pinto, A.M.; Goncalves, I.C.; Magalhaes, F.D. Graphene-based materials biocompatibility: A review. *Colloid Surf. B* **2013**, *111*, 188–202. [CrossRef] [PubMed]
12. Tjong, S.C. Structural and mechanical properties of polymer nanocomposites. *Mater. Sci. Eng. R* **2006**, *53*, 73–197. [CrossRef]
13. Xie, X.L.; Mai, Y.W.; Zhou, X.P. Dispersion and alignment of carbon nanotubes in polymer matrix: A review. *Mater. Sci. Eng. R* **2005**, *49*, 89–112. [CrossRef]
14. Pinto, A.M.; Martins, J.; Moreira, J.A.; Mendes, A.M.; Magalhaes, F.D. Dispersion of graphene nanoplatelets in poly(vinyl acetate) latex and effect on adhesive bond strength. *Polym. Int.* **2013**, *62*, 928–935. [CrossRef]
15. Pinto, A.M.; Goncalves, C.; Sousa, D.M.; Ferreira, A.R.; Moreira, J.A.; Goncalves, I.C.; Magalhaes, F.D. Smaller particle size and higher oxidation improves biocompatibility of graphene-based materials. *Carbon* **2016**, *99*, 318–329. [CrossRef]
16. Pinto, A.M.; Moreira, J.A.; Magalhaes, F.D.; Goncalves, I.C. Polymer surface adsorption as a strategy to improve the biocompatibility of graphene nanoplatelets. *Colloid Surf. B* **2016**, *146*, 818–824. [CrossRef] [PubMed]
17. Ge, C.C.; Li, Y.; Yin, J.J.; Liu, Y.; Wang, L.M.; Zhao, Y.L.; Chen, C.Y. The contributions of metal impurities and tube structure to the toxicity of carbon nanotube materials. *NPG Asia Mater.* **2012**, *4*, e32. [CrossRef]
18. Liu, X.; Guo, L.; Morris, D.; Kane, A.B.; Hurt, R.H. Targeted removal of bioavailable metal as a detoxification strategy for carbon nanotubes. *Carbon* **2008**, *46*, 489–500. [CrossRef] [PubMed]
19. Liu, Y.; Zhao, Y.L.; Sun, B.Y.; Chen, C.Y. Understanding the toxicity of carbon nanotubes. *Acc. Chem. Res.* **2013**, *46*, 702–713. [CrossRef] [PubMed]
20. Tejral, G.; Panyala, N.R.; Havel, J. Carbon nanotubes: Toxicological impact on human health and environment. *J. Appl. Biomed.* **2009**, *7*, 1–13.
21. Das, T.K.; Prusty, S. Graphene-based polymer composites and their applications. *Polym. Plast. Technol.* **2013**, *52*, 319–331. [CrossRef]
22. Huang, X.; Qi, X.Y.; Boey, F.; Zhang, H. Graphene-based composites. *Chem. Soc. Rev.* **2012**, *41*, 666–686. [CrossRef] [PubMed]
23. Kotov, N.A. Materials science: Carbon sheet solutions. *Nature* **2006**, *442*, 254–255. [CrossRef] [PubMed]

24. Kuilla, T.; Bhadra, S.; Yao, D.H.; Kim, N.H.; Bose, S.; Lee, J.H. Recent advances in graphene based polymer composites. *Prog. Polym. Sci.* **2010**, *35*, 1350–1375. [CrossRef]

25. Gupta, B.; Revagade, N.; Hilborn, J. Poly(lactic acid) fiber: An overview. *Prog. Polym. Sci.* **2007**, *32*, 455–482. [CrossRef]

26. Hu, Y.Z.; Daoud, W.A.; Cheuk, K.K.L.; Lin, C.S.K. Newly developed techniques on polycondensation, ring-opening polymerization and polymer modification: Focus on poly(lactic acid). *Materials* **2016**, *9*, 133. [CrossRef]

27. Lim, L.T.; Auras, R.; Rubino, M. Processing technologies for poly(lactic acid). *Prog. Polym. Sci.* **2008**, *33*, 820–852. [CrossRef]

28. Nampoothiri, K.M.; Nair, N.R.; John, R.P. An overview of the recent developments in polylactide (PLA) research. *Bioresour. Technol.* **2010**, *101*, 8493–8501. [CrossRef] [PubMed]

29. Rasal, R.M.; Janorkar, A.V.; Hirt, D.E. Poly(lactic acid) modifications. *Prog. Polym. Sci.* **2010**, *35*, 338–356. [CrossRef]

30. Pang, X.A.; Zhuang, X.L.; Tang, Z.H.; Chen, X.S. Polylactic acid (PLA): Research, development and industrialization. *Biotechnol. J.* **2010**, *5*, 1125–1136. [CrossRef] [PubMed]

31. Allen, M.J.; Tung, V.C.; Kaner, R.B. Honeycomb carbon: A review of graphene. *Chem. Rev.* **2010**, *110*, 132–145. [CrossRef] [PubMed]

32. Balandin, A.A. Thermal properties of graphene and nanostructured carbon materials. *Nat. Mater.* **2011**, *10*, 569–581. [CrossRef] [PubMed]

33. Baughman, R.H.; Zakhidov, A.A.; de Heer, W.A. Carbon nanotubes—The route toward applications. *Science* **2002**, *297*, 787–792. [CrossRef] [PubMed]

34. Dreyer, D.R.; Park, S.; Bielawski, C.W.; Ruoff, R.S. The chemistry of graphene oxide. *Chem. Soc. Rev.* **2010**, *39*, 228–240. [CrossRef] [PubMed]

35. Lin, X.H.; Gai, J.G. Synthesis and applications of large-area single-layer graphene. *RSC Adv.* **2016**, *6*, 17818–17844. [CrossRef]

36. Nguyen, V.H. Recent advances in experimental basic research on graphene and graphene-based nanostructures. *Adv. Nat. Sci. Nanosci.* **2016**, *7*. [CrossRef]

37. Novoselov, K.S.; Fal'ko, V.I.; Colombo, L.; Gellert, P.R.; Schwab, M.G.; Kim, K. A roadmap for graphene. *Nature* **2012**, *490*, 192–200. [CrossRef] [PubMed]

38. Park, S.; Ruoff, R.S. Chemical methods for the production of graphenes. *Nat. Nanotechnol.* **2009**, *4*, 217–224. [CrossRef] [PubMed]

39. Rao, C.N.R.; Sood, A.K.; Subrahmanyam, K.S.; Govindaraj, A. Graphene: The new two-dimensional nanomaterial. *Angew. Chem. Int. Ed.* **2009**, *48*, 7752–7777. [CrossRef] [PubMed]

40. Singh, V.; Joung, D.; Zhai, L.; Das, S.; Khondaker, S.I.; Seal, S. Graphene based materials: Past, present and future. *Prog. Mater. Sci.* **2011**, *56*, 1178–1271. [CrossRef]

41. Tasis, D.; Tagmatarchis, N.; Bianco, A.; Prato, M. Chemistry of carbon nanotubes. *Chem. Rev.* **2006**, *106*, 1105–1136. [CrossRef] [PubMed]

42. Thostenson, E.T.; Ren, Z.F.; Chou, T.W. Advances in the science and technology of carbon nanotubes and their composites: A review. *Compos. Sci. Technol.* **2001**, *61*, 1899–1912. [CrossRef]

43. Yang, F.; Wang, X.; Li, M.H.; Liu, X.Y.; Zhao, X.L.; Zhang, D.Q.; Zhang, Y.; Yang, J.; Li, Y. Templated synthesis of single-walled carbon nanotubes with specific structure. *Acc. Chem. Res.* **2016**, *49*, 606–615. [CrossRef] [PubMed]

44. Yeung, C.S.; Tian, W.Q.; Liu, L.V.; Wang, Y.A. Chemistry of single-walled carbon nanotubes. *J. Comput. Theor. Nanosci.* **2009**, *6*, 1213–1235. [CrossRef]

45. Zhang, F.; Hou, P.X.; Liu, C.; Cheng, H.M. Epitaxial growth of single-wall carbon nanotubes. *Carbon* **2016**, *102*, 181–197. [CrossRef]

46. Zhu, Y.W.; Murali, S.; Cai, W.W.; Li, X.S.; Suk, J.W.; Potts, J.R.; Ruoff, R.S. Graphene and graphene oxide: Synthesis, properties, and applications. *Adv. Mater.* **2010**, *22*, 3906–3924. [CrossRef] [PubMed]

47. Datta, R.; Henry, M. Lactic acid: Recent advances in products, processes and technologies—A review. *J. Chem. Technol. Biotechnol.* **2006**, *81*, 1119–1129. [CrossRef]

48. Jamshidian, M.; Tehrany, E.A.; Imran, M.; Jacquot, M.; Desobry, S. Poly-lactic acid: Production, applications, nanocomposites, and release studies. *Compr. Rev. Food Sci. Food Saf.* **2010**, *9*, 552–571. [CrossRef]

49. Garlotta, D. A literature review of poly(lactic acid). *J. Polym. Environ.* **2001**, *9*, 63–84. [CrossRef]

50. Auras, R.; Lim, L.-T.; Selke, S.E.M.; Tsuji, H. *Poly (lactic) acid: Synthesis, Structures, Properties, Processing, and Applications*; John Wiley and Sons: San Francisco, CA, USA, 2010.

51. Drumright, R.E.; Gruber, P.R.; Henton, D.E. Polylactic acid technology. *Adv. Mater.* **2000**, *12*, 1841–1846. [CrossRef]

52. Shogren, R.L.; Doane, W.M.; Garlotta, D.; Lawton, J.W.; Willett, J.L. Biodegradation of starch/polylactic acid/poly(hydroxyester-ether) composite bars in soil. *Polym. Degrad. Stab.* **2003**, *79*, 405–411. [CrossRef]

53. Zeng, J.B.; Li, K.A.; Du, A.K. Compatibilization strategies in poly(lactic acid)-based blends. *RSC Adv.* **2015**, *5*, 32546–32565. [CrossRef]

54. Semba, T.; Kitagawa, K.; Ishiaku, U.S.; Hamada, H. The effect of crosslinking on the mechanical properties of polylactic acid/polycaprolactone blends. *J. Appl. Polym. Sci.* **2006**, *101*, 1816–1825. [CrossRef]

55. Wang, H.; Sun, X.Z.; Seib, P. Mechanical properties of poly(lactic acid) and wheat starch blends with methylenediphenyl diisocyanate. *J. Appl. Polym. Sci.* **2002**, *84*, 1257–1262. [CrossRef]

56. Balakrishnan, H.; Hassan, A.; Wahit, M.U.; Yussuf, A.A.; Razak, S.B.A. Novel toughened polylactic acid nanocomposite: Mechanical, thermal and morphological properties. *Mater. Des.* **2010**, *31*, 3289–3298. [CrossRef]

57. Broz, M.E.; VanderHart, D.L.; Washburn, N.R. Structure and mechanical properties of poly(D,L-lactic acid)/poly(epsilon-caprolactone) blends. *Biomaterials* **2003**, *24*, 4181–4190. [CrossRef]

58. Abdelwahab, M.A.; Flynn, A.; Chiou, B.S.; Imam, S.; Orts, W.; Chiellini, E. Thermal, mechanical and morphological characterization of plasticized pla-phb blends. *Polym. Degrad. Stab.* **2012**, *97*, 1822–1828. [CrossRef]

59. Yew, G.H.; Yusof, A.M.M.; Ishak, Z.A.M.; Ishiaku, U.S. Water absorption and enzymatic degradation of poly(lactic acid)/rice starch composites. *Polym. Degrad. Stab.* **2005**, *90*, 488–500. [CrossRef]

60. Pellis, A.; Acero, E.H.; Ferrario, V.; Ribitsch, D.; Guebitz, G.M.; Gardossi, L. The closure of the cycle: Enzymatic synthesis and functionalization of bio-based polyesters. *Trends Biotechnol.* **2016**, *34*, 316–328. [CrossRef] [PubMed]

61. Gumel, A.M.; Annuar, M.S.M.; Heidelberg, T. Current application of controlled degradation processes in polymer modification and functionalization. *J. Appl. Polym. Sci.* **2013**, *129*, 3079–3088. [CrossRef]

62. Hoveizi, E.; Nabiuni, M.; Parivar, K.; Rajabi-Zeleti, S.; Tavakol, S. Functionalisation and surface modification of electrospun polylactic acid scaffold for tissue engineering. *Cell Biol. Int.* **2014**, *38*, 41–49. [CrossRef] [PubMed]

63. Kucharczyk, P.; Poljansek, I.; Sedlarik, V.; Kasparkova, V.; Salakova, A.; Drbohlav, J.; Cvelbar, U.; Saha, P. Functionalization of polylactic acid through direct melt polycondensation in the presence of tricarboxylic acid. *J. Appl. Polym. Sci.* **2011**, *122*, 1275–1285. [CrossRef]

64. Yuan, X.B.; Kang, C.S.; Zhao, Y.H.; Gu, M.Q.; Pu, P.Y.; Tian, N.J.; Sheng, J. Surface multi-functionalization of poly(lactic acid) nanoparticles and c6 glioma cell targeting in vivo. *Chin. J. Polym. Sci.* **2009**, *27*, 231–239. [CrossRef]

65. Iwatake, A.; Nogi, M.; Yano, H. Cellulose nanofiber-reinforced polylactic acid. *Compos. Sci. Technol.* **2008**, *68*, 2103–2106. [CrossRef]

66. Weiss, J.; McClements, D.J.; Takhistov, P. Functional materials in food nanotechnology. *J. Food Sci.* **2007**, *59*, 274–275. [CrossRef]

67. Oksman, K.; Mathew, A.P.; Bondeson, D.; Kvien, I. Manufacturing process of cellulose whiskers/polylactic acid nanocomposites. *Compos. Sci. Technol.* **2006**, *66*, 2776–2784. [CrossRef]

68. Ray, S.S.; Maiti, P.; Okamoto, M.; Yamada, K.; Ueda, K. New polylactide/layered silicate nanocomposites. 1. Preparation, characterization, and properties. *Macromolecules* **2002**, *35*, 3104–3110.

69. Ray, S.S.; Yamada, K.; Okamoto, M.; Ogami, A.; Ueda, K. New polylactide/layered silicate nanocomposites. 3. High-performance biodegradable materials. *Chem. Mater.* **2003**, *15*, 1456–1465.

70. Yu, L.; Dean, K.; Li, L. Polymer blends and composites from renewable resources. *Prog. Polym. Sci.* **2006**, *31*, 576–602. [CrossRef]

71. Das, K.; Ray, D.; Banerjee, I.; Bandyopadhyay, N.R.; Sengupta, S.; Mohanty, A.K.; Misra, M. Crystalline morphology of PLA/clay nanocomposite films and its correlation with other properties. *J. Appl. Polym. Sci.* **2010**, *118*, 143–151. [CrossRef]

72. Bitinis, N.; Sanz, A.; Nogales, A.; Verdejo, R.; Lopez-Manchado, M.A.; Ezquerra, T.A. Deformation mechanisms in polylactic acid/natural rubber/organoclay bionanocomposites as revealed by synchrotron X-ray scattering. *Soft Matter* **2012**, *8*, 8990–8997. [CrossRef]

73. Nofar, M.; Tabatabaei, A.; Park, C.B. Effects of nano-/micro-sized additives on the crystallization behaviors of pla and pla/CO_2 mixtures. *Polymer* **2013**, *54*, 2382–2391. [CrossRef]

74. Keshtkar, M.; Nofar, M.; Park, C.B.; Carreau, P.J. Extruded PLA/clay nanocomposite foams blown with supercritical CO_2. *Polymer* **2014**, *55*, 4077–4090. [CrossRef]

75. Ayana, B.; Suin, S.; Khatua, B.B. Highly exfoliated eco-friendly thermoplastic starch (TPS)/poly(lactic acid)(PLA)/clay nanocomposites using unmodified nanoclay. *Carbohydr. Polym.* **2014**, *110*, 430–439.

76. Singh, S.; Ghosh, A.K.; Maiti, S.N.; Raha, S.; Gupta, R.K.; Bhattacharya, S. Morphology and rheological behavior of polylactic acid/clay nanocomposites. *Polym. Eng. Sci.* **2012**, *52*, 225–232. [CrossRef]

77. Hapuarachchi, T.D.; Peijs, T. Multiwalled carbon nanotubes and sepiolite nanoclays as flame retardants for polylactide and its natural fiber reinforced composites. *Compos. Part A* **2010**, *41*, 954–963. [CrossRef]

78. Busolo, M.A.; Fernandez, P.; Ocio, M.J.; Lagaron, J.M. Novel silver-based nanoclay as an antimicrobial in polylactic acid food packaging coatings. *Food Addit. Contam.* **2010**, *27*, 1617–1626. [CrossRef] [PubMed]

79. Meng, Q.K.; Hetzer, M.; De Kee, D. Pla/clay/wood nanocomposites: Nanoclay effects on mechanical and thermal properties. *J. Compos. Mater.* **2011**, *45*, 1145–1158. [CrossRef]

80. As'habi, L.; Jafari, S.H.; Khonakdar, H.A.; Boldt, R.; Wagenknecht, U.; Heinrich, G. Tuning the processability, morphology and biodegradability of clay incorporated PLA/LLDPE blends via selective localization of nanoclay induced by melt mixing sequence. *Express Polym. Lett.* **2013**, *7*, 21–39. [CrossRef]

81. Lai, S.M.; Hsieh, Y.T. Preparation and properties of polylactic acid (PLA)/silica nanocomposites. *J. Macromol. Sci. B* **2016**, *55*, 211–228. [CrossRef]

82. Basilissi, L.; Di Silvestro, G.; Farina, H.; Ortenzi, M.A. Synthesis and characterization of pla nanocomposites containing nanosilica modified with different organosilanes II: Effect of the organosilanes on the properties of nanocomposites: Thermal characterization. *J. Appl. Polym. Sci.* **2013**, *128*, 3057–3063. [CrossRef]

83. Mooney, E.; Mackle, J.N.; Blond, D.J.P.; O'Cearbhaill, E.; Shaw, G.; Blau, W.J.; Barry, F.P.; Barron, V.; Murphy, J.M. The electrical stimulation of carbon nanotubes to provide a cardiomimetic cue to mscs. *Biomaterials* **2012**, *33*, 6132–6139. [CrossRef] [PubMed]

84. Obarzanek-Fojt, M.; Elbs-Glatz, Y.; Lizundia, E.; Diener, L.; Sarasua, J.R.; Bruinink, A. From implantation to degradation—Are poly (L-lactide)/multiwall carbon nanotube composite materials really cytocompatible? *Nanomed. Nanotechnol.* **2014**, *10*, 1041–1051. [CrossRef] [PubMed]

85. Gorrasi, G.; Milone, C.; Piperopoulos, E.; Lanza, M.; Sorrentino, A. Hybrid clay mineral-carbon nanotube-PLA nanocomposite films. Preparation and photodegradation effect on their mechanical, thermal and electrical properties. *Appl. Clay Sci.* **2013**, *71*, 49–54. [CrossRef]

86. Supronowicz, P.R.; Ajayan, P.M.; Ullmann, K.R.; Arulanandam, B.P.; Metzger, D.W.; Bizios, R. Novel current-conducting composite substrates for exposing osteoblasts to alternating current stimulation. *J. Biomed. Mater. Res.* **2002**, *59*, 499–506. [CrossRef] [PubMed]

87. Kumar, B.; Castro, M.; Feller, J.F. Poly(lactic acid)-multi-wall carbon nanotube conductive biopolymer nanocomposite vapour sensors. *Sens. Actuators B* **2012**, *161*, 621–628. [CrossRef]

88. Chiu, W.M.; Chang, Y.A.; Kuo, H.Y.; Lin, M.H.; Wen, H.C. A study of carbon nanotubes/biodegradable plastic polylactic acid composites. *J. Appl. Polym. Sci.* **2008**, *108*, 3024–3030. [CrossRef]

89. Novoselov, K.S.; Jiang, D.; Schedin, F.; Booth, T.J.; Khotkevich, V.V.; Morozov, S.V.; Geim, A.K. Two-dimensional atomic crystals. *Proc. Natl. Acad. Sci. USA* **2005**, *102*, 10451–10453. [CrossRef] [PubMed]

90. Avouris, P.; Dimitrakopoulos, C. Graphene: Synthesis and applications. *Mater. Today* **2012**, *15*, 86–97. [CrossRef]

91. Schwierz, F. Electronics industry-compatible graphene transistors. *Nature* **2011**, *472*, 41–42. [CrossRef] [PubMed]

92. Schwierz, F. Graphene transistors. *Nat. Nanotechnol.* **2010**, *5*, 487–496. [CrossRef] [PubMed]

93. Schwierz, F. Graphene transistors: Status, prospects, and problems. *Proc. IEEE* **2013**, *101*, 1567–1584. [CrossRef]

94. Avouris, P.; Chen, Z.H.; Perebeinos, V. Carbon-based electronics. *Nat. Nanotechnol.* **2007**, *2*, 605–615. [CrossRef] [PubMed]

95. Bao, Q.L.; Loh, K.P. Graphene photonics, plasmonics, and broadband optoelectronic devices. *ACS Nano* **2012**, *6*, 3677–3694. [CrossRef] [PubMed]

96. Wang, H.L.; Liang, Y.Y.; Sanchez, H.; Yang, Y.; Cui, L.F.; Cui, Y.; Dai, H.J. Graphene-based hybrid nanomaterials for energy storage applications. *Abstr. Pap. Am. Chem. S* **2011**, *241*, 2983–2994.

97. Pumera, M. Graphene-based nanomaterials for energy storage. *Energy Environ. Sci.* **2011**, *4*, 668–674. [CrossRef]

98. Radovic, L.R.; Mora-Vilches, C.; Salgado-Casanova, A.J.A. Catalysis: An old but new challenge for graphene-based materials. *Chin. J. Catal.* **2014**, *35*, 792–797. [CrossRef]

99. Machado, B.F.; Serp, P. Graphene-based materials for catalysis. *Catal. Sci. Technol.* **2012**, *2*, 54–75. [CrossRef]

100. Dikin, D.A.; Stankovich, S.; Zimney, E.J.; Piner, R.D.; Dommett, G.H.B.; Evmenenko, G.; Nguyen, S.T.; Ruoff, R.S. Preparation and characterization of graphene oxide paper. *Nature* **2007**, *448*, 457–460. [CrossRef] [PubMed]

101. Bunch, J.S.; Verbridge, S.S.; Alden, J.S.; van der Zande, A.M.; Parpia, J.M.; Craighead, H.G.; McEuen, P.L. Impermeable atomic membranes from graphene sheets. *Nano Lett.* **2008**, *8*, 2458–2462. [CrossRef] [PubMed]

102. Katsnelson, M.I. Graphene: Carbon in two dimensions. *Mater. Today* **2007**, *10*, 20–27. [CrossRef]

103. Cui, Y.B.; Kundalwal, S.I.; Kumar, S. Gas barrier performance of graphene/polymer nanocomposites. *Carbon* **2016**, *98*, 313–333. [CrossRef]

104. Kim, H.; Abdala, A.A.; Macosko, C.W. Graphene/polymer nanocomposites. *Macromolecules* **2010**, *43*, 6515–6530. [CrossRef]

105. Feng, L.Z.; Liu, Z.A. Graphene in biomedicine: Opportunities and challenges. *Nanomedicine* **2011**, *6*, 317–324. [CrossRef] [PubMed]

106. Lu, C.H.; Yang, H.H.; Zhu, C.L.; Chen, X.; Chen, G.N. A graphene platform for sensing biomolecules. *Angew. Chem. Int. Ed.* **2009**, *48*, 4785–4787. [CrossRef] [PubMed]

107. Kuila, T.; Bose, S.; Khanra, P.; Mishra, A.K.; Kim, N.H.; Lee, J.H. Recent advances in graphene-based biosensors. *Biosens. Bioelectron.* **2011**, *26*, 4637–4648. [CrossRef] [PubMed]

108. Shao, J.J.; Zheng, D.Y.; Li, Z.J.; Yang, Q.H. Top-down fabrication of two-dimensional nanomaterials: Controllable liquid phase exfoliation. *New Carbon Mater.* **2016**, *31*, 97–114.

109. Tang, L.B.; Li, X.M.; Ji, R.B.; Teng, K.S.; Tai, G.; Ye, J.; Wei, C.S.; Lau, S.P. Bottom-up synthesis of large-scale graphene oxide nanosheets. *J. Mater. Chem.* **2012**, *22*, 5676–5683. [CrossRef]

110. Zhang, Y.; Zhang, L.Y.; Zhou, C.W. Review of chemical vapor deposition of graphene and related applications. *Acc. Chem. Res.* **2013**, *46*, 2329–2339. [CrossRef] [PubMed]

111. Edwards, R.S.; Coleman, K.S. Graphene synthesis: Relationship to applications. *Nanoscale* **2013**, *5*, 38–51. [CrossRef] [PubMed]

112. Ciesielski, A.; Samori, P. Graphene via sonication assisted liquid-phase exfoliation. *Chem. Soc. Rev.* **2014**, *43*, 381–398. [CrossRef] [PubMed]

113. Vivekanand Prajapati, P.K.S. Arunabha Banik. Carbon nanotubes and its applications. *Int. J. Pharm. Sci. Res.* **2010**, *3*, 1099–1107.

114. Terrones, M. Science and technology of the twenty-first century: Synthesis, properties and applications of carbon nanotubes. *Annu. Rev. Mater. Res.* **2003**, *33*, 419–501. [CrossRef]

115. De Volder, M.F.L.; Tawfick, S.H.; Baughman, R.H.; Hart, A.J. Carbon nanotubes: Present and future commercial applications. *Science* **2013**, *339*, 535–539. [CrossRef] [PubMed]

116. Mamalis, A.G.; Voglander, L.O.G.; Markopoulos, A. Nanotechnology and nanostructured materials: Trends in carbon nanotubes. *Precis. Eng.* **2004**, *28*, 16–30. [CrossRef]

117. Zhang, Q.; Huang, J.Q.; Zhao, M.Q.; Qian, W.Z.; Wei, F. Carbon nanotube mass production: Principles and processes. *Chemsuschem* **2011**, *4*, 864–889. [CrossRef] [PubMed]

118. Aqel, A.; Abou El-Nour, K.M.M.; Ammar, R.A.A.; Al-Warthan, A. Carbon nanotubes, science and technology part (i) structure, synthesis and characterisation. *Arab. J. Chem.* **2012**, *5*, 1–23. [CrossRef]

119. Kumar, M.; Ando, Y. Chemical vapor deposition of carbon nanotubes: A review on growth mechanism and mass production. *J. Nanosci. Nanotechnol.* **2010**, *10*, 3739–3758. [CrossRef] [PubMed]

120. Chen, X.P.; Zhang, L.L.; Chen, S.S. Large area cvd growth of graphene. *Synth. Met.* **2015**, *210*, 95–108. [CrossRef]

121. Saito, N.; Usui, Y.; Aoki, K.; Narita, N.; Shimizu, M.; Hara, K.; Ogiwara, N.; Nakamura, K.; Ishigaki, N.; Kato, H.; et al. Carbon nanotubes: Biomaterial applications. *Chem. Soc. Rev.* **2009**, *38*, 1897–1903. [CrossRef] [PubMed]

122. Dalton, A.B.; Collins, S.; Razal, J.; Munoz, E.; Ebron, V.H.; Kim, B.G.; Coleman, J.N.; Ferraris, J.P.; Baughman, R.H. Continuous carbon nanotube composite fibers: Properties, potential applications, and problems. *J. Mater. Chem.* **2004**, *14*, 1–3. [CrossRef]

123. Wang, J. Carbon-nanotube based electrochemical biosensors: A review. *Electroanal* **2005**, *17*, 7–14. [CrossRef]

124. Darkrim, F.L.; Malbrunot, P.; Tartaglia, G.P. Review of hydrogen storage by adsorption in carbon nanotubes. *Int. J. Hydrogen Energy* **2002**, *27*, 193–202. [CrossRef]

125. Bonard, J.M.; Kind, H.; Stockli, T.; Nilsson, L.A. Field emission from carbon nanotubes: The first five years. *Solid State Electron.* **2001**, *45*, 893–914. [CrossRef]

126. Paradise, M.; Goswami, T. Carbon nanotubes—Production and industrial applications. *Mater. Des.* **2007**, *28*, 1477–1489. [CrossRef]

127. Malarkey, E.B.; Parpura, V. Applications of carbon nanotubes in neurobiology. *Neurodegener. Dis.* **2007**, *4*, 292–299. [CrossRef] [PubMed]

128. Lacerda, L.; Bianco, A.; Prato, M.; Kostarelos, K. Carbon nanotubes as nanomedicines: From toxicology to pharmacology. *Adv. Drug Deliv. Rev.* **2006**, *58*, 1460–1470. [CrossRef] [PubMed]

129. Riehemann, K. Nanotoxicity: How the body develops a way to reduce the toxicity of carbon nanotubes. *Small* **2012**, *8*, 1970–1972. [CrossRef] [PubMed]

130. Ren, W.C.; Cheng, H.M. The global growth of graphene. *Nat. Nanotechnol.* **2014**, *9*, 726–730. [CrossRef] [PubMed]

131. Si, Y.; Samulski, E.T. Synthesis of water soluble graphene. *Nano Lett.* **2008**, *8*, 1679–1682. [CrossRef] [PubMed]

132. Delogu, F.; Gorrasi, G.; Sorrentino, A. Fabrication of polymer nanocomposites via ball milling: Present status and future perspectives. *Prog. Mater. Sci.* **2017**, *86*, 75–126. [CrossRef]

133. Moniruzzaman, M.; Winey, K.I. Polymer nanocomposites containing carbon nanotubes. *Macromolecules* **2006**, *39*, 5194–5205. [CrossRef]

134. Tait, M.; Pegoretti, A.; Dorigato, A.; Kalaitzidou, K. The effect of filler type and content and the manufacturing process on the performance of multifunctional carbon/poly-lactide composites. *Carbon* **2011**, *49*, 4280–4290. [CrossRef]

135. Pinto, A.M.; Cabral, J.; Tanaka, D.A.P.; Mendes, A.M.; Magalhaes, F.D. Effect of incorporation of graphene oxide and graphene nanoplatelets on mechanical and gas permeability properties of poly(lactic acid) films. *Polym. Int.* **2013**, *62*, 33–40. [CrossRef]

136. Du, J.H.; Cheng, H.M. The fabrication, properties, and uses of graphene/polymer composites. *Macromol. Chem. Phys.* **2012**, *213*, 1060–1077. [CrossRef]

137. Huang, X.; Yin, Z.Y.; Wu, S.X.; Qi, X.Y.; He, Q.Y.; Zhang, Q.C.; Yan, Q.Y.; Boey, F.; Zhang, H. Graphene-based materials: Synthesis, characterization, properties, and applications. *Small* **2011**, *7*, 1876–1902. [CrossRef] [PubMed]

138. Moon, S.I.; Jin, F.; Lee, C.; Tsutsumi, S.; Hyon, S.H. Novel carbon nanotube/poly(L-lactic acid) nanocomposites; their modulus, thermal stability, and electrical conductivity. *Macromol. Symp.* **2005**, *224*, 287–295. [CrossRef]

139. He, L.H.; Sun, J.; Wang, X.X.; Fan, X.H.; Zhao, Q.L.; Cai, L.F.; Song, R.; Ma, Z.; Huang, W. Unzipped multiwalled carbon nanotubes-incorporated poly(L-lactide) nanocomposites with enhanced interface and hydrolytic degradation. *Mater. Chem. Phys.* **2012**, *134*, 1059–1066. [CrossRef]

140. Seligra, P.G.; Nuevo, F.; Lamanna, M.; Fama, L. Covalent grafting of carbon nanotubes to pla in order to improve compatibility. *Compos. Part. B Eng.* **2013**, *46*, 61–68. [CrossRef]

141. Yoon, J.T.; Jeong, Y.G.; Lee, S.C.; Min, B.G. Influences of poly(lactic acid)-grafted carbon nanotube on thermal, mechanical, and electrical properties of poly(lactic acid). *Polym. Adv. Technol.* **2009**, *20*, 631–638. [CrossRef]

142. Kim, H.S.; Chae, Y.S.; Park, B.H.; Yoon, J.S.; Kang, M.; Jin, H.J. Thermal and electrical conductivity of poly(L-lactide)/multiwalled carbon nanotube nanocomposites. *Curr. Appl. Phys.* **2008**, *8*, 803–806. [CrossRef]

143. Yoon, J.T.; Lee, S.C.; Jeong, Y.G. Effects of grafted chain length on mechanical and electrical properties of nanocomposites containing polylactide-grafted carbon nanotubes. *Compos. Sci. Technol.* **2010**, *70*, 776–782. [CrossRef]

144. Chiu, W.M.; Kuo, H.Y.; Tsai, P.A.; Wu, J.H. Preparation and properties of poly (lactic acid) nanocomposites filled with functionalized single-walled carbon nanotubes. *J. Polym. Environ.* **2013**, *21*, 350–358. [CrossRef]

145. Kong, Y.X.; Yuan, J.; Wang, Z.M.; Qiu, J. Study on the preparation and properties of aligned carbon nanotubes/polylactide composite fibers. *Polym. Compos.* **2012**, *33*, 1613–1619. [CrossRef]

146. Chrissafis, K.; Paraskevopoulos, K.M.; Jannakoudakis, A.; Beslikas, T.; Bikiaris, D. Oxidized multiwalled carbon nanotubes as effective reinforcement and thermal stability agents of poly(lactic acid) ligaments. *J. Appl. Polym. Sci.* **2010**, *118*, 2712–2721. [CrossRef]

147. McCullen, S.D.; Stano, K.L.; Stevens, D.R.; Roberts, W.A.; Monteiro-Riviere, N.A.; Clarke, L.I.; Gorga, R.E. Development, optimization, and characterization of electrospun poly(lactic acid) nanofibers containing multi-walled carbon nanotubes. *J. Appl. Polym. Sci.* **2007**, *105*, 1668–1678. [CrossRef]

148. Vaisman, L.; Wagner, H.D.; Marom, G. The role of surfactants in dispersion of carbon nanotubes. *Adv. Colloid Interface Sci.* **2006**, *128*, 37–46. [CrossRef] [PubMed]

149. Pinto, A.M.; Moreira, S.; Goncalves, I.C.; Gama, F.M.; Mendes, A.M.; Magalhaes, F.D. Biocompatibility of poly(lactic acid) with incorporated graphene-based materials. *Colloid Surf. B* **2013**, *104*, 229–238. [CrossRef] [PubMed]

150. Li, W.X.; Xu, Z.W.; Chen, L.; Shan, M.J.; Tian, X.; Yang, C.Y.; Lv, H.M.; Qian, X.M. A facile method to produce graphene oxide-*g*-poly(L-lactic acid) as an promising reinforcement for plla nanocomposites. *Chem. Eng. J.* **2014**, *237*, 291–299. [CrossRef]

151. Li, W.X.; Shi, C.B.; Shan, M.J.; Guo, Q.W.; Xu, Z.W.; Wang, Z.; Yang, C.Y.; Mai, W.; Niu, J.R. Influence of silanized low-dimensional carbon nanofillers on mechanical, thermomechanical, and crystallization behaviors of poly(L-lactic acid) composites—A comparative study. *J. Appl. Polym. Sci.* **2013**, *130*, 1194–1202. [CrossRef]

152. Cao, Y.W.; Feng, J.C.; Wu, P.Y. Preparation of organically dispersible graphene nanosheet powders through a lyophilization method and their poly(lactic acid) composites. *Carbon* **2010**, *48*, 3834–3839. [CrossRef]

153. Wang, H.S.; Qiu, Z.B. Crystallization behaviors of biodegradable poly(L-lactic acid)/graphene oxide nanocomposites from the amorphous state. *Thermochim. Acta* **2011**, *526*, 229–236. [CrossRef]

154. Tong, X.Z.; Song, F.; Li, M.Q.; Wang, X.L.; Chin, I.J.; Wang, Y.Z. Fabrication of graphene/polylactide nanocomposites with improved properties. *Compos. Sci. Technol.* **2013**, *88*, 33–38. [CrossRef]

155. Shen, Y.X.; Jing, T.; Ren, W.J.; Zhang, J.W.; Jiang, Z.G.; Yu, Z.Z.; Dasari, A. Chemical and thermal reduction of graphene oxide and its electrically conductive polylactic acid nanocomposites. *Compos. Sci. Technol.* **2012**, *72*, 1430–1435. [CrossRef]

156. Huang, H.D.; Ren, P.G.; Xu, J.Z.; Xu, L.; Zhong, G.J.; Hsiao, B.S.; Li, Z.M. Improved barrier properties of poly(lactic acid) with randomly dispersed graphene oxide nanosheets. *J. Membr. Sci.* **2014**, *464*, 110–118. [CrossRef]

157. Wu, D.F.; Cheng, Y.X.; Feng, S.H.; Yao, Z.; Zhang, M. Crystallization behavior of polylactide/graphene composites. *Ind. Eng. Chem. Res.* **2013**, *52*, 6731–6739. [CrossRef]

158. Sisti, L.; Belcari, J.; Mazzocchetti, L.; Totaro, G.; Vannini, M.; Giorgini, L.; Zucchelli, A.; Celli, A. Multicomponent reinforcing system for poly(butylene succinate): Composites containing poly(L-lactide) electrospun mats loaded with graphene. *Polym. Test.* **2016**, *50*, 283–291. [CrossRef]

159. Verdejo, R.; Bernal, M.M.; Romasanta, L.J.; Lopez-Manchado, M.A. Graphene filled polymer nanocomposites. *J. Mater. Chem.* **2011**, *21*, 3301–3310. [CrossRef]

160. Lin, W.Y.; Shih, Y.F.; Lin, C.H.; Lee, C.C.; Yu, Y.H. The preparation of multi-walled carbon nanotube/poly(lactic acid) composites with excellent conductivity. *J. Taiwan Inst. Chem. E* **2013**, *44*, 489–496. [CrossRef]

161. Barrau, S.; Vanmansart, C.; Moreau, M.; Addad, A.; Stoclet, G.; Lefebvre, J.M.; Seguela, R. Crystallization behavior of carbon nanotube-polylactide nanocomposites. *Macromolecules* **2011**, *44*, 6496–6502. [CrossRef]

162. Villmow, T.; Potschke, P.; Pegel, S.; Haussler, L.; Kretzschmar, B. Influence of twin-screw extrusion conditions on the dispersion of multi-walled carbon nanotubes in a poly(lactic acid) matrix. *Polymer* **2008**, *49*, 3500–3509. [CrossRef]

163. Kuan, C.F.; Kuan, H.C.; Ma, C.C.M.; Chen, C.H. Mechanical and electrical properties of multi-wall carbon nanotube/poly(lactic acid) composites. *J. Phys. Chem. Solids* **2008**, *69*, 1395–1398. [CrossRef]

164. Lei, L.; Qiu, J.H.; Sakai, E. Preparing conductive poly(lactic acid) (PLA) with poly(methyl methacrylate) (PMMA) functionalized graphene (PFG) by admicellar polymerization. *Chem. Eng. J.* **2012**, *209*, 20–27. [CrossRef]

165. Gorrasi, G.; Sorrentino, A. Photo-oxidative stabilization of carbon nanotubes on polylactic acid. *Polym. Degrad. Stab.* **2013**, *98*, 963–971. [CrossRef]

166. Ali, A.M.; Ahmad, S.H. Mechanical characterization and morphology of polylactic acid/liquid natural rubber filled with multi walled carbon nanotubes. *AIP Conf. Proc.* **2013**, *1571*, 83–89.

167. Chieng, B.W.; Ibrahim, N.A.; Yunus, W.M.Z.W.; Hussein, M.Z. Poly(lactic acid)/poly(ethylene glycol) polymer nanocomposites: Effects of graphene nanoplatelets. *Polymers* **2014**, *6*, 93–104. [CrossRef]

168. Bao, C.L.; Song, L.; Xing, W.Y.; Yuan, B.H.; Wilkie, C.A.; Huang, J.L.; Guo, Y.Q.; Hu, Y. Preparation of graphene by pressurized oxidation and multiplex reduction and its polymer nanocomposites by masterbatch-based melt blending. *J. Mater. Chem.* **2012**, *22*, 6088–6096. [CrossRef]

169. Ali, A.M.; Ahmad, S.H. Effect of processing parameter and filler content on tensile properties of multi-walled carbon nanotubes reinforced polylactic acid nanocomposite. In Proceedings of the 2012 National Physics Conference, PERFIK 2012, Bukit Tinggi, Malaysia, 19–21 Novembwr 2012; pp. 254–259.

170. Kim, S.Y.; Shin, K.S.; Lee, S.H.; Kim, K.W.; Youn, J.R. Unique crystallization behavior of multi-walled carbon nanotube filled poly(lactic acid). *Fibers Polym.* **2010**, *11*, 1018–1023. [CrossRef]

171. Murariu, M.; Dechief, A.L.; Bonnaud, L.; Paint, Y.; Gallos, A.; Fontaine, G.; Bourbigot, S.; Dubois, P. The production and properties of polylactide composites filled with expanded graphite. *Polym. Degrad. Stab.* **2010**, *95*, 889–900. [CrossRef]

172. Hassouna, F.; Laachachi, A.; Chapron, D.; El Mouedden, Y.; Toniazzo, V.; Ruch, D. Development of new approach based on raman spectroscopy to study the dispersion of expanded graphite in poly(lactide). *Polym. Degrad. Stab.* **2011**, *96*, 2040–2047. [CrossRef]

173. Potts, J.R.; Dreyer, D.R.; Bielawski, C.W.; Ruoff, R.S. Graphene-based polymer nanocomposites. *Polymer* **2011**, *52*, 5–25. [CrossRef]

174. Kim, H.; Kobayashi, S.; AbdurRahim, M.A.; Zhang, M.L.J.; Khusainova, A.; Hillmyer, M.A.; Abdala, A.A.; Macosko, C.W. Graphene/polyethylene nanocomposites: Effect of polyethylene functionalization and blending methods. *Polymer* **2011**, *52*, 1837–1846. [CrossRef]

175. Brzeziński, M.; Biela, T. Polylactide nanocomposites with functional carbon nanotubes: A focused review. *Mater. Lett.* **2014**, 244–250. [CrossRef]

176. Yang, J.H.; Lin, S.H.; Lee, Y.D. Preparation and characterization of poly(L-lactide)-graphene composites using the in situ ring-opening polymerization of plla with graphene as the initiator. *J. Mater. Chem.* **2012**, *22*, 10805–10815. [CrossRef]

177. Pramoda, K.P.; Koh, C.B.; Hazrat, H.; He, C.B. Performance enhancement of polylactide by nanoblending with poss and graphene oxide. *Polym. Compos.* **2014**, *35*, 118–126. [CrossRef]

178. Li, Q.H.; Zhou, Q.H.; Deng, D.; Yu, Q.Z.; Gu, L.; Gong, K.D.; Xu, K.H. Enhanced thermal and electrical properties of poly (D,L-lactide)/multi-walled carbon nanotubes composites by in-situ polymerization. *Trans. Nonferr. Met. Soc. China* **2013**, *23*, 1421–1427. [CrossRef]

179. Yoon, O.J.; Sohn, I.Y.; Kim, D.J.; Lee, N.E. Enhancement of thermomechanical properties of poly(D,L-lactic-*co*-glycolic acid) and graphene oxide composite films for scaffolds. *Macromol. Res.* **2012**, *20*, 789–794. [CrossRef]

180. Pinto, A.M.; Goncalves, C.; Goncalves, I.C.; Magalhaes, F.D. Effect of biodegradation on thermo-mechanical properties and biocompatibility of poly(lactic acid)/graphene nanoplatelets composites. *Eur. Polym. J.* **2016**, *85*, 431–444. [CrossRef]

181. Desa, M.S.Z.M.; Hassan, A.; Arsad, A. The effect of natural rubber toughening on mechanical properties of poly(lactic acid)/multiwalled carbon nanotube nanocomposite. *Adv. Mater. Res.* **2013**, *747*, 639–642. [CrossRef]

182. Goncalves, C.; Pinto, A.; Machado, A.V.; Moreira, J.A.; Gonçalves, I.C.; Magalhães, F.D. Biocompatible reinforcement of poly(lactic acid) with graphene nanoplatelets. *Polym. Compos.* **2016**. [CrossRef]

183. Shao, S.J.; Zhou, S.B.; Li, L.; Li, J.R.; Luo, C.; Wang, J.X.; Li, X.H.; Weng, J. Osteoblast function on electrically conductive electrospun pla/mwcnts nanofibers. *Biomaterials* **2011**, *32*, 2821–2833. [CrossRef] [PubMed]

184. Potschke, P.; Andres, T.; Villmow, T.; Pegel, S.; Brunig, H.; Kobashi, K.; Fischer, D.; Haussler, L. Liquid sensing properties of fibers prepared by melt spinning from poly(lactic acid) containing multi-walled carbon nanotubes. *Compos. Sci. Technol.* **2010**, *70*, 343–349. [CrossRef]

185. Ma, H.B.; Su, W.X.; Tai, Z.X.; Sun, D.F.; Yan, X.B.; Liu, B.; Xue, Q.J. Preparation and cytocompatibility of polylactic acid/hydroxyapatite/graphene oxide nanocomposite fibrous membrane. *Chin. Sci. Bull.* **2012**, *57*, 3051–3058. [CrossRef]

186. Magrez, A.; Kasas, S.; Salicio, V.; Pasquier, N.; Seo, J.W.; Celio, M.; Catsicas, S.; Schwaller, B.; Forro, L. Cellular toxicity of carbon-based nanomaterials. *Nano Lett.* **2006**, *6*, 1121–1125. [CrossRef] [PubMed]

187. McCullen, S.D.; Stevens, D.R.; Roberts, W.A.; Clarke, L.I.; Bernacki, S.H.; Gorga, R.E.; Loboa, E.G. Characterization of electrospun nanocomposite scaffolds and biocompatibility with adipose-derived human mesenchymal stem cells. *Int. J. Nanomed.* **2007**, *2*, 253–263.

188. Sherrell, P.C.; Thompson, B.C.; Wassei, J.K.; Gelmi, A.A.; Higgins, M.J.; Kaner, R.B.; Wallace, G.G. Maintaining cytocompatibility of biopolymers through a graphene layer for electrical stimulation of nerve cells. *Adv. Funct. Mater.* **2014**, *24*, 769–776. [CrossRef]

189. Li, Z.Q.; Zhao, X.W.; Ye, L.; Coates, P.; Caton-Rose, F.; Martyn, M. Structure and blood compatibility of highly oriented PLA/mwnts composites produced by solid hot drawing. *J. Biomater. Appl.* **2014**, *28*, 978–989. [CrossRef] [PubMed]

190. An, X.L.; Ma, H.B.; Liu, B.; Wang, J.Z. Graphene oxide reinforced polylactic acid/polyurethane antibacterial composites. *J. Nanomater.* **2013**, *2013*. [CrossRef]

191. Yoon, O.J.; Jung, C.Y.; Sohn, I.Y.; Kim, H.J.; Hong, B.; Jhon, M.S.; Lee, N.E. Nanocomposite nanofibers of poly(D,L-lactic-co-glycolic acid) and graphene oxide nanosheets. *Compos. Part A* **2011**, *42*, 1978–1984. [CrossRef]

192. Kanczler, J.M.; Mirmalek-Sani, S.H.; Hanley, N.A.; Ivanov, A.L.; Barry, J.J.A.; Upton, C.; Shakesheff, K.M.; Howdle, S.M.; Antonov, E.N.; Bagratashvili, V.N.; et al. Biocompatibility and osteogenic potential of human fetal femur-derived cells on surface selective laser sintered scaffolds. *Acta Biomater.* **2009**, *5*, 2063–2071. [CrossRef] [PubMed]

193. Pinto, V.C.; Costa-Almeida, R.; Rodrigues, I.; Guardão, L.; Soares, R.; Guedes, R.M. Biocompatibility of PLA/GNP and PLA/CNT-cooh nanocomposites. *J. Biomed. Mater. Res. Part A* **2017**, *105A*, 2182–2190. [CrossRef] [PubMed]

polymers

MDPI

Review

Thermal Conductivity of Graphene-Polymer Composites: Mechanisms, Properties, and Applications

An Li, Cong Zhang and Yang-Fei Zhang *

Department of Materials Science and Engineering, College of Engineering, Peking University, Beijing 100871, China; lian1993@pku.edu.cn (A.L.); 1601214778@pku.edu.cn (C.Z.)
* Correspondence: zhangyangfei@pku.edu.cn; Tel.: +86-10-6275-9815

Received: 5 August 2017; Accepted: 7 September 2017; Published: 15 September 2017

Abstract: With the integration and miniaturization of electronic devices, thermal management has become a crucial issue that strongly affects their performance, reliability, and lifetime. One of the current interests in polymer-based composites is thermal conductive composites that dissipate the thermal energy produced by electronic, optoelectronic, and photonic devices and systems. Ultrahigh thermal conductivity makes graphene the most promising filler for thermal conductive composites. This article reviews the mechanisms of thermal conduction, the recent advances, and the influencing factors on graphene-polymer composites (GPC). In the end, we also discuss the applications of GPC in thermal engineering. This article summarizes the research on graphene-polymer thermal conductive composites in recent years and provides guidance on the preparation of composites with high thermal conductivity.

Keywords: graphene; polymer composites; thermal conductivity; mechanisms; properties; applications

1. Introduction

Thermal management has become a crucial issue in the modern electronics industry as electronic devices have become more integrated and miniaturized. The power required for some processor modules can reach 250 W in a high-performance computer, leading to heat loads as large as 1 kW in this system [1]. If the heat can-not be dissipated promptly, the lifetime and the efficiency of the system could be reduced, or even breakdown. In this situation, materials with high thermal conductivity are strongly needed to dissipate the heat and solve the problem [2].

Polymers have a lot of advantages, such as being lightweight, low cost, easy to process, and exhibiting good corrosion resistance. However, most polymers are heat insulators and have a thermal conductivity between 0.1 and 0.5 W m^{-1} K^{-1} [3], which is due to their amorphous state. There are three kinds of carriers in solids to transport energy: phonons, electrons, and photons [4]. Phonons are quantized modes of vibration in a rigid crystal lattice, which is the fundamental mechanism of heat conduction in most polymers. Polymers in amorphous state are usually considered to have lots of defects that contribute to numerous phonon scatting, leading to a low thermal conductivity [5].

In past years, a lot of works have studied thermal conductive polymer-based composites. Many different materials with high thermal conductivity have been used as fillers to improve the thermal conductivity of composites, such as boron nitride (BN) [6–9], carbon nanotubes (CNTs) [10–14], aluminum oxide [15–17], diamond [18–21], and graphene [22,23].

Graphene has attracted great attention because of its unique two dimensional (2D) structure and novel properties, such as the zero-gap band structure, high electron mobility, and high thermal conductivity [24]. Balandin and his co-workers reported a measurement of the thermal conductivity of suspended single-layer graphene around 5000 W m^{-1} K^{-1}, which was one of the highest thermal conductivities of currently known materials [25]. Although there are lots of reviews on the thermal conductivity of polymer-based composites, system summaries on thermal conductive graphene-polymer composites are rare [2–4]. In this situation, it is necessary to review the advances in the thermal conductivity of graphene-polymer composites.

In this article, we review the advances in thermal conductivity of graphene-polymer composites in recent years. Special attention is given to the mechanism, the properties, and the influence factors of graphene-polymer composites. Additionally, we discuss the applications of thermal conductive graphene-polymer composites.

2. Thermal Conductive Mechanisms

2.1. Thermal Conductive Mechanisms in Graphene

In solid materials, heat is carried by phonons and electrons [26]. In metals, thermal conductivity is due to free carriers of electrons. Copper is a good thermal conductor with a thermal conductivity of 400 W m^{-1} K^{-1} at room temperature, and the attribution from phonons is limited to 1–2% of the total [27]. The thermal conductivity of graphene is attributed to phonons and electrons because of its metallic property [28]. However, the contribution of electrons to the thermal conductivity of graphene is relatively rare. In general, it is believed that the thermal conductivity of graphene is mainly accomplished by phonons [27]. Figure 1 is a schematic of heat conduction in crystalline materials, which can also be applied to graphene [29]. When one side of the crystal lattice makes contact with the heat source, heat conducts to the first layer atoms in the form of vibrations. Due to the dense packing of atoms in the lattice and the strong chemical bonds between them, the vibrations of the first layer atoms quickly spread to the neighboring atoms, and the neighboring atoms pass the vibrations to the other neighboring atoms, which results in rapid heat transfer in crystalline materials. In graphene, which has the ideal structure, all of the carbon atoms are fixed by a covalent bond to a layer. When some of the atoms in the graphene come into contact with the heat source and begin to vibrate, the vibrations will quickly pass to the surrounding atoms by the strong force of the covalent bond. In other words, the heat transfers from one position to another in graphene. In some studies, the researchers believe that the heat in graphene is transferred by the form of phonon waves, and some researchers have detected and proved this speculation [30–32]. In fact, most of the graphene used to manufacture the thermal conductive composites is multilayer graphene, such as graphene nanosheets and graphene nanoplatelets. When one of the layers in multilayer graphene begins to vibrate, due to the weak force of the van der Waals force between each layer, vibrations are difficult to pass on to the adjacent graphene layers. That is, heat is difficult to transfer through the interlayer of graphene. As a result, anisotropic heat conduction exists in the multilayer graphene. This phenomenon has been proved by many researches [25,33–36].

2.2. Thermal Conductive Mechanisms in Polymers

Thermal conduction through a polymer is a complicated process, which is influenced by many parameters like crystallinity, temperature, orientation of the macromolecules, and so on [37–39]. Phonons are usually considered to be thermal carriers in polymers because there is a mere free electron [40]. Burger and her colleagues discussed the mechanism of heat transfer in an amorphous polymer and described it using a schematic diagram, which is presented in Figure 2 [29]. When the surface of the polymer makes contact with the heat source, heat transfers to the first atom of the molecular chain in the form of a vibration, then the nearest atom, and then the next. Heat will not propagate as a wave, like in graphene, but diffuse slower. Heat transfer in a molecular chain will also cause the disordered vibration and rotation of atoms, which significantly reduces the thermal conductivity of the polymer. A good conductor has a complete lattice structure, and atoms accumulate closely. When heat reaches the first atom, it will quickly transfer to the last one. However, heat transfer in a bad conductor causes the vibration and rotation of atoms, which will significantly reduce the thermal conductivity [29].

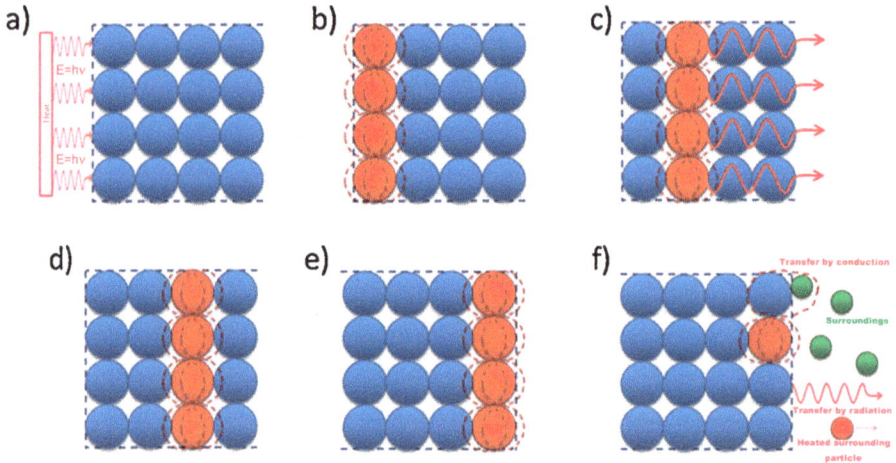

Figure 1. The schematic of thermal conductance in a crystalline material [29]. Copyright (2016), with permission from Elsevier.

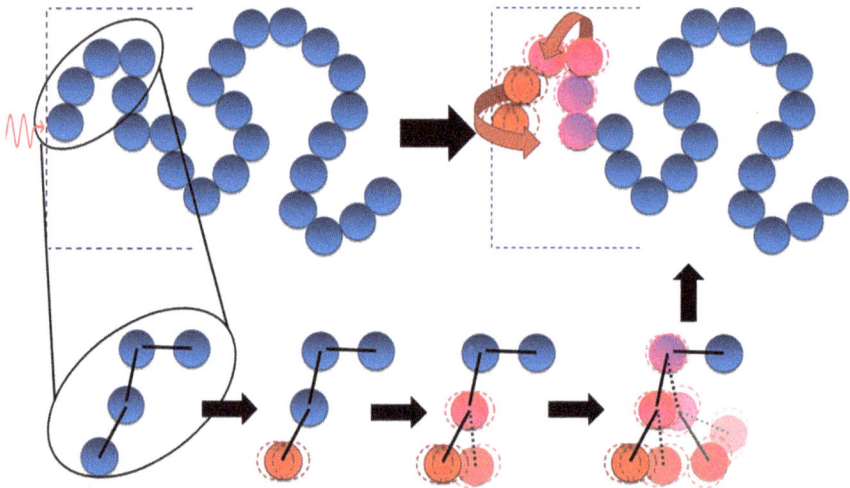

Figure 2. Schematic of thermal conductive mechanisms in polymer [29]. Copyright (2016), with permission from Elsevier.

2.3. Thermal Conductive Mechanisms in Graphene-Polymer Composites

The thermal conductive mechanism of graphene in polymers is more complex. In general, graphene has a very high specific surface area. When being added in a polymer, large numbers of interfaces are produced [41]. These interfaces will lead to phonon scattering and introduce ultrahigh interfacial thermal resistance. Therefore, it is difficult for heat to transfer through the graphene-polymer interface [42]. There is much research discussing the thermal conductive mechanisms in the interface of graphene-polymer composites [2–4,29]. Since mismatches between graphene and the polymer exist, the interface will result in phonon scattering and hinder the heat transfer [43]. For example, supposing that during the same time of Δt, the heat transfers from one side of the graphene to the other. But in the polymer, the heat passes over a very short distance attributed to phonon scattering. When the loading of the filler is below the percolation threshold, the fillers cannot connect to each other to form a thermal conduction pathway. In this case, the interfacial thermal resistance of graphene and the polymer will be the main factor determining the thermal conductivity of the composite. Surface modification of the graphene has been proved to be an applicable method to enhance graphene-polymer interface interaction, and an efficient technique to decrease interfacial thermal resistance. In a composite, graphene acts as a highly thermal conductive channel, while the modified surface affords covalent and non-covalent bonding with the molecular chains of the polymer matrix, which will facilitate the phonon transfer from the graphene to the polymer and also promote the phonon transfer from the polymer to the graphene [44]. In many studies, researchers have considered that the molecular chains of polymer and the molecular chains on the surface of graphene can intertwine with each other and form an interlayer. This interlayer will decrease the interfacial phonon scattering and minimize the interface thermal resistance by intertwined molecular chains [45,46]. However, when the loading is above the percolation threshold, the heat in the composite mainly transfers through the thermal conduction pathway, due to the high thermal conductivity of graphene. Figure 3 shows the case of graphene acting as an efficient thermal conduction channel. In the course of time Δt, the heat could transfer over a longer distance in graphene than the polymer matrix. When composites make contact with the heat source, heat transfers though graphene very quickly, which will increase the thermal conductivity. Increasing the number of thermal pathways and reducing the thermal resistance between graphene and the graphene-polymer interface are recommended steps for preparing a composite with high thermal conductivity [3].

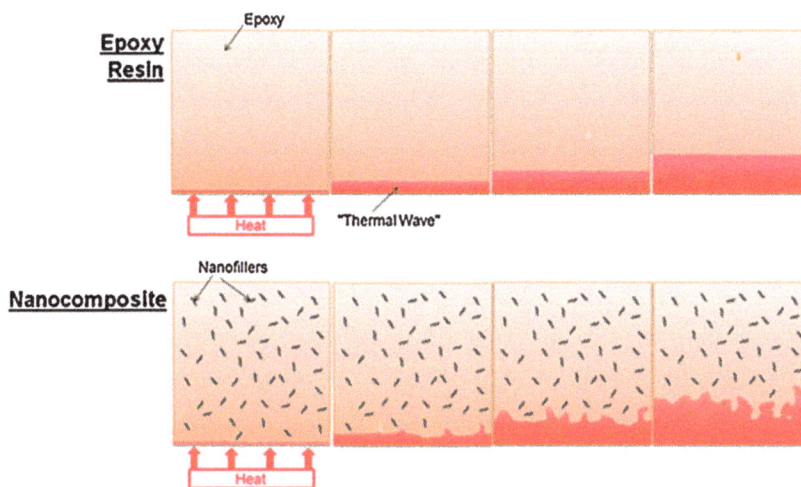

Figure 3. Thermal conductivity by graphene in a graphene-polymer composite [29]. Copyright (2016), with permission from Elsevier.

3. Recent Advances in Thermal Conductivity of Graphene-Polymer Composites

In recent years, more studies have been using graphene and its derived materials to prepare thermal conductive composite materials. The morphology of graphene in the polymer matrix significantly affects the thermal conductivity of the composites [2]. In this section, we review the advances in thermal conductivity of graphene-polymer composites. To review them more systematically, this section is divided into two parts, according to the different morphology of graphene in the polymer matrix. In the first part, the random dispersion of graphene in yjr polymer matrix is discussed. Random dispersion refers to the addition of graphene to the matrix, which is performed by a simple method, such as agitation, sonication, and blending, etc. Besides, there is no special method employed to control the orientation of graphene in matrix. In the second part, we discuss graphene with a specific orientation in the polymer matrix. This refers to the unusual structures of graphene in the polymer matrix, including the orientation, three-dimension structure (3D), and separate structure, etc. The term "graphene-related materials" is used to refer to the materials associated with graphene, which have different names in different literatures. These include graphene nanosheets, graphene nanoplatelets, graphene sheets, graphene flakes, graphene film, reduced graphene oxides, and graphene foam, etc.

3.1. Graphene with Random Orientation in the Polymer Matrix

Graphene with a random orientation in the polymer matrix can be manufactured by many methods, such as solution mixing, melt mixing, and in-situ polymerization, etc. [47–52]. Table 1 lists the thermal conductivity, thermal conductivity enhancement (TCE) per wt %, preparation methods, and surface preparation methods of graphene with a random orientation. The thermal conductivity enhancement is measured by a term of TCE per wt %, which refers to the enhancement of thermal conductivity by per weight content of graphene in composites [53]. In order to find the most effective methods to enhance the thermal conductivity of composites, we compared the TCE per wt % of every composite shown in Table 1. The results are shown in Figure 4. From Figure 4, we can see that graphene is an efficient filler for enhancing the thermal conductivity of the polymer matrix. The TCE per wt % of graphene is around 50%, which means several percent of graphene can significantly increase the thermal conductivity of the composite. But the TCE per wt % of unmodified graphene is difficult to exceed 100%. However, when using graphene modified by covalent or noncovalent bonds, the TCE per wt % can be very close to 100%. When using a titanate coupling agent to modify graphene, the TCE per wt % is as high as 357.8%. The researchers believe the interfacial force between graphene and polymer has been enhanced by surface modification. This enhancement could reduce the interfacial thermal resistance and disperse graphene more uniformly [54].

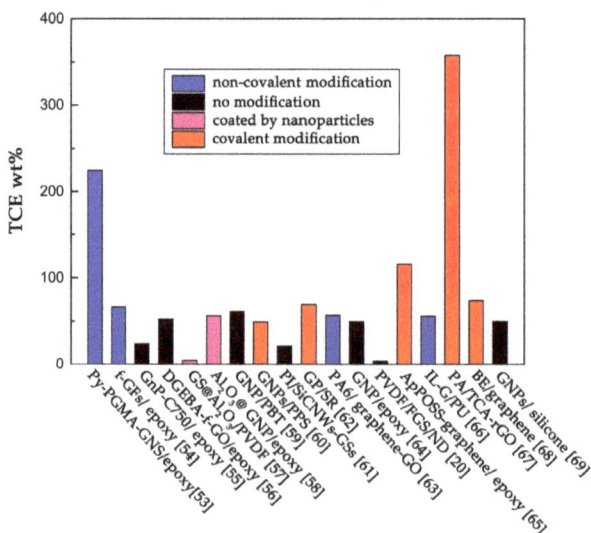

Figure 4. TCE wt % of composites in Table 1.

Table 1. Thermal conductivity of polymer composites filled with graphene and graphene-related materials with a random orientation.

Sample	Graphene content (wt %)	Thermal conductivity (W m^{-1} k^{-1})	Thermal conductivity enhancement (TCE) per wt %	Preparation method	Surface preparation methods
Py-PGMA-GNS/epoxy [55]	3.8	1.91	225%	In-situ polymerization	Non-covalent modification
f-GFs/epoxy [56]	10	1.53	66.5%	In-situ polymerization	Non-covalent modification
GnP-C750/epoxy [57]	5	0.45	23.8%	In-situ polymerization	no
DGEBA-f-GO/epoxy [58]	4.64 [1]	0.72	52.3%	In-situ polymerization	no
GS@Al2O3/PVDF [59]	40	0.586	4.8%	solution mixing	Coated by alumina nanoparticals
Al2O3@GNP/epoxy [60]	12	1.49	56.4%	solution mixing	Coated by alumina
GNP/PBT [61]	20	1.98	61%	In-situ polymerization	no
GNPs/PPS [62]	37.8 [1]	4.414	49%	melt mixing	Covalent modification
PI/SiCNWs-GSs [63]	7	0.577	21%	solution mixing	no
GP/SR [54]	0.72	0.3	69.4%	mechanical blending	Covalent modification
PA6/graphene-GO [64]	10	2.14	56.9%	In-situ polymerization	Non-covalent modification
GNP/epoxy [65]	25	2.67	49.4%	solution mixing	no
PVDF/FGS/ND [20]	45	0.66	3.9%	solution mixing	no
ApPOSS-graphene/epoxy [66]	0.5	0.348	115.8%	solution mixing	Covalent modification
IL-G/PU [67]	0.608	0.3012	55.9%	In-situ polymerization	Non-covalent modification
PA/TCA-rGO [68]	5	5.1	357.8%	melt mixing	Covalent modification
BE/graphene [69]	2.5	0.542	73.7%	solution mixing	Covalent modification
GNPs/silicone [70]	16	~2.6	49.7%	In-situ polymerization	no

[1] volume fraction was converted into weight fraction.

3.2. Graphene with Specific Orientation in the Polymer Matrix

There are a variety of specific orientations of graphene in the polymer matrix, such as orientation, segregated structure, 3D structure, and so on [53,71–75]. Specific orientations of graphene give special properties to the composites. The thermal properties of recent studies in graphene-polymer composites are listed in Table 2, and the enhancements of each composite are compared in Figure 5. From Figure 5, it seems that the orientation and 3D structure are more efficient structures for improving the thermal conductivity of composites. By comparing Figures 4 and 5, we also find that graphene with a specific orientation is more efficient than that with a random orientation. The researchers believe that this is mainly because that graphene plays the role of the thermal conduction pathway in the polymer matrix, and the heat transfers through the graphene pathway preferentially [76]. The purpose of orientation, a segregated structure, and 3D structure is establishing the thermal pathway in the polymer matrix, which can transfer heat more efficiently [77–80].

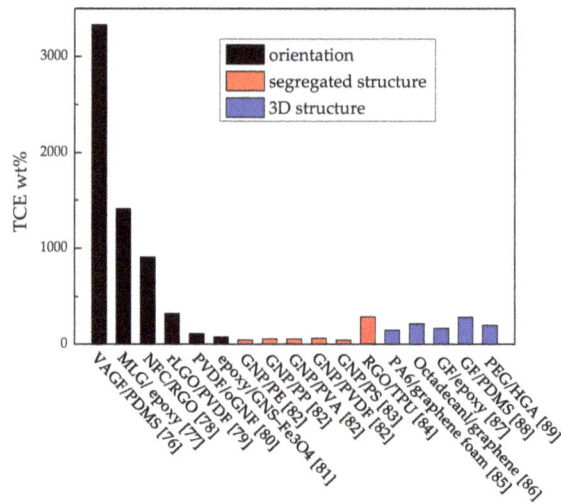

Figure 5. TCE wt % of composites in Table 2.

Table 2. Thermal conductivity of polymer composites filled with graphene and graphene-related materials with a specific orientation.

Sample	Graphene content (wt %)	Thermal conductivity (W m^{-1} k^{-1})	Thermal conductivity enhancement (TCE) per wt %	Specific orientation of graphene
VAGF/PDMS [76]	92.3	614.85	3329%	orientation
MLG/epoxy [77]	11.8	33.54	1412.7%	orientation
NFC/RGO [81]	1	12.6	910%	orientation
rLGO/PVDF [82]	27.2	19.5	~323.5%	orientation
PVDF/oGNF [83]	~36.8	~10	~113.2%	orientation
epoxy/GNS–Fe$_3$O$_4$ [84]	~1.74	~0.6	~79.9%	orientation
GNP/PE [85]	10	1.84	45.7%	segregated structure
GNP/PP [85]	10	1.53	59.5%	segregated structure
GNP/PVA [85]	10	1.43	58%	segregated structure
GNP/PVDF [85]	10	1.47	67.3%	segregated structure

Table 2. *Cont.*

Sample	Graphene content (wt %)	Thermal conductivity (W m^{-1} k^{-1})	Thermal conductivity enhancement (TCE) per wt %	Specific orientation of graphene
GNP/PS [86]	~9.2	~0.9	43.3%	segregated structure
RGO/TPU [87]	1.04	0.8	288%	segregated structure
PA6/graphene foam [78]	2	0.847	150%	3D structure
Octadecanl/graphene [88]	12	5.92	216%	3D structure
GF/epoxy [89]	5	1.52	170%	3D structure
GF/PDMS [90]	0.7	0.56	285%	3D structure
PEG/HGA [79]	1.8	1.43	200.6%	3D structure

4. Influence Factors on Thermal Conductivity of Graphene-Polymer Composites

There are many factors affecting the thermal conductivity of graphene-polymer composites, such as the defects on graphene, the orientation of graphene in the polymer, the graphene loading, and the surface modification, etc. [3–5]. In this section, we mainly review the influence of the characteristics of graphene (such as the defect, morphology, number of layers, and size), the loading of graphene, the orientation of graphene in the polymer matrix, and the interface between graphene and the polymer on the thermal conductivity.

4.1. The Characteristics of Graphene

The characteristics of graphene have a great influence on the thermal conductivity of graphene-polymer composites [80,91–93]. Hoda et al. investigated the thermal conductivity of graphene as a function of the density of defects. Graphene was suspended over ~7.5 µm size square holes and the optothermal Raman technique was employed to measure the thermal conductivity of graphene in air. They found that the thermal conductivity of suspended graphene decreased from ~1.8 × 10^3 W m^{-1} K^{-1} to ~4.0 × 10^2 W m^{-1} K^{-1} near room temperature as the density of defects changed from 2.0 × 10^{10} cm^{-2} to 1.8 × 10^{11} cm^{-2} [94]. Xin et al. employed a high temperature to obtain defect-free graphene and investigated the thermal conductivities of polymer composites filled with graphene of different defect contents. The graphene annealing at 2200 °C had the least amount of defects, and the composite filled with it had the highest thermal conductivity, reaching 3.55 W m^{-1} K^{-1}. This is because the high-temperature annealing heals defects and removes oxygen functional groups on graphene, thus reducing the phonon scattering centers [95]. The morphology of graphene also has an influence on the thermal conductivity of composites. Chu et al. pointed out that when using graphene nanoplates with more wrinkles as a filler, the composites will exhibit lower thermal conductivity. The reason for this is that the waviness of GNPs significantly affects the intrinsic characteristics of GNPs (such as thermal conductivity, aspect ratio) and the interfacial phonon coupling behavior between GNPs and polymers [96]. Kim et al. investigated the effects of the graphene layer and size on the thermal conductivity of composites and found that the thermal conductivity across the graphene/epoxy interface increases when increasing the number of graphene layers [97]. Kim et al. prepared composites filled with graphene of varied sizes and thicknesses. A similar result is found that a larger size and thickness of the graphene nanoplatelets results in an effective improvement in the thermal conductivity and heat dissipation ability of the composite [98].

4.2. The Loading of Graphene

The loading of graphene exerts a significant effect on the electrical and thermal conductivity of the composites [22,99,100]. It is found that there is a critical loading (percolation threshold) of graphene when the conductive composite is prepared. When the loading exceeds this value, the electrical conductivity of the composite material is improved significantly. However, it is difficult to determine whether there is a percolation threshold phenomenon in thermally conductive composites.

Khan et al. researched the thermal conductivity of graphene sheets-epoxy composites. The thermal conductivity increases with increasing graphene loading and there is no percolation threshold [101]. Fazel investigated the thermal conductivity of graphene/1-octadecanol (stearyl alcohol) composites and reported a similar finding [22]. Michael et al. found strong evidence for the existence of a thermal percolation threshold in graphene nanoplatelets (GnPs)-polymer composites. Below the percolation threshold (loading < 0.17), the polymer mediates between adjacent GnPs and the GnP cannot make sufficient contact, resulting in gaps. Above the percolation threshold (loading > 0.17), there is a sharp rise in the thermal conductivity, which means that direct GnP-GnP contacts have been formed [23]. Li et al. also found a similar phenomenon in graphene-epoxy composite [102].

4.3. The Orientation of Graphene in the Polymer Matrix

Many researchers believe that graphene with a specific orientation, like orientation and 3D structure, is much better than graphene with a random orientation when preparing thermal conductive composites [76–79,81–90]. Zhang et al. poured polydimethylsiloxane (PDMS) into a vertically aligned graphene film (VAGF) to manufacture a high-orientation graphene-polymer composite. The thermal conductivity of this composite was as high as 614.85 $Wm^{-1} K^{-1}$, which is higher than copper at room temperature. It is claimed that this dramatic enhancement is attributed to the rapid and effective heat-transfer path formed by orientated graphene [76]. Zhao et al. prepared a GF/PDMS composite with a thermal conductivity of 0.56 $W m^{-1} K^{-1}$, 20% higher than that of GS/PDMS composite at the loading of 0.7 wt %. They believe that the unique interconnected structure of GF acts as an efficient thermal pathway in the polymer matrix [90]. Figure 6. is the schematic of thermal conductance in a polymer, oriented graphene/polymer composite, and 3D graphene/polymer composite.

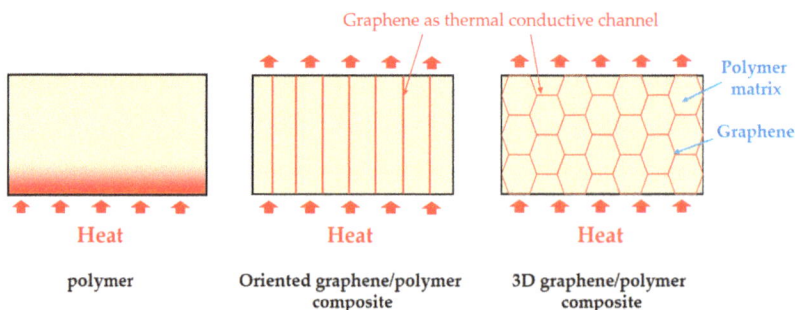

Figure 6. Schematic of thermal conductance in a polymer, oriented graphene/polymer composite, and 3D graphene/polymer composite.

4.4. The Interface between Graphene and the Polymer

It is considered that the interface between graphene and the polymer plays an important role in thermal conductive composites. Since phonons are the main form of thermal conductance in graphene-polymer composites, bad coupling in vibration modes at the graphene-polymer interface will generate huge interfacial thermal resistance. Chemical bonding between graphene and the polymer can efficiently decrease the phonon scattering at the interface and reduce interfacial thermal resistance [69]. Gao et al. investigated the influence of surface-grafted polymer chains on the thermal conductivity of a graphene-polyamide-6,6 nanocomposite. It was found that the through-plane interfacial thermal conductivity is proportional to the grafting density. Meanwhile, it first rises and then saturates as the grafting length increases. However, the in-plane thermal conductivity of graphene decreases rapidly as the grafting density increases. There is a maximum thermal conductivity of the composite because of these two competing factors [103]. Wang et al. studied the interfacial thermal resistance for polymer composites reinforced by various covalently functionalized graphene using molecular dynamics simulations. Among the various functional groups, like methyl, phenyl, butyl, formyl, carboxyl, amines, and hydroxyl, butyl is found to be the most effective one in reducing the interfacial thermal resistance [104]. Eslami et al. investigated the heat transport between graphene and polyamide-6,6 oligomers. They found that well-organized (chain stretching) polymer layers between the graphene show an interesting anisotropic heat conduction. The heat conduction in the parallel direction to the graphene surface is higher than that in the perpendicular direction [105,106].

5. Applications of Graphene-Polymer Composites in Thermal Engineering

Nowadays, with the improving demand in emerging industries, thermal conductive materials with novel properties are widely required [107–109]. Compared with other thermal conductive materials (metal, ceramics, carbon-related materials), polymer matrix composites have many outstanding properties, such as being lightweight and easy to process, and exhibiting good corrosion resistance and vibration damping, etc. In this section, some emerging applications of graphene-polymer composites are listed, such as electronic packaging, batteries, and energy storage.

5.1. Electronic Packaging

In the electronic industry, thermal management has been a serious challenge because of the miniaturization and functionalization of electronic devices. To control the temperature of all components in devices, an effective thermal conductive path must be used [47]. Thermal interface materials (TIM) are used to provide an effective heat conduction path between the two solid surfaces due to their ability to conform to rough surfaces and high thermal conductivity [110]. Figure 7 is a schematic diagram of TIM [101]. The international technology roadmap for high-performance chips at 14 nm is a power density greater than 100 W/cm^2 and junction-to-ambient thermal resistance of less than 0.2 °C/W [2]. There is a need for TIM to dissipate heat when the chip is operating. However, the thermal conductivity of commercial TIM is relatively low as most of them are less than 5 W m^{-1} K^{-1}. Employing graphene to prepare a thermal conductive material as TIM has been attracting a lot of attention.

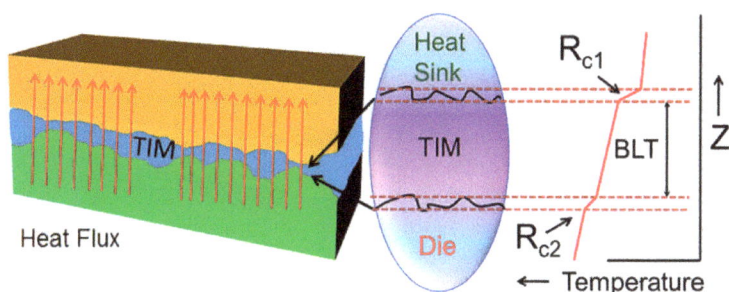

Figure 7. Schematic illustrating the action of thermal interface material, which fills the gaps between two contacting surfaces and conducts the heat produced by electronic drives [101]. (Copyright (2012) the American Chemical Society).

5.2. Thermal Energy Storage

The effort towards thermal energy storage has been intensified over the past years. Thermal conductivity is an important parameter in thermal energy storage materials, which significantly influences the rate of heat storage and extraction [111]. Therefore, graphene and graphene derivatives are used as thermal conductive carriers to improve the thermal conductivity of thermal energy storage materials. Ji et al. embedded continuous ultrathin-graphite foams (UGFs) in phase change materials to manufacture a composite, and improved the thermal conductivity by 18 times [112]. Mehrali et al.

prepared phase change materials by the vacuum impregnation of paraffin within graphene oxide, and the maximum energy storage value was 64.89 kJ/kg [113]. A phase change material consisting of graphene aerogel and octadecanoic acid was produced by Zhong et al. When the loading of GA reached 20 vol %, the thermal conductivity of this composite achieved 2.635 W/m^{-1} K^{-1}, which is about 14 times that of the OA [114].

5.3. Batteries

As batteries have become more powerful in recent years, thermal management has turned into a special issue in the battery system. When a battery is used at a high charging/discharging rate, the rate of heat generation may exceed the rate of heat dissipation. In this situation, the battery may be inefficient or even catch fire. A thermal management system is needed to maintain the battery pack at an optimum temperature. Khan et al. incorporated 8 wt % graphene nanoflake in polyacrylonitrile (PAN) fiber separators, and the thermal conductivity increased from 3.5 to 8.5 W m^{-1} K^{-1}. They think Lithium-ion batteries have become the major source of power for portable electronic devices. Separators are one of the major components of these batteries, and the improvement of thermal conductivity in separators is an option for long-lasting Li-ion battery fabrications [115]. Hallaj et al. presented a novel thermal management system and investigated it for electric vehicle applications. They think it is important to manage the heat in a battery under both cold and hot conditions [116].

6. Conclusions

In this paper, we have reviewed the graphene-polymer thermal conductive material in recent years. The thermal conductive mechanisms in graphene, polymers, and their composites have been discussed. The recent advances on thermal conductivity of graphene-polymer composites have also been reviewed. Furthermore, we have discussed the factors influencing the thermal conductivity of graphene-polymer composites, such as the characteristics, the loading, the orientation of graphene, and the interface. Finally, the applications of thermal conductive graphene-polymer composites have been demonstrated. This review reveals the relationship between thermal conductive mechanisms and properties and also provides guidance on the preparation of composites with high thermal conductivity.

Acknowledgments: This work is supported by the National Natural Science Foundation of China (No. 11202005).

Author Contributions: Yang-Fei Zhang and An Li. conceived and designed the content of the paper. An Li, Cong Zhang and Yang-Fei Zhang wrote the paper.

Conflicts of Interest: The authors declare no conflict of interest.

Abbreviations

Py-PGMA-GNS/epoxy	Pyrene-end poly(glycidyl methacrylate)-graphene nanosheet/epoxy composite
f-GFs/epoxy	Non-covalently functionalized graphene flakes/epoxy composite
GnP-C750/epoxy	Graphene nanoplatelets (sizes < 1 μm)/epoxy composite
DGEBA-f-GO/epoxy	Diglycidyl ether of bisphenol-A functionalized graphene oxide/epoxy composite
GS@Al$_2$O$_3$/PVDF	Alumina-coated graphene sheet/poly(vinylidene fluoride) composite
Al$_2$O$_3$@GNP/epoxy	Alumina nanoparticles decorated graphene nanoplatelets/epoxy composite
GNP/PBT	Graphene nanoplatelet/polybutylene terephthalate composite
GNPs/PPS	Graphene nanoplatelets/polyphenylene sulfide composite
PI/SiCNWs-GSs	Polyimide/SiC nanowires grown on graphene sheets composite
GP/SR	Graphene/silicone rubber
PA$_6$/graphene-GO	Polyamide-6/graphene-graphene oxide composite
GNP/epoxy	Graphene nanoplatelets/epoxy composite
PVDF/FGS/ND	Poly(vinylidene fluoride)/functionalized graphene sheets/nanodiamonds composite

ApPOSS-graphene/epoxy	Aminopropylisobutyl polyhedral oligomeric silsesquioxane grafted graphene/epoxy composite
IL-G/PU	1-allyl-methylimidazolium chloride ionic liquid modified graphene/polyurethane composite
PA/TCA-rGO	Titanate coupling agent modified reduced graphene/polyamide composite
BE/graphene	Bio-based polyester/graphene composite
GNPs/silicone	Graphene nanoplatelets/silicone composite
VAGF/PDMS	Vertically aligned graphene film/polydimethylsiloxane
MLG/epoxy	Multilayer graphene/epoxy composite
NFC/RGO	Nanofibrillated cellulose/epoxy composite
rLGO/PVDF	Highly self-aligned large-area reduced graphene oxide/poly (vinylidene fluoride-co-hexafluoropropylene) composite
PVDF/oGNF	Oriented graphene nanoflake/poly(vinylidene fluoride) (PVDF) composite
epoxy/GNS–Fe_3O_4	Epoxy/graphene nanosheets-Fe_3O_4
GNP/PE	Graphene nanoplatelets/polyethylene
GNP/PP	Graphene nanoplatelets/polypropylene
GNP/PVA	Graphene nanoplatelets/poly(vinyl alcohol)
GNP/PVDF	Graphene nanoplatelets/poly(vinylideneuoride)
GNP/PS	Graphene nanoplates/polystyrene; RGO/TPU
RGO/TPU	Reduced graphene oxide/thermoplastic polyurethane
PA_6/graphene foam	Polyamide-6/graphene foam
GF/epoxy	Graphene foams/epoxy
GF/PDMS	Graphene foam/polydimethylsiloxane
PEG/HGA	Polyethylene glycol/hybrid graphene aerogels

References

1. Moore, A.L.; Shi, L. Emerging challenges and materials for thermal management of electronics. *Mater. Today* **2014**, *17*, 163–174. [CrossRef]
2. Chen, H.; Ginzburg, V.V.; Yang, J.; Yang, Y.; Liu, W.; Huang, Y.; Du, L.; Chen, B. Thermal Conductivity of Polymer-Based Composites: Fundamentals and Applications. *Prog. Polym. Sci.* **2016**, *59*, 41–85. [CrossRef]
3. Huang, X.; Jiang, P.; Tanaka, T. A review of dielectric polymer composites with high thermal conductivity. *IEEE Electr. Insul. Mag.* **2011**, *27*, 8–16. [CrossRef]
4. Han, Z.; Fina, A. Thermal conductivity of carbon nanotubes and their polymer nanocomposites: A review. *Prog. Polym. Sci.* **2011**, *36*, 914–944. [CrossRef]
5. Agari, Y.; Ueda, A.; Omura, Y.; Nagai, S. Thermal diffusivity and conductivity of PMMA/PC blends. *Polymer* **1997**, *38*, 801–807. [CrossRef]
6. Hu, J.; Huang, Y.; Yao, Y.; Pan, G.; Sun, J.; Zeng, X.; Sun, R.; Xu, J.B.; Song, B.; Wong, C.P. A Polymer Composite with Improved Thermal Conductivity by Constructing Hierarchically Ordered Three-Dimensional Interconnected Network of BN. *ACS Appl. Mater. Interfaces* **2017**, 13544–13553. [CrossRef] [PubMed]
7. Lim, H.S.; Jin, W.O.; Kim, S.Y.; Yoo, M.J.; Park, S.D.; Lee, W.S. Anisotropically Alignable Magnetic Boron Nitride Platelets Decorated with Iron Oxide Nanoparticles. *Chem. Mater.* **2013**, *25*, 3315–3319. [CrossRef]
8. Yao, Y.; Zeng, X.; Wang, F.; Sun, R.; Xu, J.; Wong, C.P. Significant Enhancement of Thermal Conductivity in Bioinspired Freestanding Boron Nitride Papers Filled with Graphene Oxide. *Chem. Mater.* **2016**, *80*, 1357–1359. [CrossRef]
9. Fang, H.; Bai, S.L.; Wong, C.P. Thermal, mechanical and dielectric properties of flexible BN foam and BN nanosheets reinforced polymer composites for electronic packaging application. *Compos. A Appl. Sci. Manuf.* **2017**, *100*, 71–80. [CrossRef]
10. Yu, A.; Ramesh, P.; Sun, X.; Bekyarova, E.; Itkis, M.E.; Haddon, R.C. Enhanced Thermal Conductivity in a Hybrid Graphite Nanoplatelet—Carbon Nanotube Filler for Epoxy Composites. *Adv. Mater.* **2010**, *20*, 4740–4744. [CrossRef]
11. Moisala, A.; Li, Q.; Kinloch, I.A.; Windle, A.H. Thermal and electrical conductivity of single- and multi-walled carbon nanotube-epoxy composites. *Compos. Sci. Technol.* **2006**, *66*, 1285–1288. [CrossRef]

12. Yang, S.Y.; Ma, C.C.M.; Teng, C.C.; Huang, Y.W.; Liao, S.H.; Huang, Y.L.; Tien, H.W.; Lee, T.M.; Chiou, K.C. Effect of functionalized carbon nanotubes on the thermal conductivity of epoxy composites. *Carbon* **2010**, *48*, 592–603. [CrossRef]

13. Song, Y.S.; Youn, J.R. Evaluation of effective thermal conductivity for carbon nanotube/polymer composites using control volume finite element method. *Carbon* **2006**, *44*, 710–717. [CrossRef]

14. Nan, C.W.; Liu, G.; Lin, Y.; Li, M. Interface effect on thermal conductivity of carbon nanotube composites. *Appl. Phys. Lett.* **2004**, *85*, 3549–3551. [CrossRef]

15. Choi, S.; Kim, J. Thermal conductivity of epoxy composites with a binary-particle system of aluminum oxide and aluminum nitride fillers. *Compos. B Eng.* **2013**, *51*, 140–147. [CrossRef]

16. Li, B.; Li, R.; Xie, Y. Properties and effect of preparation method of thermally conductive polypropylene/aluminum oxide composite. *J. Mater. Sci.* **2017**, *52*, 2524–2533. [CrossRef]

17. Im, H.; Kim, J. Enhancement of the thermal conductivity of aluminum oxide–epoxy terminated poly(dimethyl siloxane) with a metal oxide containing polysiloxane. *J. Mater. Sci.* **2011**, *46*, 6571–6580. [CrossRef]

18. Kidalov, S.V.; Shakhov, F.M. Thermal Conductivity of Diamond Composites. *Materials* **2009**, *2*, 2467–2495. [CrossRef]

19. Cho, H.B.; Konno, A.; Fujihara, T.; Suzuki, T.; Tanaka, S.; Jiang, W.; Suematsu, H.; Niihara, K.; Nakayama, T. Self-assemblies of linearly aligned diamond fillers in polysiloxane/diamond composite films with enhanced thermal conductivity. *Compos. Sci. Technol.* **2012**, *72*, 112–118. [CrossRef]

20. Yu, J.; Qian, R.; Jiang, P. Enhanced thermal conductivity for PVDF composites with a hybrid functionalized graphene sheet-nanodiamond filler. *Fibers Polym.* **2013**, *14*, 1317–1323. [CrossRef]

21. Saw, W.P.S.; Mariatti, M. Properties of synthetic diamond and graphene nanoplatelet-filled epoxy thin film composites for electronic applications. *J. Mater. Sci. Mater. Electron.* **2012**, *23*, 817–824. [CrossRef]

22. Shtein, M.; Nadiv, R.; Buzaglo, M.; Kahil, K.; Regev, O. Thermally Conductive Graphene-Polymer Composites: Size, Percolation, and Synergy Effects. *Chem. Mater.* **2015**, *27*, 2100–2106. [CrossRef]

23. Yavari, F.; Fard, H.R.; Pashayi, K.; Rafiee, M.A.; Zamiri, A.; Yu, Z.; Ozisik, R.; Borcatasciuc, T.; Koratkar, N. Enhanced Thermal Conductivity in a Nanostructured Phase Change Composite due to Low Concentration Graphene Additives. *J. Phys. Chem. C* **2011**, *115*, 8753–8758. [CrossRef]

24. Geim, A.K. Graphene: status and prospects. *Science* **2009**, *324*, 1530–1534. [CrossRef] [PubMed]

25. Balandin, A.A.; Ghosh, S.; Bao, W.; Calizo, I.; Teweldebrhan, D.; Miao, F.; Lau, C.N. Superior thermal conductivity of single-layer graphene. *Nano Lett.* **2008**, *8*, 902–907. [CrossRef] [PubMed]

26. Maldovan, M. Sound and heat revolutions in phononics. *Nature* **2013**, *503*, 209–217. [CrossRef] [PubMed]

27. Balandin, A.A. Thermal properties of graphene and nanostructured carbon materials. *Nat. Mater.* **2011**, *10*, 569. [CrossRef] [PubMed]

28. Pu, H.H.; Rhim, S.H.; Hirschmugl, C.J.; Gajdardziska-Josifovska, M.; Weinert, M.; Chen, J.H. Anisotropic thermal conductivity of semiconducting graphene monoxide. *Appl. Phys. Lett.* **2013**, *102*, 569–581. [CrossRef]

29. Burger, N.; Laachachi, A.; Ferriol, M.; Lutz, M.; Toniazzo, V.; Ruch, D. Review of thermal conductivity in composites: Mechanisms, parameters and theory. *Prog. Polym. Sci.* **2016**, *61*, 1–28. [CrossRef]

30. Narula, R.; Bonini, N.; Marzari, N.; Reich, S. Dominant phonon wave vectors and strain-induced splitting of the 2D Raman mode of graphene. *Phys. Rev. B* **2012**, *85*, 115451. [CrossRef]

31. Yao, W.; Cao, B. Triggering wave-domain heat conduction in graphene. *Phys. Lett. A* **2016**, *380*, 2105–2110. [CrossRef]

32. Yao, W.; Cao, B. Thermal wave propagation in graphene studied by molecular dynamics simulations. *Chin. Sci. Bull.* **2014**, *59*, 3495–3503. [CrossRef]

33. Ghosh, S.; Calizo, I.; Teweldebrhan, D.; Pokatilov, E.P. Extremely high thermal conductivity of graphene: Prospects for thermal management applications in nanoelectronic circuits. *Appl. Phys. Lett.* **2008**, *92*, 151911–151913. [CrossRef]

34. Zhixin, G.; Dier, Z.; Xingao, G. Thermal conductivity of graphene nanoribbons. *Appl. Phys. Lett.* **2009**, *95*, 163103. [CrossRef]

35. Nika, D.L.; Ghosh, S.; Pokatilov, E.P.; Balandin, A.A. Lattice thermal conductivity of graphene flakes: Comparison with bulk graphite. *Appl. Phys. Lett.* **2009**, *94*, 203103. [CrossRef]

36. Faugeras, C.; Faugeras, B.; Orlita, M.; Potemski, M.; Nair, R.R.; Geim, A.K. Thermal conductivity of graphene in corbino membrane geometry. *ACS Nano* **2010**, *4*, 1889–1892. [CrossRef] [PubMed]

37. Rossinsky, E.; Müllerplathe, F. Anisotropy of the thermal conductivity in a crystalline polymer: Reverse nonequilibrium molecular dynamics simulation of the delta phase of syndiotactic polystyrene. *J. Chem. Phys.* **2009**, *130*, 134905. [CrossRef] [PubMed]

38. Choy, C.L.; Greig, D. The low-temperature thermal conductivity of a semi-crystalline polymer, polyethylene terephthalate. *J. Phys. C Solid State Phys.* **1975**, *8*, 3121–3130. [CrossRef]

39. Choy, C.L.; Chen, F.C.; Luk, W.H. Thermal conductivity of oriented crystalline polymers. *J. Polym. Sci. Polym. Phys.* **1980**, *18*, 1187–1207. [CrossRef]

40. Lee, J.H.; Koh, C.Y.; Singer, J.P.; Jeon, S.J.; Maldovan, M.; Stein, O.; Thomas, E.L. 25th anniversary article: ordered polymer structures for the engineering of photons and phonons. *Adv. Mater.* **2014**, *26*, 532–569. [CrossRef] [PubMed]

41. Xu, P.; Loomis, J.; Bradshaw, R.D.; Panchapakesan, B. Load transfer and mechanical properties of chemically reduced graphene reinforcements in polymer composites. *Nanotechnology* **2012**, *23*, 3847–3856. [CrossRef] [PubMed]

42. Luo, T.; Lloyd, J.R. Enhancement of Thermal Energy Transport Across Graphene/Graphite and Polymer Interfaces: A Molecular Dynamics Study. *Adv. Funct. Mater.* **2012**, *22*, 2495–2502. [CrossRef]

43. Liu, Y.; Huang, J.; Yang, B.; Sumpter, B.G.; Qiao, R. Duality of the interfacial thermal conductance in graphene-based nanocomposites. *Carbon* **2014**, *75*, 169–177. [CrossRef]

44. Kuila, T.; Bose, S.; Hong, C.E.; Uddin, M.E.; Khanra, P.; Kim, N.H.; Lee, J.H. Preparation of functionalized graphene/linear low density polyethylene composites by a solution mixing method. *Carbon* **2011**, *49*, 1033–1037. [CrossRef]

45. Garzón, C.; Palza, H. Electrical behavior of polypropylene composites melt mixed with carbon-based particles: Effect of the kind of particle and annealing process. *Compos. Sci. Technol.* **2014**, *99*, 117–123. [CrossRef]

46. Hu, Z.; Li, N.; Li, J.; Zhang, C.; Song, Y.; Li, X.; Wu, G.; Xie, F.; Huang, Y. Facile preparation of poly(*p*-phenylene benzobisoxazole)/graphene composite films via one-pot in situ polymerization. *Polymer* **2015**, *71*, 8–14. [CrossRef]

47. Yu, A.; Ramesh, P.; Itkis, M.E.; Elena Bekyarova, A.; Haddon, R.C. Graphite Nanoplatelet−Epoxy Composite Thermal Interface Materials. *J. Phys. Chem. C* **2007**, *111*, 7565–7569. [CrossRef]

48. Irwin, P.C.; Cao, P.; Bansal, A.; Schadler, L.S. Thermal and mechanical properties of polyimide nanocomposites. In Proceedings of the 2003 Annual Report Conference on Electrical Insulation and Dielectric Phenomena, Albuquerque, NM, USA, 19–22 October 2003; pp. 120–123. [CrossRef]

49. Min, C.; Yu, D.; Cao, J.; Wang, G.; Feng, L. A graphite nanoplatelet/epoxy composite with high dielectric constant and high thermal conductivity. *Carbon* **2013**, *55*, 116–125. [CrossRef]

50. Al-Saygh, A.; Ponnamma, D.; Almaadeed, M.; Poornima, V.P.; Karim, A.; Hassan, M. Flexible Pressure Sensor Based on PVDF Nanocomposites Containing Reduced Graphene Oxide-Titania Hybrid Nanolayers. *Polymers* **2017**, *9*, 33. [CrossRef]

51. Li, Y.; Lian, H.; Hu, Y.; Chang, W.; Cui, X.; Liu, Y. Enhancement in Mechanical and Shape Memory Properties for Liquid Crystalline Polyurethane Strengthened by Graphene Oxide. *Polymers* **2016**, *8*, 236. [CrossRef]

52. Kim, D.S.; Dhand, V.; Rhee, K.Y.; Park, S.J. Study on the Effect of Silanization and Improvement in the Tensile Behavior of Graphene-Chitosan-Composite. *Polymers* **2015**, *7*, 527–551. [CrossRef]

53. Kim, H.; Macosko, C.W. Processing-property relationships of polycarbonate/graphene composites. *Polymer* **2009**, *50*, 3797–3809. [CrossRef]

54. Tian, L.; Wang, Y.; Li, Z.; Mei, H.; Shang, Y. The thermal conductivity-dependant drag reduction mechanism of water droplets controlled by graphene/silicone rubber composites. *Exp. Therm. Fluid Sci.* **2017**, *85*, 363–369. [CrossRef]

55. Teng, C.C.; Ma, C.C.M.; Lu, C.H.; Yang, S.Y.; Lee, S.H.; Hsiao, M.C.; Yen, M.Y.; Chiou, K.C.; Lee, T.M. Thermal conductivity and structure of non-covalent functionalized graphene/epoxy composites. *Carbon* **2011**, *49*, 5107–5116. [CrossRef]

56. Song, S.H.; Park, K.H.; Kim, B.H.; Choi, Y.W.; Jun, G.H.; Lee, D.J.; Kong, B.S.; Paik, K.W.; Jeon, S. Enhanced thermal conductivity of epoxy-graphene composites by using non-oxidized graphene flakes with non-covalent functionalization. *Adv. Mater.* **2013**, *25*, 732–737. [CrossRef] [PubMed]

57. Wang, F.; Drzal, L.T.; Yan, Q.; Huang, Z. Mechanical properties and thermal conductivity of graphene nanoplatelet/epoxy composites. *J. Mater. Sci.* **2015**, *50*, 1082–1093. [CrossRef]

58. Wan, Y.; Tang, L.; Gong, L.; Yan, D.; Li, Y.; Wu, L.; Jiang, J.; Lai, G. Grafting of epoxy chains onto graphene oxide for epoxy composites with improved mechanical and thermal properties. *Carbon* **2014**, *69*, 467–480. [CrossRef]

59. Qian, R.; Yu, J.; Wu, C.; Zhai, X.; Jiang, P. Alumina-coated graphene sheet hybrids for electrically insulating polymer composites with high thermal conductivity. *RSC Adv.* **2013**, *3*, 17373–17379. [CrossRef]

60. Sun, R.; Yao, H.; Zhang, H.B.; Li, Y.; Mai, Y.W.; Yu, Z.Z. Decoration of defect-free graphene nanoplatelets with alumina for thermally conductive and electrically insulating epoxy composites. *Compos. Sci. Technol.* **2016**, *137*, 16–23. [CrossRef]

61. Kim, S.Y.; Ye, J.N.; Yu, J. Thermal conductivity of graphene nanoplatelets filled composites fabricated by solvent-free processing for the excellent filler dispersion and a theoretical approach for the composites containing the geometrized fillers. *Compos. A Appl. Sci. Manuf.* **2015**, *69*, 219–225. [CrossRef]

62. Gu, J.; Xie, C.; Li, H.; Dang, J.; Geng, W.; Zhang, Q. Thermal percolation behavior of graphene nanoplatelets/polyphenylene sulfide thermal conductivity composites. *Polym. Compos.* **2014**, *35*, 1087–1092. [CrossRef]

63. Dai, W.; Yu, J.; Liu, Z.; Wang, Y.; Song, Y.; Lyu, J.; Bai, H.; Nishimura, K.; Jiang, N. Enhanced thermal conductivity and retained electrical insulation for polyimide composites with SiC nanowires grown on graphene hybrid fillers. *Compos. A Appl. Sci. Manuf.* **2015**, *76*, 73–81. [CrossRef]

64. Chen, J.; Chen, X.; Meng, F.; Li, D.; Tian, X.; Wang, Z.; Zhou, Z. Super-high thermal conductivity of polyamide-6/graphene-graphene oxide composites through in situ polymerization. *High Perform. Polym.* **2016**, 585–594. [CrossRef]

65. Guo, W.; Chen, G. Fabrication of graphene/epoxy resin composites with much enhanced thermal conductivity via ball milling technique. *J. Appl. Polym. Sci.* **2014**, *131*, 338–347. [CrossRef]

66. Zong, P.; Fu, J.; Chen, L.; Yin, J.; Dong, X.; Yuan, S.; Shi, L.Y.; Deng, W. Effect of Aminopropylisobutyl Polyhedral Oligomeric Silsesquioxane Functionalized Graphene on the Thermal Conductivity and Electrical Insulation Properties of Epoxy Composites. *RSC Adv.* **2016**, *6*, 10498–10506. [CrossRef]

67. Ma, W.S.; Li, W.; Fang, Y.; Wang, S.F. Non-covalently modified reduced graphene oxide/polyurethane nanocomposites with good mechanical and thermal properties. *J. Mater. Sci.* **2014**, *49*, 562–571. [CrossRef]

68. Cho, E.C.; Huang, J.H.; Li, C.P.; Chang-Jian, C.W.; Lee, K.C.; Hsiao, Y.S.; Huang, J.H. Graphene-based thermoplastic composites and their application for LED thermal management. *Carbon* **2016**, *102*, 66–73. [CrossRef]

69. Tang, Z.; Kang, H.; Shen, Z.; Guo, B.; Zhang, L.; Jia, D. Grafting of Polyester onto Graphene for Electrically and Thermally Conductive Composites. *Macromolecules* **2012**, *45*, 3444–3451. [CrossRef]

70. Varenik, M.; Nadiv, R.; Levy, I.; Vasilyev, G.; Regev, O. Breaking through the Solid/Liquid Processability Barrier: Thermal Conductivity and Rheology in Hybrid Graphene-Graphite Polymer Composites. *ACS Appl. Mater. Interfaces* **2017**, 7556–7564. [CrossRef] [PubMed]

71. Ling, J.; Zhai, W.; Feng, W.; Shen, B.; Zhang, J.; Zheng, W.G. Facile Preparation of Lightweight Microcellular Polyetherimide/Graphene Composite Foams for Electromagnetic Interference Shielding. *ACS Appl. Mater. Interfaces* **2013**, *5*, 2677–2684. [CrossRef] [PubMed]

72. Li, M.; Gao, C.; Hu, H.; Zhao, Z. Electrical conductivity of thermally reduced graphene oxide/polymer composites with a segregated structure. *Carbon* **2013**, *65*, 371–373. [CrossRef]

73. Hu, H.; Zhang, G.; Xiao, L.; Wang, H.; Zhang, Q.; Zhao, Z. Preparation and electrical conductivity of graphene/ultrahigh molecular weight polyethylene composites with a segregated structure. *Carbon* **2012**, *50*, 4596–4599. [CrossRef]

74. Lin, Y.; Liu, S.; Liu, L. A new approach to construct three dimensional segregated graphene structures in rubber composites for enhanced conductive, mechanical and barrier properties. *J. Mater. Chem. C* **2016**, *4*, 2353–2358. [CrossRef]

75. Wang, M.; Duan, X.; Xu, Y.; Duan, X. Functional Three-Dimensional Graphene/Polymer Composites. *ACS Nano* **2016**, *10*, 7231. [CrossRef] [PubMed]

76. Zhang, Y.F.; Han, D.; Zhao, Y.H.; Bai, S.L. High-performance thermal interface materials consisting of vertically aligned graphene film and polymer. *Carbon* **2016**, *109*, 552–557. [CrossRef]

77. Li, Q.; Guo, Y.; Li, W.; Qiu, S.; Zhu, C.; Wei, X.; Chen, M.; Liu, C.; Liao, S.; Gong, Y. Ultrahigh Thermal Conductivity of Assembled Aligned Multilayer Graphene/Epoxy Composite. *Chem. Mater.* **2014**, *26*, 4459–4465. [CrossRef]

78. Li, X.; Shao, L.; Song, N.; Shi, L.; Ding, P. Enhanced thermal-conductive and anti-dripping properties of polyamide composites by 3D graphene structures at low filler content. *Compos. A Appl. Sci. Manuf.* **2016**, *88*, 305–314. [CrossRef]

79. Yang, J.; Qi, G.Q.; Liu, Y.; Bao, R.Y.; Liu, Z.Y.; Yang, W.; Xie, B.H.; Yang, M.B. Hybrid graphene aerogels/phase change material composites: Thermal conductivity, shape-stabilization and light-to-thermal energy storage. *Carbon* **2016**, *100*, 693–702. [CrossRef]

80. Liu, D.; Yang, P.; Yuan, X.; Guo, J.; Liao, N. The defect location effect on thermal conductivity of graphene nanoribbons based on molecular dynamics. *Phys. Lett. A* **2015**, *379*, 810–814. [CrossRef]

81. Song, N.; Jiao, D.; Cui, S.; Hou, X.; Ding, P.; Shi, L. Highly Anisotropic Thermal Conductivity of Layer-by-Layer Assembled Nanofibrillated Cellulose/Graphene Nanosheets Hybrid Films for Thermal Management. *ACS Appl. Mater. Interfaces* **2017**, *9*, 2924. [CrossRef] [PubMed]

82. Kumar, P.; Yu, S.; Shahzad, F.; Hong, S.M.; Kim, Y.H.; Chong, M.K. Ultrahigh electrically and thermally conductive self-aligned graphene/polymer composites using large-area reduced graphene oxides. *Carbon* **2016**, *101*, 120–128. [CrossRef]

83. Jung, H.; Yu, S.; Bae, N.S.; Cho, S.M.; Kim, R.H.; Cho, S.H.; Hwang, I.; Jeong, B.; Ji, S.R.; Hwang, J.Y. High through-plane thermal conduction of graphene nanoflake filled polymer composites melt-processed in an L-shape kinked tube. *ACS Appl. Mater. Interfaces* **2015**, *7*, 15256. [CrossRef] [PubMed]

84. Yan, H.; Tang, Y.; Long, W.; Li, Y. Enhanced thermal conductivity in polymer composites with aligned graphene nanosheets. *J. Mater. Sci.* **2014**, *49*, 5256–5264. [CrossRef]

85. Alam, F.E.; Dai, W.; Yang, M.; Du, S.; Li, X.; Yu, J.; Jiang, N.; Lin, C.T. In situ formation of a cellular graphene framework in thermoplastic composites leading to superior thermal conductivity. *J. Mater. Chem. A* **2017**, 6164–6169. [CrossRef]

86. Wu, K.; Lei, C.; Huang, R.; Yang, W.; Chai, S.; Geng, C.; Chen, F.; Fu, Q. Design and Preparation of a Unique Segregated Double Network with Excellent Thermal Conductive Property. *ACS Appl. Mater. Interfaces* **2017**, *9*, 7637–7647. [CrossRef] [PubMed]

87. Li, A.; Zhang, C.; Zhang, Y.F. RGO/TPU composite with a segregated structure as thermal interface material. *Compos. A Appl. Sci. Manuf.* **2017**, 108–114. [CrossRef]

88. Yang, J.; Li, X.; Han, S.; Zhang, Y.; Min, P.; Koratkar, N.; Yu, Z.Z. Air-dried, high-density graphene hybrid aerogels for phase change composites with exceptional thermal conductivity and shape stability. *J. Mater. Chem. A* **2016**, 18067–18074. [CrossRef]

89. Liu, Z.; Shen, D.; Yu, J.; Dai, W.; Li, C.; Du, S.; Jiang, N.; Li, H.; Lin, C.T. Exceptionally high thermal and electrical conductivity of three-dimensional graphene-foam-based polymer composites. *RSC Adv.* **2016**, *6*, 22364–22369. [CrossRef]

90. Zhao, Y.H.; Wu, Z.K.; Bai, S.L. Study on thermal properties of graphene foam/graphene sheets filled polymer composites. *Compos. A Appl. Sci. Manuf.* **2015**, *72*, 200–206. [CrossRef]

91. Zabihi, Z.; Araghi, H. Effect of functional groups on thermal conductivity of graphene/paraffin nanocomposite. *Phys. Lett. A* **2016**, *380*, 3828–3831. [CrossRef]

92. Adamyan, V.; Zavalniuk, V. Lattice thermal conductivity of graphene with conventionally isotopic defects. *J. Phys. Condens. Matter.* **2012**, *24*, 415401. [CrossRef] [PubMed]

93. Kim, J.Y.; Lee, J.H.; Grossman, J.C. Thermal transport in functionalized graphene. *ACS Nano* **2012**, *6*, 9050–9057. [CrossRef] [PubMed]

94. Malekpour, H.; Ramnani, P.; Srinivasan, S.; Balasubramanian, G.; Nika, D.L.; Mulchandani, A.; Lake, R.K.; Balandin, A.A. Thermal conductivity of graphene with defects induced by electron beam irradiation. *Nanoscale* **2016**, *8*, 14608–14616. [CrossRef] [PubMed]

95. Xin, G.; Sun, H.; Scott, S.M.; Yao, T.; Lu, F.; Shao, D.; Hu, T.; Wang, G.; Ran, G.; Lian, J. Advanced Phase Change Composite by Thermally Annealed Defect-Free Graphene for Thermal Energy Storage. *ACS Appl. Mater. Interfaces* **2014**, *6*, 15262–15271. [CrossRef] [PubMed]

96. Chu, K.; Li, W.S.; Dong, H. Role of graphene waviness on the thermal conductivity of graphene composites. *Appl. Phys. A* **2013**, *111*, 221–225. [CrossRef]

97. Shen, X.; Wang, Z.; Wu, Y.; Liu, X.; He, Y.B.; Kim, J.K. Multilayer Graphene Enables Higher Efficiency in Improving Thermal Conductivities of Graphene/Epoxy Composites. *Nano Lett.* **2016**, *16*, 3585–3593. [CrossRef] [PubMed]

98. Kim, H.S.; Bae, H.S.; Yu, J.; Kim, S.Y. Thermal conductivity of polymer composites with the geometrical characteristics of graphene nanoplatelets. *Sci. Rep.* **2016**, *6*, 26825. [CrossRef] [PubMed]

99. Zhan, Y.; Lavorgna, M.; Buonocore, G.; Xia, H. Enhancing electrical conductivity of rubber composites by constructing interconnected network of self-assembled graphene with latex mixing. *J. Mater. Chem.* **2012**, *22*, 10464–10468. [CrossRef]

100. Chakraborty, I.; Bodurtha, K.J.; Heeder, N.J.; Godfrin, M.P.; Tripathi, A.; Hurt, R.H.; Shukla, A.; Bose, A. Massive electrical conductivity enhancement of multilayer graphene/polystyrene composites using a nonconductive filler. *ACS Appl. Mater. Interfaces* **2014**, *6*, 16472–16475. [CrossRef] [PubMed]

101. Khan, M.F.S.; Alexander, A.B. Graphene–Multilayer Graphene Nanocomposites as Highly Efficient Thermal Interface Materials. *Naon Lett.* **2012**, *12*, 861–867. [CrossRef]

102. Li, A.; Zhang, C.; Zhang, Y.F. Graphene nanosheets-filled epoxy composites prepared by a fast dispersion method. *J. Appl. Polym. Sci.* **2017**, *134*, 45152. [CrossRef]

103. Gao, Y.; Müllerplathe, F. Increasing the Thermal Conductivity of Graphene-Polyamide-6,6 Nanocomposites by Surface-Grafted Polymer Chains: Calculation with Molecular Dynamics and Effective-Medium Approximation. *J. Phys. Chem. B* **2016**, *120*, 1336–1346. [CrossRef] [PubMed]

104. Wang, Y.; Zhan, H.F.; Xiang, Y.; Yang, C.; Wang, C.M.; Zhang, Y.Y. Effect of Covalent Functionalization on Thermal Transport across Graphene–Polymer Interfaces. *J. Phys. Chem. C* **2015**, *119*, 12731–12738. [CrossRef]

105. Eslami, H.; Mohammadzadeh, L.; Mehdipour, N. Anisotropic heat transport in nanoconfined polyamide-6,6 oligomers: Atomistic reverse nonequilibrium molecular dynamics simulation. *J. Chem. Phys.* **2012**, *136*, 104901. [CrossRef] [PubMed]

106. Eslami, H.; Mohammadzadeh, L.; Mehdipour, N. Reverse nonequilibrium molecular dynamics simulation of thermal conductivity in nanoconfined polyamide-6,6. *J. Chem. Phys.* **2011**, *135*, 064703. [CrossRef] [PubMed]

107. Prasher, R.S.; Matayabas, J.C. Thermal contact resistance of cured gel polymeric thermal interface material. *IEEE Trans. Compon. Packag. Technol.* **2004**, *27*, 702–709. [CrossRef]

108. Liu, J.; Michel, B.; Rencz, M.; Tantolin, C.; Sarno, C.; Miessner, R.; Schuett, K.V.; Tang, X.; Demoustier, S.; Ziaei, A. Recent progress of thermal interface material research—An overview. In Proceedings of the 14th International Workshop on Thermal Inveatigation of ICs and Systems (THERMINIC 2008), Rome, Italy, 24–26 September 2008; pp. 156–162. [CrossRef]

109. And, M.M.; Winey, K.I. Polymer Nanocomposites Containing Carbon Nanotubes. *Macromolecules* **2006**, *39*, 543–545. [CrossRef]

110. Xu, J.; Munari, A.; Dalton, E.; Mathewson, A. Silver nanowire array-polymer composite as thermal interface material. *J. Appl. Phys.* **2009**, *106*, 124310. [CrossRef]

111. Alkan, C.; Sarı, A.; Karaipekli, A. Preparation, thermal properties and thermal reliability of microencapsulated n-eicosane as novel phase change material for thermal energy storage. *Energy Convers. Manag.* **2011**, *52*, 687–692. [CrossRef]

112. Ji, H.; Sellan, D.P.; Pettes, M.T.; Kong, X.; Ji, J.; Shi, L.; Ruoff, R.S. Enhanced thermal conductivity of phase change materials with ultrathin-graphite foams for thermal energy storage. *Energy Environ. Sci.* **2014**, *7*, 1185–1192. [CrossRef]

113. Mehrali, M.; Latibari, S.T.; Mehrali, M.; Metselaar, H.S.M.; Silakhori, M. Shape-stabilized phase change materials with high thermal conductivity based on paraffin/graphene oxide composite. *Energy Convers. Manag.* **2013**, *67*, 275–282. [CrossRef]

114. Zhong, Y.; Zhou, M.; Huang, F.; Lin, T.; Wan, D. Effect of graphene aerogel on thermal behavior of phase change materials for thermal management. *Sol. Energy Mater. Sol. Cells* **2013**, *113*, 195–200. [CrossRef]

115. Khan, W.S.; Asmatulu, R.; Rodriguez, V.; Ceylan, M. Enhancing thermal and ionic conductivities of electrospun PAN and PMMA nanofibers by graphene nanoflake additions for battery-separator applications. *Int. J. Energy Res.* **2015**, *38*, 2044–2051. [CrossRef]

116. Hallaj, S.A.; Selman, J.R. ChemInform Abstract: A Novel Thermal Management System for Electric Vehicle Batteries Using Phase-Change Material. *Cheminform* **2001**, *32*, 3231–3236. [CrossRef]

Review

Intercalation Polymerization Approach for Preparing Graphene/Polymer Composites

Yifan Guo [1], Fuxi Peng [1], Huagao Wang [1], Fei Huang [1], Fanbin Meng [1,*], David Hui [2] and Zuowan Zhou [1,*]

[1] Key Laboratory of Advanced Technologies of Materials (Ministry of Education), School of Materials Science and Engineering, Southwest Jiaotong University, Chengdu 610031, China; yfguo@my.swjtu.edu.cn (Y.G.); fuxipeng@my.swjtu.edu.cn (F.P.); huagaowang@163.com (H.W.); fhuang0137@163.com (F.H.)
[2] Department of Mechanical Engineering, University of New Orleans, New Orleans, LA 70148, USA; dhui@uno.edu
*   Correspondence: mengfanbin@swjtu.edu.cn (F.M.); zwzhou@swjtu.edu.cn (Z.Z.); Tel.: +86-028-8760-0454

Received: 5 December 2017; Accepted: 5 January 2018; Published: 10 January 2018

Abstract: The rapid development of society has promoted increasing demand for various polymer materials. A large variety of efforts have been applied in order for graphene strengthened polymer composites to satisfy different requirements. Graphene/polymer composites synthesized by traditional strategies display some striking defects, like weak interfacial interaction and agglomeration of graphene, leading to poor improvement in performance. Furthermore, the creation of pre-prepared graphene while being necessary always involves troublesome processes. Among the various preparation strategies, an appealing approach relies on intercalation and polymerization in the interlayer of graphite and has attracted researchers' attention due to its reliable, fast and simple synthesis. In this review, we introduce an intercalation polymerization strategy to graphene/polymer composites by the intercalation of molecules/ions into graphite interlayers, as well as subsequent polymerization. The key point for regulating intercalation polymerization is tuning the structure of graphite and intercalants for better interaction. Potential applications of the resulting graphene/polymer composites, including electrical conductivity, electromagnetic absorption, mechanical properties and thermal conductivity, are also reviewed. Furthermore, the shortcomings, challenges and prospects of intercalation polymerization are discussed, which will be helpful to researchers working in related fields.

Keywords: graphene/polymer composites; intercalation of graphite; exfoliation intercalation polymerization; interaction

1. Introduction

Graphene, a single atom-thick sheet composed of sp^2-hybridized carbon, has received considerable attention since its first fabrication through mechanical exfoliation in 2004 [1]. Favored for its unique two-dimensional structure and extraordinary electrical [2,3], thermal [4,5], mechanical properties [6–9], graphene is widely researched in energy storage and conversion, spintronic devices, photonics and optoelectronics and other kinds of materials. In recent years, the hybridization or composites based on graphene and its derivatives has attracted much interest in physics, chemistry and materials domains. Among this research, the introduction of graphene in polymer significantly increases Young's modulus [10–12], and electrical [13–15] or thermal conductivity [16–18] of polymer composites, particularly at low volume fractions (<1 wt %). Moreover, some special properties of the polymer composites such as shape memory [19–21], chemiluminescence [22] and microwave absorption [23–26] may emerge, resulting from the interaction between graphene and polymer.

Melting blend and solution mixing are the most economically attractive and scalable methods for prepared graphene/polymer composites [27–31]. However, agglomerate pre-prepared graphene is always hard to disperse in polymer melt or solution because of their high viscosity. Moreover, interfacial interactions between the graphene and polymer matrix are weak, resulting in low enhancement of polymer properties. In situ polymerization after the dispersion of graphene in a monomer is another way to synthesize graphene/polymer composites [32,33]. On the one hand, a particular monomer can be used to disperse graphene in the system [34]. On the other hand, dispersed graphene layers act as the hard template of polymerization, leading to strong intercalation between graphene and polymer [35,36].

Pre-prepared graphene is needed when the aforementioned methods are used to process the graphene/polymer composites. According to recent reports, bottom-up approaches, including chemical vapor deposition (CVD) [37,38] and epitaxial growth [39], are widely used to produce high-quality graphene. Although large crystal domain, specific layer graphene can be synthesized via tuning of carbon source and growth conditions, the high cost and low yield of these methods associated with difficulties in exfoliating graphene from the substrate limit their application in industrial production. Therefore, most of the graphene used in the further processes is produced by exfoliation of natural graphite (NG) or highly ordered pyrolytic graphite, named top-down approaches. Among these approaches, dry exfoliation performed by using mechanical, electrostatic or electromagnetic forces can result in grain boundary-free graphene [1]. However, these approaches are impractical for large-scale applications. The thickness and size of graphene layers can hardly be controlled, and it is thus unsuitable for use in composite preparation. By comparison, sonication-assisted liquid phase exfoliation in reasonable solvents has been considered as one of the most promising routes for the mass production of low-cost and high-quality graphene [40–42]. However, the long time required for sonication and the further purification process (always involving ultracentrifugation) may limit the production period when applied in industry-scale production. For the reduction of graphene oxide (GO), the synthesis of graphite oxide (GtO) always involves successive oxidative treatments containing a strong acid and oxidant. Only in recent years have some efforts been made to avoid of using such environmentally damaging substances [43,44]. Moreover, chemical oxidation always introduces an oxygen-containing functional group in the basal plane or edges, acting as active sites for further modification and functional applications such as biosensing, catalytic, electromagnetic waves absorption, supercapacitors etc. [45] On the other hand, oxidation of graphene leads to damage of the basal plane, thus degrading some properties relying on the perfect crystalline structure (typically tensile strength and electrical or thermal conductivity) [44]. But the π–π conjugate can be partially recovered relying on the reduction degree of GO [46]. Furthermore, even if high-quality graphene could be produced on a large scale, the pre-prepared graphene powder or concentrated slurry is always difficult to disperse uniformly whether in a polymer matrix or monomer solution. This usually results in a limited performance improvement in graphene/polymer composites.

In recent years, the exfoliation of graphite intercalated compounds (GIC) has been deemed another interesting approach to realizing the exfoliation of graphite to the graphene layer. GIC is formed by insertion of particular atomic or molecular layers between the layers of graphite. The weak Van der Waals interaction, a distance of 3.35 Å, and an abundant π-electron cloud between graphite layers ensure the intercalation process of alkali metal [47,48], sulphuric [49] and some metal chloride [50,51]. The graphene layer can be then easily exfoliated with the assistant of mechanical or heat treatment [52]. This approach has gained attention due to easily available raw materials, its simple operation and high-quality products. Furthermore, industry-scale production can be expected to be based on this method.

Inspired by the exfoliation of GIC, in situ intercalation polymerization using organic monomers was recently proposed for synthesizing a graphene/polymer composite in a one-step process [53]. Benefiting from an abundant π-electron cloud, different kinds of monomer cation can penetrate into the interlayer of graphite and subsequently polymerize in the gap of the planes. While the intercalants weaken the inter-planar bonding, polymerization then separates the layers from the intergallery, resulting in the formation of graphene/polymer composites. Interest in the reliable, fast and simple synthesis means that intercalation polymerization has gained more attention in the strategies of graphene/polymer composites preparation. Therefore, how to intercalate molecules/ions/clusters into graphite, and how to conduct polymerization in the graphite interlayers, are now research topics. In this review, we discuss the recent progress of the intercalation of graphite, including inorganic-GIC mostly synthesized by two-zone vapor transport and electrochemistry methods; and organic intercalating compounds synthesized by electrochemistry, cation exchange or chemical methods. Furthermore, polymerization conducted in graphite interlayers, which can be divided into monomers initiated by pre-intercalated compounds and polymerization of intercalative monomers (in situ intercalation polymerization), are reviewed here. Some regular results, shortcomings, challenges, and prospects of intercalation methods and interlayer polymerization are also suggested. Potential applications of graphene/polymer composites prepared by intercalation polymerization, including electrical conductivity, electromagnetic absorption, mechanical properties and thermal conductivity, are introduced, which will be helpful to people working in related fields.

2. Intercalation of Graphite

2.1. Traditional Graphite Intercalated Compounds (GIC)

The capture of organic monomers in the interlayer of graphite is the prerequisite for intercalation polymerization and the consequent exfoliation of graphene. Therefore, the intercalation of molecules, ions or clusters is one of the key issues in the process. In fact, the intercalation of graphite has been researched for more than one hundred years since the first synthesis of GIC reported by Schaffäutl (1841). Owing to the layered structure, natural graphite provides shelter for guest molecules with subnanometer interlayer distance. While graphite can act as an electron donor or acceptor based on the reaction conditions [54], hundreds of kinds of atomic and molecular layers with various physical/chemical characteristics, have been intercalated into the interlayer space of graphite host material to form GIC [55].

GIC can be generally classified in terms of a "stage index n", where n means the number of graphite layers between two adjacent intercalant layers. As shown in Figure 1, for example, GIC with stage of 1 indicates that 1 graphene layer is covered by adjacent intercalant layers. What should be mentioned here is that the intercalant layers can be more than 1 atom thick. Since most intercalants are inorganic, the formed GIC are generally classified according to the electrons that are donors or acceptors of intercalants. The most widely used donor intercalants are alkali metals [48]. Other donors like alkaline earth metals [56,57] and lanthanides [58–60] can also be used to synthesize donor GIC. When it comes to electron acceptor intercalants, a very large variety of compounds have been prepared using Lewis acid intercalants such as halogen [61], metal chlorides [50,51,62], bromides [63], fluorides [64] and oxyhalides [55], acidic oxides such as SO_3, and strong Brønsted acids [65,66] such as H_2SO_4 or HNO_3. The dominant method for synthesizing GIC is the two-zone vapour transport method [67–69]. Intercalation in intercalants that are molten [62,70] or in solution [71] can also obtain GIC. Apparently, the intercalation process is dominated by the donor–acceptor interaction between host graphite and guest intercalants. Another way to achieve intercalation is to utilize electrochemical reactions. The graphite can act as either an anodic electrode or cathodic electrode depending on the electrophile or nucleophile of intercalants [72–76]. It is worth mentioning that if the graphite was applied as an anode, the lithium ion can penetrate into the graphite layer and recombine with electrons in the intergallery to form stable intercalation compounds [77]. This process has been developed to commercialize lithium-ion batteries [78,79] and further improved in aluminum-based batteries [80] when an aluminum foil anode and ionic liquid electrolyte are used.

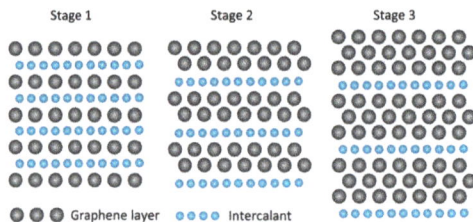

Figure 1. Schematic illustration of graphite intercalated compounds (GIC). Adapted with permission from [81]. Copyright © 2012 Elsevier.

2.2. Organic Intercalating Compounds

2.2.1. Electrochemical Methods

Similarly, organic molecules or ions can also achieve intercalation, but only a few studies have been reported. The intercalation of organic molecules by electrochemical methods is always regarded as a side effect of lithium-ion battery charging. Ionic liquids like *N*-butyl-*N*-methylpyrrolidinium bis(trifluoromethanesulfonyl)imide (Pyr$_{14}$TFSI) or its smaller derivative, and other carbonates like propylene carbonate (PC), dimethyl sulphoxide (DMSO) and dimethylformamide (DMF)s are typically applied as an electrolyte for a lithium-ion battery. When charging, the *N*-butyl-*N*-methylpyrrolidinium cation (Pyr$_{14}^+$) [82,83] or PC [84] or other electrolyte molecules [85–88] can easily co-intercalate into the graphite anode with lithium ions. Besides the lithium-ion battery system, intercalation of organic molecules/ions by electrochemical methods mostly exist in the co-intercalation phenomenon with AlCl$_4^-$, PF$_6^-$, ClO$_4^-$ et al. [80,89–91]. Palermo et al. [91] reported that acetonitrile can co-intercalate with ClO$_4^-$, and ClO$_4^-$ and is indispensable in the intercalation process. This process involves intercalation of the large and negatively charged ClO$_4^-$ through grain boundaries or defect sites of a graphite anode, which favor the further penetration of the smaller, uncharged acetonitrile molecules. What should be mentioned here is that although most research focuses on organic molecules for co-intercalation, sporadic investigations indicate that organic ions can singly penetrate the interlayer of graphite in a special environment, for examples, dual-graphite cells [82,83] schemed in Figure 2. When Pyr$_{14}$TFSI and lithium bis(trifluoromethanesulfonyl)imide (LiTFSI) were applied as an electrolyte, the graphite anode can be intercalated by bis(trifluoromethanesulfonyl)imide anion (TFSI$^-$) individually in the charging process.

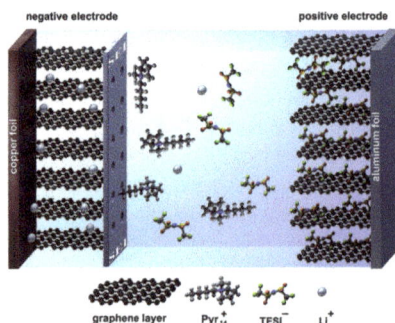

Figure 2. Schematic illustration of a dual-graphite cell with no effective solid electrolyte interphase layer at the graphite anode during the charge process. The negative graphite electrode suffers from exfoliation reactions caused by co-intercalation of the relatively large Pyr$_{14}^+$ cations [82]. Published by The Royal Society of Chemistry.

2.2.2. Cation Exchange Methods

Cation exchange is another effective method for the intercalation of organic compound [92,93]. This idea follows a similar mechanism to the intercalation of montmorillonite [94], but unfortunately, no cation lives in the intergallery of pure graphite. Therefore, graphite should be pre-treated to ensure enough cation in its interlayers. Lerner et al. [95] used GIC as raw material, and the Na-ethylenediamine complex in the interlayers can be easily displaced by tetrabutylammonium ion (TBA^+) in DMF through a cation-exchange reaction. Moreover, cation exchange can also perform in the electrochemical process. While Li^+ have intercalated into the graphite cathode in charging, positively charged TBA^+ can penetrate into the graphite lattice by cation exchange with the intercalated lithium ions [96]. However, electrodecomposition of the intercalated TBA^+ appears in this reaction, and thus it is hard to obtain a stable TBA^+ intercalated compound.

2.2.3. Chemical Methods

Organic molecules can also directly intercalate into graphite layers by chemical methods, but this method always involve co-intercalation with alkali metal cations [71,97]. Metallic Li, Na, or K together with 1,2-diaminopropane (1,2-DAP) can realize co-intercalation with the protection of inert gas, but this process always takes a long time (1–3 days) [71]. The resulting compounds show different orientations of 1,2-DAP in the interlays, depending on the co-intercalated alkali metal.

However, the intercalation of pure organic molecules by chemical methods is far more difficult than co-intercalation with the help of alkali metals. Limited research has been done to successfully synthesize organic GIC using only graphite and organic intercalants. Although it is hard to form an organic layer in the graphite gallery, a limited number of organic molecules can still intercalate into graphite by π–π or cation–π intercalation between intercalants and graphite. Naphthalene, which consists of a fused pair of benzene rings, can penetrate into the edge of graphite, without further intercalation, acting as a "molecular wedge" [98]. This result was confirmed by the slight shift and obvious intensity decrease of the (002) plane of graphite in an X-ray diffraction (XRD) pattern. Similar results were obtained for the intercalation of cationic aniline (denoted as ANi^+) [53] and caprolactam onium ion (denoted as CL^+) [18], although the major driving force for intercalation is cation–π intercalation rather than π–π intercalation.

The intercalation of organic molecules into the graphite crystal is intrinsically impeded by the interlay's Van der Waals interaction. Therefore, weakening of the inter-plane interaction would significantly facilitate the intercalation process. The most widely used method is oxidation of graphite. As schemed in Figure 3, natural graphite oxidized by low-concentration $KMnO_4$ at relatively higher temperature can lead to edge-selectively oxidized graphite (EOG) with low-degree oxidation. Long-chain tetradecyl-ammonium cation ($C_{14}N^+$) can then spontaneously intercalate into graphite, forming an integrated $C_{14}N^+$ layer in the graphite gallery [99]; in other words, intercalation compounds. If there is a higher oxidation degree for graphite, it may transform into graphite oxide with a larger distance and weaker interaction between graphite planes, making it easier to capture more and larger molecules, for example, tetraalkylammonium ions (TAA^+) [100], alcohol [101] or even polymers like poly(vinyl alcohol) (PVA) [102], poly(diallyldimethylammonium chloride) (PDDA) [103] and poly(vinyl acetate) (PVAc) [104] etc.

Despite the difficulty in forming an organic layer, the intercalation of special molecules into expanded graphite (EG) or natural graphite has been confirmed, as mentioned above. Basically, the driving force for intercalation was firstly due to π–π intercalation between intercalants and graphite. This idea is proved by the fact that naphthalene and aniline (ANi), both of which possess benzene rings, can intercalate into graphite layers [53,98]. As shown in Figure 4, first-principle simulation of the intercalation of ANi molecule into bilayer graphene was performed by Zhou et al. [53]. The positive formation energy of 2.01 eV proved its energetically favorable reaction. Meanwhile, it was noticed that the cationic ANi would be easier to intercalate into the graphite layers as ANi^+ obtains higher formation energy of 2.81 eV, and experimental data further confirmed the simulation results. It seems that the cation–π interaction between the intercalary cation and the graphite interlayer of the π-electron is another important force for intercalation. This theory was soon authenticated by the further study on the intercalation of CL⁺ [18]. By comntrast with ANi⁺, CL⁺ do not have a benzene-like structure, and thus there is no π–π intercalation between CL⁺ and graphite. Consequently, the intercalation force is almost all attributed to the cation–π interaction. Moreover, research also indicates that the adsorption of cation on the graphite surface can significantly decrease the interaction between the graphite layers [18], facilitating the succeeding intercalation of organic cation.

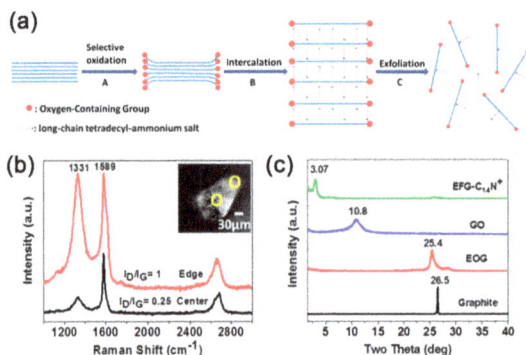

Figure 3. (**a**) Schematic illustration of the intercalation of edge-selective oxidized graphite (EOG); (**b**) micro-Raman spectra measured at the edge and on the basal plane of EOG; and (**c**) X-ray diffraction (XRD) of graphite, EOG, graphene oxide (GO) and EOG-C₁₄N⁺ intercalated compound. Adapted with permission from [99]. Copyright © 2013 Springer Nature.

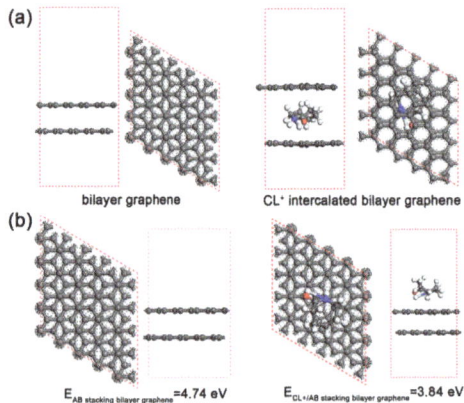

Figure 4. (**a**) The geometric structures of bilayer graphene and CL⁺ intercalated bilayer graphene; and (**b**) the calculated interlayer binding energy of the AB stacking bilayer graphene and CL⁺ absorbed bilayer graphene. Adapted with permission from [18]. Copyright © 2017 Elsevier.

3. Polymerization in the Interlayers of Graphite

3.1. Intercalation Polymerization Methods

Polymerization in the interlayers of graphite can be generally divided into two strategies as illustrated in Figure 5: polymerization initiated by pre-intercalated compounds and polymerization initiated after the intercalation of monomers (in situ intercalation polymerization).

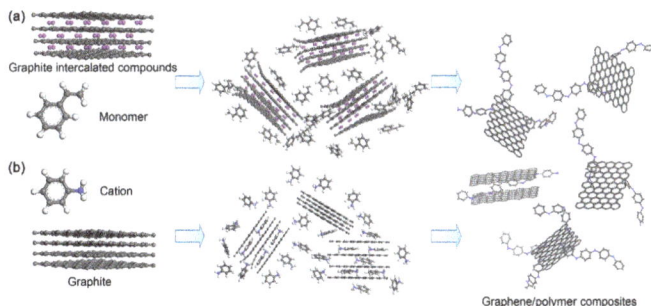

Figure 5. Schematic of the two kinds of polymerization in the interlayer of graphite: (**a**) polymerization initiated by pre-intercalated compounds; and (**b**) polymerization initiated after intercalation of monomers (in situ intercalation polymerization).

3.1.1. Polymerization Initiated by Pre-Intercalated Compounds

For this situation, GIC is always used as pre-intercalated compounds. When graphite is intercalated by alkali metals, an electron cloud of the alkali metal tends to migrate to graphite, thus forming an ionic compound [55]. Then, anionic polymerization of vinyl or epoxide monomers can be initiated by the negatively charged graphite layer of the alkali metal–GIC [105,106]. However, limited by the interlayer distance of GIC, monomers are hard to absorb into the interlayer of graphite in solution for further polymerization [107]. Instead, unsaturated hydrocarbon vapor such as styrene or isoprene were used to penetrate the interlayer galleries of potassium intercalated graphite, and then these underwent anionic polymerization, leading to the gradual expansion of the distance between graphite layers and the final exfoliation of graphite nanosheets [108,109]. It should be noted that the stage of alkali metal–GIC seems to be important for controlling the intercalation polymerization. For example, when KC_{24} (stage 2 potassium intercalated graphite) is used as the initiator, the reaction rate of intercalation polymerization can be several times faster than that of KC_8 (stage 1 potassium intercalated graphite) [108]. However, KC_8 exhibits much more effective exfoliation of graphite layers, while the products obtained from higher-stage GIC are mixed with un-exfoliated graphite [105].

Besides the intercalation polymerization initiated by alkali metal–GIC, some interesting work has been reported to synthesize polymer functionalized graphene nanoribbons (GNRs) using multiwalled carbon nanotubes (MWCNTs) as raw material [110]. In an analogy to the intercalation chemistry of graphite, the intercalation of potassium vapor or solvent-stabilized potassium cations into MWCNTs can lead to an expansion of the d-space between MWCNT layers, causing the MWCNTs to partially or fully split [111–113]. Thus, the fissures are functioned with aryl anions and their associated metal cations and converted into edge-negatively charged macroinitiators for the subsequent anionic polymerization of vinyl monomers [110]. This strategy can be described in Figure 6. Furthermore, the active carboanionic edge of unzipped MWCNTs can be further functioned by *N*-vinylformamide to act as nucleophilic agents and initiate a polymerization of epoxy resin (Figure 7) [106]. Thus, GNR functioned with different kinds of polymers can be synthesized following this idea [114–116]. Since the active carboanionic site mostly appears at the edges of GNR, it would always result in site-selective polymerization. Therefore, this strategy leads to polymer functionalized edges of graphene nanoribbon, but the basal planes can still remain sp^2-hybridized carbon [110].

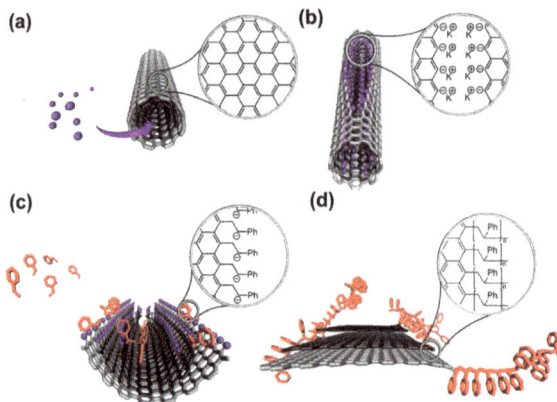

Figure 6. Reaction scheme of multiwalled carbon nanotubes (MWCNTs) unzipping and edge-selective in situ polymerization of vinyl monomers. (**a**) Intercalation of MWCNTs by potassium naphthalenide; (**b**) formation of longitudinal fissure in the walls; (**c**) polymerization of styrene assists in exfoliation of MWCNTs; (**d**) polymer functionalized GNRs. Adapted with permission from [110]. Copyright © 2013 American Chemical Society.

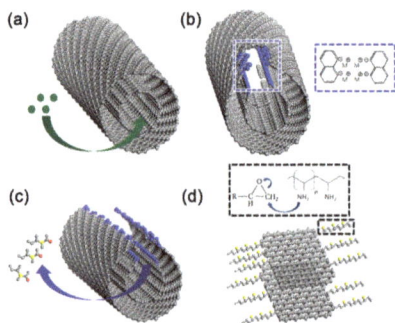

Figure 7. Reaction scheme of unzipping and edge-functioned MWCNTs for initiating polymerization of the epoxy resin. (**a**) intercalation of MWCNTs by alkali metal; (**b**) longitudinal unzipping and formation of carbanions, stabilized by cation; (**c**) in situ functionalization of unzipped MWCNTs by *N*-vinylformamide; (**d**) edge-functioned MWCNTs formation upon *N*-vinylformamide hydrolysis. The polymerization reaction is marked by dashed box. Adapted with permission from [106]. Copyright © 2016 Elsevier.

3.1.2. In Situ Intercalation Polymerization

As mentioned, some kinds of organic molecules can intercalate into graphite by π–π or cation–π intercalation between intercalants and graphite interlayers. Although a limited amount of molecules can penetrate into the interlayer of graphite, these polymerizable monomers can be initiated by subsequently added initiators. Zhou et al. performed the polymerization of aniline confined in graphite layers, resulting in graphene/polyaniline (PANi) hybrids by a one-step in situ intercalation polymerization [53]. An alogous method was then applied to prepare polypyrrole (PPy) or polyamide-6 (PA-6)/graphite nanoflake composites, confirming the universality of in situ intercalation polymerization [18,117]. This strategy is summarized in Figure 8. Monomer cations absorb on the surface of graphite to decrease the interaction between graphite layers, which facilies the following intercalation of monomer cations by π–π or cation–π intercalation. As more cationic complexes insert into the layers, the graphite interlayer space turns to a larger space and thus further weakens the intercalation between interlaminations. After initiating the polymerization, monomer cations confined in graphite interlayers grow into polymer chains gradually. A large amount of heat would be generated in this process, involving the movement of long-chain molecules. These effects lead to a violent separation of graphite and exfoliate into graphene. Furthermore, the exfoliated graphene is pasted and stabilized by the onsite synthesized polymer molecules to prevent its agglomeration.

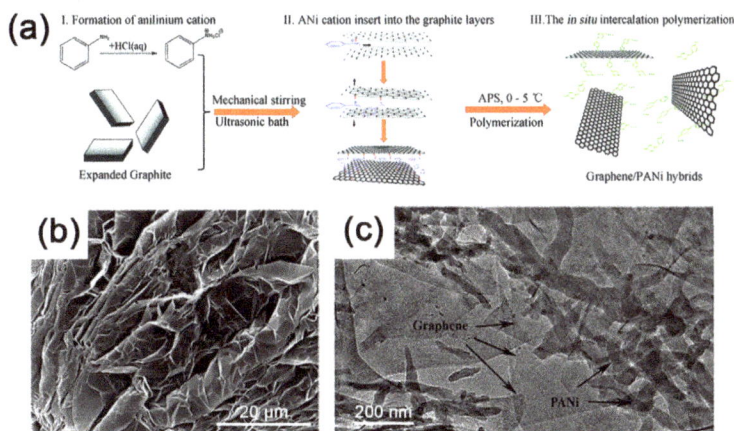

Figure 8. (**a**) Schematic for the in situ intercalation polymerization of ANi$^+$ into EG to synthesize graphene/polyaniline hybrids; (**b**) scanning electron microscope (SEM) image of expanded graphite; and (**c**) transmission electron microscope (TEM) image of graphene/polyaniline hybrids obtained by in situ intercalation polymerization. Adapted with permission from [53]. Copyright © 2014 Royal Society of Chemistry.

Since the interlayer distance of graphite is 3.35 Å, it can be thought as a natural nanoreactor, and in situ intercalation polymerization performed in the graphite interlayers can be recognized as a typical 2D-confined polymerization. Moreover, sp^2-hybridized carbon in graphite provides abundant π-electrons, leading to a special 2D electron-rich confined space. Polymer synthesized in such a unique nanoscale-confined space is partitioned from that of the surrounding bulk space. In situ polymerization in the interlayer of graphite results in the hybridization of graphene/polymer induced by the nano-confined effect and electron interaction, which may further influence the band structure of hybrids [53,117,118]. In addition, a nano-confined space always causes geometric conformational transformation or orientation of confined molecules [71], which might be used for the further study of molecular structure.

Owing to its larger interlayer distance and functioned oxygen-containing group, graphite oxide can be more easily intercalated than graphite by not only the cationic complex but also molecules like vinyl alcohol [107], vinyl acetate [104] and methyl methacrylate [119] for interlayer polymerization. Sandwich-like polymer/graphene oxide composites with highly crumpled and intercalated structures can be obtained by the in situ interlayer polymerization [107,119,120]. The extraordinary crumpled structure might be attributed to the interlayer chain movement and hybrid interactions between the polymer and graphene oxide. Besides weakening of interlayer interaction, some research indicates that the surface wettability of graphite to monomers is another critical factor for the exfoliation of graphene [119]. Chemical expanded graphite (CEG) is used for further oxidization to introduce oxygen functional groups on the graphite surface. Benefiting from its open and highly surface-accessible pore structures, diffusion resistance of the oxidizer in the interlayers of CEG significantly reduces [121]. The two-stage oxidization (as illustrated in Figure 9a) results in the spatially uniform oxidization of graphite layers (Figure 9b) which is different from traditional graphite oxide functioned mostly in the peripheral region [122,123]. In this way, the wetting capability of CEG to monomers can be improved by the uniformly grafted oxygen functional groups, and finally leads to spontaneously and uniform exfoliation of CEG into single- and few-layer graphene in graphene/polymer composites during the interlayer polymerization.

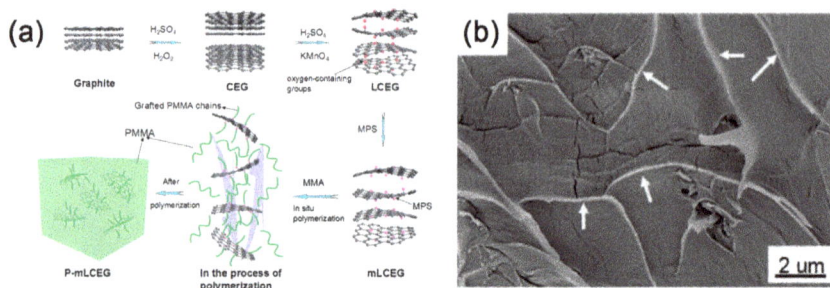

Figure 9. (**a**) Schematic of the preparation of polymethyl methacrylate (PMMA)/graphene composites by interlayer polymerization; and (**b**) SEM image of freeze-fractured cross sections of composites. Adapted with permission from [119]. Copyright © 2017 American Chemical Society.

3.2. Characterization of Intercalation Polymerization

For the characterization of intercalation polymerization, the primary consideration is to focus on the intercalation and exfoliation of graphite, and XRD is the most important test. The XRD pattern of graphite always exhibits sharp characteristic diffraction peaks at $2\theta = 26.5°$ ($d = 3.35$ Å), which are assigned to the (002) plane of graphite. The interlayer distance will be enlarged if graphite is intercalated by a guest molecule, leading to intensity decrease or disappearance of this peak. Instead, new diffraction peaks corresponding to the changed interlayer distance may appear as shown in Figure 10. Once intercalation polymerization achieves the exfoliation of graphene, these peaks will disappear due to the separation of graphite layers. Therefore, the XRD pattern can be used to effectively analyze the intercalation and exfoliation of graphite.

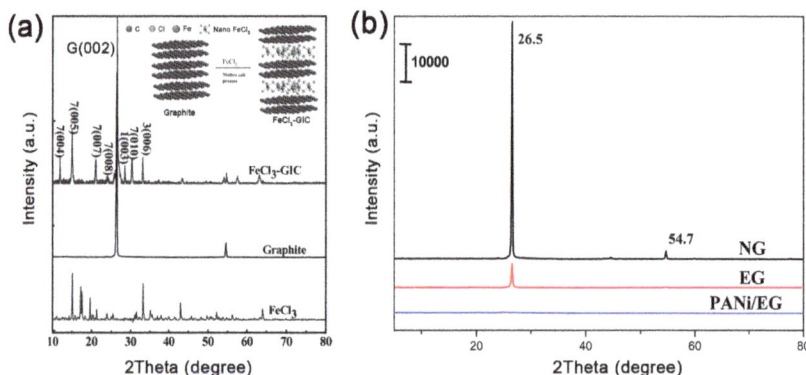

Figure 10. (**a**) XRD patterns of graphite, FeCl$_3$ and FeCl$_3$-GIC; insert is the schematic of graphite and FeCl$_3$-GIC. Adapted with permission from [62]. Copyright © 2014 Royal Society of Chemistry. (**b**) XRD patterns of natural graphite, expanded graphite (EG) and graphene/polyaniline (PANi)/EG hybrids synthesized by intercalation polymerization. Adapted with permission from [53]. Copyright © 2014 Wiley Online Library.

Exfoliated graphene can also be distinguished by morphology characterization using a scanning electron microscope (SEM), transmission electron microscope (TEM) and atomic force microscope (AFM) etc. Highly-stacked natural graphite or worm-like expanded graphite are significantly different from exfoliated single- and few-layer graphene, as depicted in Figure 11. Furthermore, high-resolution TEM and AFM can be helpful in confirming the number of graphene layers. It must be noted that sometimes the number of graphene layers calculated from thickness are not accurate due to the coated polymer on the graphene.

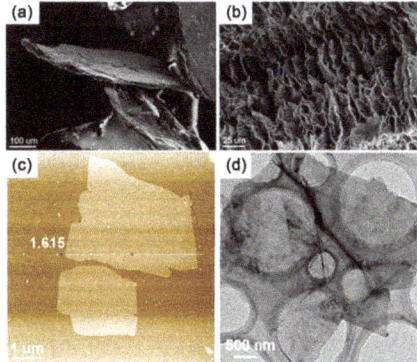

Figure 11. SEM images of (**a**) natural graphite and (**b**) expanded graphite; (**c**) atomic force microscope (AFM) image and (**d**) TEM image of PMMA/graphene composites. Adapted with permission from [119]. Copyright © 2017 American Chemical Society.

As the intercalation polymerization goes on in a typical 2D-confined space, the structure of graphene may change due to the hybridizing interactions between exfoliated graphene and synthesized polymer. Therefore, some forms of structural characterization can also be applied to analyse intercalation polymerization such as Fourier-transform–infrared (FT–IR) spectra and laser Raman spectroscopy. For example, the interaction between the N-atom in PANi and π-electrons in graphene leads to the blue shifts of C–N, C=N stretching vibrations (Figure 12a). Meanwhile, the exfoliation and hybridization of graphene influence the π-electron cloud in graphite, resulting in overlapping of the D band (at 1350 cm^{-1}) and G band (at 1580 cm^{-1}), and the disappearing of the 2D band (at 2700 cm^{-1}) in the Raman spectra (Figure 12b) [53].

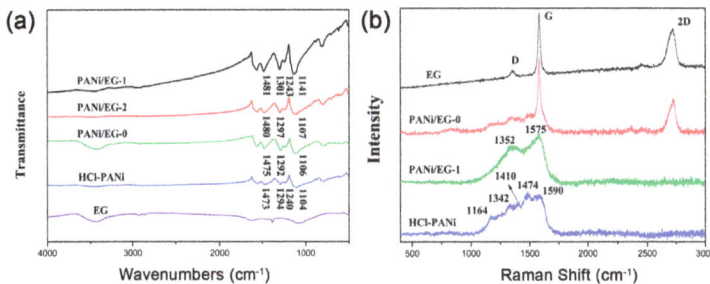

Figure 12. (**a**) Fourier-transform–infrared (FT–IR) and (**b**) Raman spectra of expanded graphite (EG), PANi and PANi/EG composites. (PANi/EG grinding mixture was denoted as PANi/EG-0, the intercalation polymerization and in situ polymerization of ANi$^+$ into 1 wt % EG was denoted as PANi/EG-1 and PANi/EG-2, respectively.) Adapted with permission from [53].Copyright © 2014 Royal Society of Chemistry.

Besides the above-mentioned methods, many other characterizations have been used to study the intercalation polymerization and synthesized composites, such as X-ray photoelectron spectroscopy (XPS), differential scanning calorimetry (DSC) and polarized optical microscopy (POM) etc. [18,119] However, some fundamental research, for example that on intercalation efficiency, are still challenging, and require further study. With the development of in situ characterization and theoretical simulation, a better understanding of intercalation polymerization can be achieved.

3.3. Influence Factors on Intercalation Polymerization

As presented above, intercalating molecules/ions/clusters into graphite, and polymerization in the graphite interlayers, are the key points for intercalation polymerization. There are many factors affecting this process. Therefore, based on literature results, we mainly review the influencing factors on intercalation polymerization from three aspects, i.e., the source of graphene, intercalant species, and process parameters of intercalation polymerization.

3.3.1. Source of Graphene

The source of graphene in resulting graphene/polymer composites is important in the intercalation polymerization. It can be divided into natural graphite, expanded graphite, modified graphite and carbon nanotube. Because of the differences in structure, their performances in intercalation and exfoliation are also different.

Natural graphite with a complete crystal structure and large planes are the first choice for preparing high-quality graphene. Most research into traditional GIC used NG as raw materials. As mentioned above, NG can be fully intercalated by alkali metal, but few studies have achieved the intercalation of organic monomers [55]. That might be due to its highly stacked layers. Thus, until now NG has only been used as an initiator for anionic polymerization after the intercalation of alkali metal. However, only thick graphite flakes are exfoliated in related reports, indicating an insufficient contact between polymerizable monomers and the initiating segment of GIC [105,108,109]. It seems that monomers can only contact the edge of the GIC without further penetrating into the interlayer galleries, leading to limited exfoliation. Potassium intercalated MWCNTs are in a similar situation when applied in intercalation polymerization. While intercalation of MWCNTs leads to a partial or full split, monomers can only polymerize at the edges of fissures without further intercalating [106,110]. Actually, when MWCNTs are used for polymerization, as shown in Figure 13a,b, the size of GNR in the resulting composites is quite small and limited by the superficial area of pristine MWCNTs [106,115].

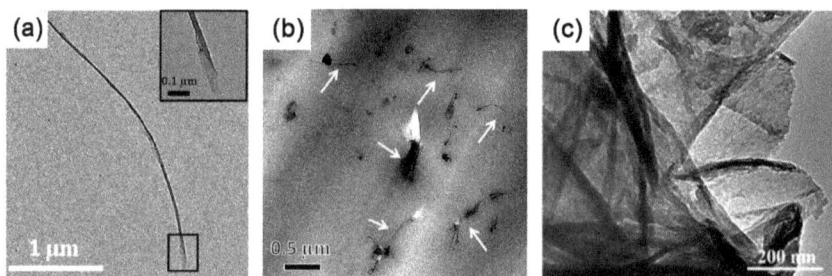

Figure 13. TEM images of (**a**) partially unzipped MWCNTs, inset is the unzipped layer; and (**b**) graphene nanoribbon (GNR)/epoxy nanocomposites. Adapted with permission from [106]. Copyright © 2016 Elsevier. (**c**) Polypyrrole (PPy)/GO synthesized by intercalation polymerization. Adapted with permission from [120]. Copyright © 2016 Elsevier.

The intercalation of organic monomers is impeded by the highly-stacking layers due to strong interlayer interaction of natural graphite. The stacked layers of NG also lead to a limited area of accessible surface for monomers. Therefore, EG with an open, highly surface-accessible pore structure (Figure 11a,b) is the best substitute for NG, which facilitates the access and intercalation of monomers [53,117]. As more monomers are able to absorb on the surface of EG due to the worm-like structure, single- or few-layered graphene with large scale (Figure 8c) can be effectively exfoliated by the subsequent polymerization [53]. However, the exfoliation of EG is insufficient at relatively higher filler loading (more than 4 wt %), indicating the limitation of utilizing the physical structure of graphite.

Comparing with NG and EG, GtO and CEG possess not only larger interlayer distance due to weakened inter-plane interaction, but also abundant functional groups including hydroxyl, carboxyl and grafted molecules. These functional groups induce strong interaction between monomers and graphite layers, thus making for effective intercalation and exfoliation (Figure 13c) [119,120]. Furthermore, the modification of graphite significantly improves the surface wettability of graphite layers to monomers, resulting in the spontaneous exfoliation of graphene. It seems that reasonable modification of graphite can be helpful in intercalation polymerization together with the highly accessible surface area of graphite layers, which inspire us to tune the structure of graphite for more efficient intercalation polymerization.

3.3.2. Intercalant Species

The choosing of intercalant is another key factor for graphite intercalation. Traditional intercalants for synthesising GIC have been systematically reviewed in ref. [55], but the organic intercalants have not been discussed before. In order for the one-component organic molecules to intercalate, their structure should be carefully considered. Intercalation of naphthalene or aniline molecules infers that π–π interactions can be utilized in this process [53,98]. However, the incomplete intercalation indicates that π–π interactions are not strong enough for sufficient intercalation. Benefiting from the strong cation–π interactions, aniline cation exhibits a more pronounced effect in intercalation and exfoliation [53]. Moreover, the successful intercalation of pyrrole cation or caprolactam onium ion further confirms the contribution of a positive charge [18,117]. Thus we can say that the intercalative process is dominated by the strong cation–π interactions between monomers and graphite, and π–π interactions may also help this process.

3.3.3. Process Parameters of Intercalation Polymerization

Feeding a ratio of monomers to graphite can significantly influence the exfoliation and dispersion of graphene in the polymer matrix. Because of its poor ability in dissolution, the addition of graphite is always less than 1 wt % of monomers, otherwise exfoliated graphene would be difficult to homogeneously disperse in the matrix [117]. But for hydrophilic CEG or GtO, their content can even be increased up to 10 wt % with only a few aggregations, as shown in Figure 14 [119].

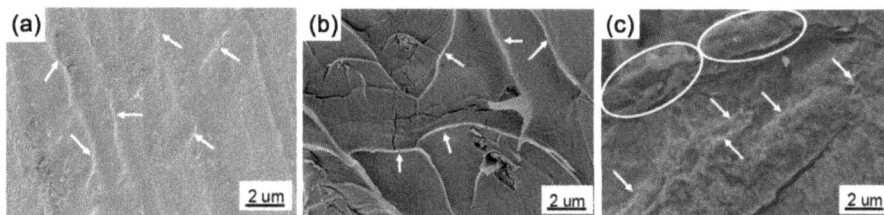

Figure 14. SEM image of freeze-fractured cross sections of PMMA/chemical expanded graphite (CEG) with CEG contents of (**a**) 1 wt %, (**b**) 4 wt %, (**c**) 10 wt %; graphene sheets are denoted by the arrows, and the ovals indicate aggregations of graphene sheets. Adapted with permission from [119]. Copyright © 2017 American Chemical Society.

Moreover, ultrasonication is necessary to help the monomers intercalating into the interlayers of graphite. With the ultrasonication-assisted intercalation, worm-like EG or stacked GtO can be separated and dispersed into flakes [53]. However, it is easy to understand that violent ultrasonication may break the complete graphite layers into small fragments. With short-duration ultrasonic exfoliation, large GO flakes (lateral size of 10–20 μm) can be obtained. Long-duration ultrasonic also results in flakes smaller than 1 μm, with more than 75% of them having a size in the range 0.1–0.4 μm [124]. Thus, a mild and reasonable power of ultrasonication is of importance in the intercalation process, which facilities high-efficient exfoliation in the polymerization.

4. Application of Graphene/Polymer Composites

Intercalation polymerization provides a new method for synthesizing graphene/polymer composites. Polymerization conducted in the 2D-confined space of graphite layers leads to graphene and polymer hybrids which can be easily distinguished from general polymers synthesized in normal environment. Strong hybridization interaction between polymer molecules and graphene can induce some amazing performance change. In this section, some emerging applications of graphene/polymer composites synthesized by intercalation polymerization are reviewed, including electrical conductivity, electromagnetic absorption, mechanical properties and thermal conductivity.

4.1. Electrical Conductivity

Graphene is widely used as nanofiller for improving the electrical conductivity of polymers and decreasing the percolation threshold, because of its large specific surface area and extraordinary electrical property. But contrary to original intentions, the agglomerate of graphene sheets in polymer composites during processing always inhibits the expected effects. In situ polymerization conducted in the interlayer of graphite not only exfoliates graphene layers, but also isolate layers by onsite synthesized polymer. For the PMMA/graphene composite synthesized by intercalation polymerization with the addition of 1.5 wt % of CEG, electrical conductivity increases about 12 orders of magnitude to 1.63×10^{-2} S/m [119]. This value is far beyond the percolation threshold, implying the good dispersion of exfoliated graphene in composites. Even more astonishing, a PMMA/graphene composite with an extremely high electrical conductivity of 1719 S/m can be obtained when 10 wt % of CEG was used in polymerization, which is one of the highest values reported for graphene/polymer composites as compared in Table 1.

Table 1. Comprehensive comparison of the electrical conductivity of graphene/polymer composites.

Material	Synthesis method	Filler content	Electrical conductivity (S/m)	Reference
PMMA/Graphene	Intercalation polymerization	4 wt % 10 wt %	17.55 1719	[119]
PMMA/rGO	In situ polymerization	3 wt %	1.5	[125]
PMMA/rGO	Aqueous mixing	2 wt %	3.7×10^{-2}	[126]
PEO/Graphene	Aqueous mixing	2 wt %	6×10^{-2}	[127]
PBT/rGO	Aqueous mixing	10 wt %	9×10^{-2}	[128]
PET/Graphene	Melt mixing	7 wt %	$\sim 10^{-4}$	[129]
PI/rGO	In situ polymerization	30 wt %	11	[130]
Epoxy/Graphene foam	Prepreg-hot press	10 wt %	230	[131]

However, interesting results are reported when conducting polymers were used for interlayer polymerization. Polyaniline/graphene hybrids synthesized by in situ intercalation polymerization display obvious decrease in electrical conductivity as compared to those of HCl doped polyaniline or expanded graphite [53]. This can mostly be attributed to the hybridizing intercalation between polyaniline molecule and graphene. While the interlayer of graphite acts as nanoreactors, the strong confined effect would occur during the confined polymerization, which behaves as electron cloud migration between graphene and polymer molecules. The hybridizing intercalation, on the one hand, reduces the doping degree of polyaniline, leading to lower carrier concentration, and, on the other hand, affects the conjugated system in graphene. Furthermore, π–π staking might also exist in graphene/polyaniline hybrids. Taken together, the electrical conductivity of the hybrids exhibits an unusual decrease when compared to pure polyaniline or expanded graphite.

4.2. Electromagnetic Wave Absorption

While digital devices and rapid development of radar detecting technology change our lifestyle, the electromagnetic waves (EM) generated also lead to the grim problem of EM interference. Thus the protection and shielding of electromagnetic radiation has been widely considered as a serious problem, and the microwave absorbing materials is desperately desired by society. As is known, impedance matching and EM-wave attenuation in the interior of materials are two principles for promoting EM-wave absorption. The former ensures as little reflection as possible at the surface of materials, and the latter leads to energy dissipation of the EM wave. Therefore, synergistic effects of the dielectric loss and magnetic loss are important for promoting EM absorption.

Intercalation polymerization has brought some obvious change in physical parameters for graphene/conductive polymers. For example, the conductivity and permittivity of the hybrids exhibit extraordinary change as compared with pure conductive polymers or graphite. A much better impedance match can be obtained for graphene/polyaniline hybrids synthesized by intercalation polymerization, facilitating the improvement of microwave absorption [53]. Besides, defects and hybridizing points induced by hybridizing interaction between polyaniline and graphene act as an extra polarization center and cause additional relaxation. As shown in Figure 15, the resulting hybrids show significant enhancement in microwave absorption, and the minimum reflection loss (RL) reached −36.9 dB with a thickness of 3.5 mm. Moreover, absorption bandwidth with RL below −10 dB is in the frequency range of 5–18 GHz, depicting a broad frequency band for the application. Furthermore, based on intercalation polymerization, our group has also developed other similar works such as graphene/polypyrrole or graphene oxide/polypyrrole hybrids for microwave absorption [117,120].

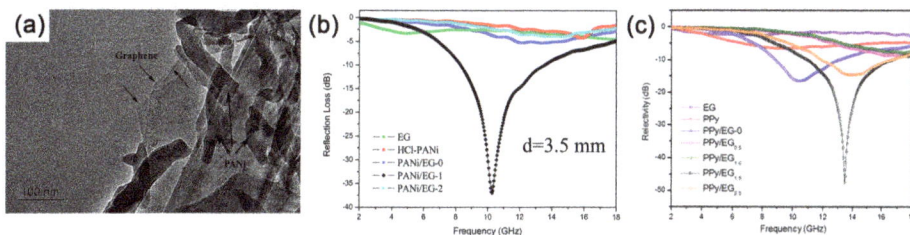

Figure 15. (a) TEM image of PANi/EG hybrids hybrids synthesized by intercalation polymerization of ANi+ into 1 wt % EG; and (b) the calculated RL in the frequency range of 2–18 GHz (PANi/EG grinding mixture was denoted as PANi/EG-0, the intercalation polymerization and in situ polymerization of ANi+ into 1 wt % EG was denoted as PANi/EG-1 and PANi/EG-2, respectively). Adapted with permission from [53]. Copyright © 2014 Royal Society of Chemistry. (c) Calculated RL of PPy/EG with a thickness of 2.7 mm in the frequency range of 2–18 GHz (hybrids with different addition of EG were denoted as PPy/EG$_x$, where x = 0, 0.5, 1.0, 1.5, 2.0 wt %). Adapted with permission from [117]. Copyright © 2015 Elsevier.

Among these hybrids, PPy/GO exhibits the best result for microwave absorption. The minimum RL reaches −58.1 dB at 12.4 GHz with a thickness of 2.96 mm, and a wide broad bandwidth (< −10 dB) of 6.2 GHz (Figure 16a) indicates its extraordinary performance among different microwave-absorbing materials [120]. For graphene/conductive polymer composites, their EM loss mainly comes from dielectric loss with almost no magnetic response. Benefiting from the strong hybridization effect, the interaction between –NH in PPy and –CO in GO introduce new unsymmetrical centers, which results in additional charge rearrangement and orbital hybridization due to electric dipole polarization. In addition, crumpled structures of PPy/GO (as shown in Figure 13c) would lead to more interfacial losses or relaxations at a higher frequency. The mechanism for the dielectric loss enhancement of PPy/GO composite is illustrated in Figure 16b. Recent work on microwave absorption of polymer composites is summarized in Table 2. It can be seen that intercalation polymerization plays a key role in the polymer composites to improve their performance in microwave absorption.

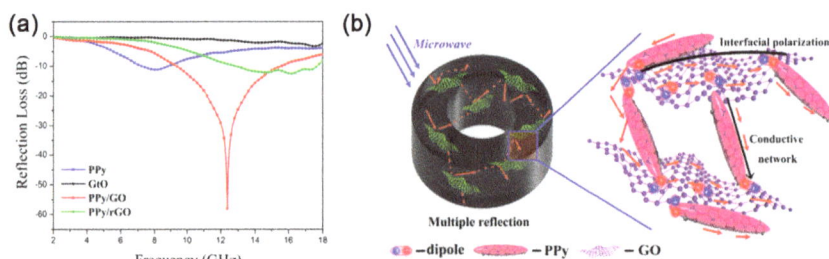

Figure 16. (**a**) The reflection loss (RL) of the samples with a thickness of 2.96 mm; and (**b**) schematics of electromagnetic waves (EM) loss-enhancement mechanism of PPy/GO. Adapted with permission from [120]. Copyright © 2016 by the Elsevier.

Table 2. EM wave absorption of different graphene/polymer composites.

Absorber	Synthesis method	Matrix	Absorber content	Thickness (mm)	RL min (dB)	RL < −10 dB bandwidth (GHz)	Reference
PPy/GO	Intercalation polymerization	Wax	30 wt %	2.96	−58.1	6.2	[120]
PANi/Graphene	Intercalation polymerization	Wax	10 wt %	3.5	−36.9	5.3	[53]
PPy/Graphene	Intercalation polymerization	Wax	10 wt %	2.7	−48.0	3.4	[117]
PANi/Graphene	In situ polymerization	Wax	25 wt %	3.04	−38.8	2.3	[132]
PEO/rGO	Aqueous mixing	PEO	2.6 vol %	1.8	−38.8	4.1	[133]
NBR/GO	Aqueous mixing	NBR	10 wt %	3	−57.0	4.5	[134]
PANi/Graphene foam	In situ polymerization	Graphene foam	-	2	−52.5	3.0	[135]
PANi/rGO	In situ polymerization	Wax	50 wt %	2	−41.4	4.2	[136]

4.3. Mechanical Properties

The mechanical properties of composites are worth expecting because of the homogeneous disperse of graphene and the strong interfacial interactions induced by in situ intercalation. When GtO is intercalated and exfoliated, the tensile strength of PVA increases from 42.3 MPa of pure PVA to 50.8 MPa with only 0.04 wt % GtO loading, and Young's modulus increases from 1477 to 2123 MPa [107]. The significant improvement of mechanical properties at such low loading of GO can be due to the uniform dispersion of exfoliated GO, the aligned GO parallel to the film and the hydrogen bonding interaction between GO and polymer chains. But, limited by the initial strength of a dilapidated GO plane and the efficiency of intercalation polymerization, the mechanical properties of PVA are difficult to improve further. Thus, stronger interfacial interactions between graphene planes and polymer matrix are expected. Therefore, uniform oxidized graphite layers are functioned by introducing polymerizable C=C bonds on the graphene surface, ensuring polymer molecules covalent grafting onto graphene in subsequent interlayer polymerization, as shown in Figure 17 [119].

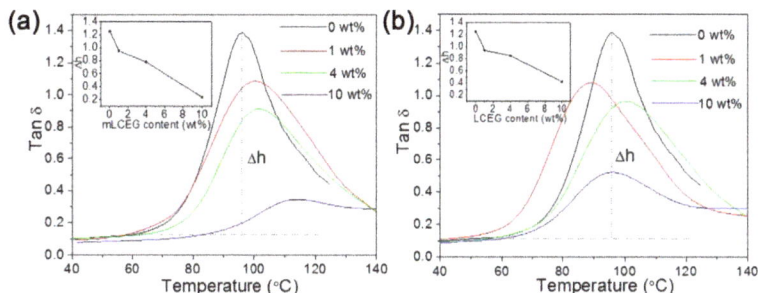

Figure 17. (**a**) Storage modulus; and (**b**) damping loss of PMMA/chemical expanded graphite (CEG) composites with different filler contents as a function of temperature. Adapted with permission from [119]. Copyright © 2017 American Chemical Society.

Covalent bonding between polymer chains and graphene planes leads to better interfacial interaction, cooperating with the good dispersion of graphene, composites exhibit a 3-fold increase in the storage modulus with 10 wt % functioned CEG [119]. As summarized in Table 3, the intercalation polymerization significantly improves the mechanical properties of composites when compared to other synthesis methods. Furthermore, gradually decreasing transition temperature and decreasing of damping loss indicates a typical restricted relaxation behavior and effective interface load transfer, which is reasonably related to the modified in situ intercalation polymerization.

Table 3. Improvement in the mechanical properties of composites synthesized by different methods.

Material	Synthesis method	Filler content	Mechanical properties relative to neat polymer (ΔE, $\Delta E'$, $\Delta\sigma_{max}$, ΔK_{IC}) *	Reference
PMMA/Graphene	Intercalation polymerization	10 wt %	$\Delta E'$ = 299% (at 45 °C)	[119]
TPU/GNR	Intercalation polymerization	0.5 wt %	ΔE = 70%, $\Delta E'$ = 175% (at −25 °C), $\Delta\sigma_{max}$ = 15%	[115]
Epoxy/GNR	Intercalation polymerization	0.15 wt %	ΔE = 11%, ΔK_{IC} = 43%	[106]
PVA/GO	Intercalation polymerization	0.04 wt %	ΔE = 43%, $\Delta\sigma_{max}$ = 20%	[107]
PMMA/rGO	In situ polymerization	2 wt %	ΔE = 13%, $\Delta\sigma_{max}$ = −41%	[137]
PMMA/Graphene	Twin screw extruding	20 wt %	ΔE = 7%, $\Delta E'$ = 22% (at 100 °C), $\Delta\sigma_{max}$ = 3%	[138]
Epoxy/rGO	Ball mill	2 wt %	ΔE = 5%, $\Delta\sigma_{max}$ = 0%, ΔK_{IC} = 50%	[139]
Epoxy/Functioalized-GO	In situ polymerization	0.5 wt %	ΔE = 16%, $\Delta\sigma_{max}$ = −75%, ΔK_{IC} = 33%	[140]
Thermoplastic polyurethane (TPU)/Graphene	Aqueous mixing	3 wt %	ΔE = 43%, $\Delta\sigma_{max}$ = −22%	[141]

* ΔE: maximum Young's modulus improvement; $\Delta E'$: maximum storage modulus improvement; $\Delta\sigma_{max}$: maximum tensile strength improvement; ΔK_{IC}: maximum fracture toughness improvement.

4.4. Thermal Conductivity

Since most polymers exhibit poor ability in conducting heat flow, graphene has long been expected to promote their thermal conductivity (TC). Similar to electrical conductivity, the dispersion of graphene in the polymer matrix is one of the key points for higher thermal conductivity. Thus, in situ intercalation polymer can be a useful method for fabricating polymers with high thermal conductivities. As depicted in Figure 18, polyamide-6/graphite nanoflakes synthesized by intercalation polymerization exhibits significant thermal conductive improvement to 2.49 W/(m·K) with 12 wt %

EG loading, as that of pure polyamide-6 is only 0.32 W/(m·K) [18,142]. Compared with composites prepared by in situ polymerization or melt mixing with EG, intercalation polymerization results in not only better dispersion of graphite nanoflake but also better interfacial connections. Generally, better compatibility always means a better phonon match between EG and the polymer matrix, further decreasing the thermal interface resistance and improving the percolation. Moreover, the thermal conductivity of PA-6 composites synthesized by the intercalation polymerization is much higher than that of most reported graphene/polymer composites (Table 4). Therefore, in situ intercalation polymerization provides a good idea for constructing highly efficient thermal conductive pathways within the matrix network.

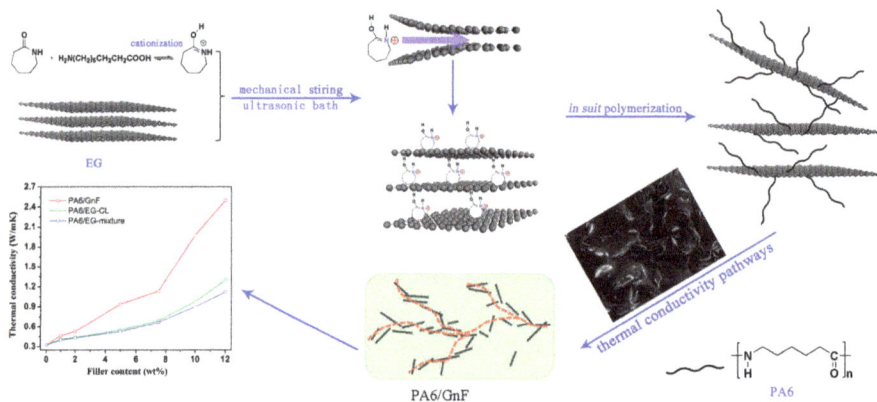

Figure 18. Schematic for the in situ intercalation polymerization of CL^+ into EG to synthesize polyamide-6/graphite nanoflakes composites. Adapted with permission from [18]. Copyright © 2017 Elsevier.

Table 4. Thermal conductivity polymer/graphene composites synthesized by different methods.

Material	Synthesis method	Filler content	TC (W/(m·K))	TC enhancement compared to neat polymer	Reference
PA-6/Graphite nanoflakes	Intercalation polymerization	12 wt %	2.49	678%	[18]
PA-6/rGO	In situ polymerization	10 wt %	0.416	112%	[143]
PA-6/Graphene foam	In situ polymerization	2 wt %	0.847	300%	[144]
PA-6/Graphene-GO	In situ polymerization	10 wt %	2.14	569%	[142]
PA-6/Graphite	Twin screw extruding	30 wt %	1.37	350%	[145]
PS/Graphite nanoflakes	Melt mixing	~9.2 wt %	0.9	398%	[146]
PP/Graphite nanoflakes	Aqueous mixing	10 wt %	1.53	595%	[147]
PVA/Graphite nanoflakes	Aqueous mixing	10 wt %	1.43	580%	[147]
PBT/Graphite nanoflakes	In situ polymerization	20 wt %	1.98	1320%	[148]

4.5. Other Applications

Except for the above applications, graphene/polymer composites synthesized by intercalation polymerization have also been used in other fields like sensing, electrochemical supercapacitor and gas barriers. For examples, PVA/GO synthesized by intercalation polymerization can form an optically transparent, flexible film with much lower water vapor permeability than neat PVA, as shown in Figure 19a,b [107]. Similar results are reported for thermoplastic polyurethane (TPU)/GNR composites. Nitrogen gas effective diffusivity decreased by 3 orders of magnitude with only 0.5 wt % GNRs (Figure 19c) [115]. Some other applications of synthesized graphene/polymer composites are summarized in Table 5. Although a few researches, these works give a sight for expanding the application fields of intercalation polymerization.

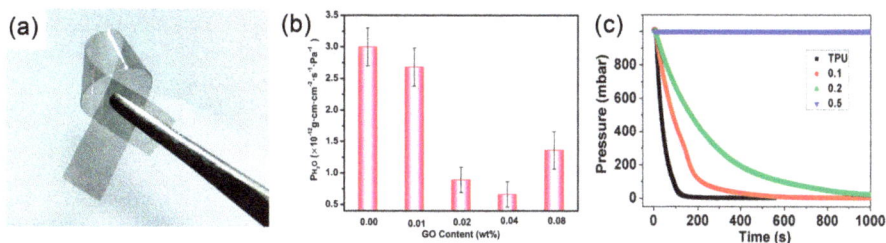

Figure 19. (a) Digital image of poly(vinyl alcohol (PVA)/GO composites film synthesized by intercalation polymerization; and (b) its water vapor permeability. Adapted with permission from [107]. Copyright © 2016 Royal Society of Chemistry. (c) Pressure drop curves of thermoplastic polyurethane (TPU)/GNR composites. Adapted with permission from [115]. Copyright © 2013 by the American Chemical Society.

Table 5. Other applications of graphene/polymer composites synthesized by intercalation polymerization.

Application	Material	Description	Reference
Sensing of serotonin	PLA/GO	Electrochemical detection with high concentration range (0.1–100.0 μM) and low detection limit (0.08 μM, where $s/n = 3$)	[149]
Sensing of methanol	PANi/GO	High sensitivity ($\Delta R/R_0 = 20.9$–37) for methanol vapor (100–500 ppm) *	[150]
Electrochemical supercapacitor	PANi/GO	High specific capacitance of 543.75 F/g and reversible electrochemical response up to 150th repeated cycles	[151]
Water vapor barrier	PVA/GO	Water vapor permeability declines about 5-fold to 0.66×10^{-12} g·cm·(cm^2·s·Pa)$^{-1}$ by adding 0.04 wt % GO	[107]
Nitrogen gas barrier	TPU/GNR	Nitrogen gas effective diffusivity decreased by 3 orders of magnitude with only 0.5 wt % GNRs.	[115]

* $\Delta R/R_0 = (R - R_0)/R_0$, where, R_0 and R are the initial resistance of sensor in the air and in target gas, respectively.

5. Conclusions and Outlook

Based on the above generalizations about intercalation polymerization, it can be concluded that the intercalation chemistry of graphite and subsequent interlayer polymerization have attracted increasing attention, and research of intercalation polymerization and the resulting composites has indeed become attractive. The presented review has highlighted recent developments relating to intercalation, polymerization and the performance of the as-synthesized graphene/polymer composites.

For intercalation polymerization, what is important is the interaction between organic monomers and graphite interlayers. If the interaction is not strong enough, monomers cannot penetrate into the deep intergallery for sufficient exfoliation, which leads to only thick graphite flakes or edge-functioned layers. In situ intercalation polymerization successfully disperses graphene in synthesized polymer composites. However, the intercalation efficiency of monomers is still too low to form GIC, thus limiting the content of graphene in the matrix. Moreover, once organic monomer-GIC is successfully synthesized, the layer number of exfoliated graphene will be theoretically controllable. Therefore, improving the intercalation efficiency becomes a serious issue for intercalation polymerization, and tuning the interaction between monomers and graphite can be an effective way of doing this. What we can do to tune the interaction is to carefully design the structure of intercalative monomers and graphite. Cation–π interactions play an essential role in the intercalation process, and therefore cationic monomers or oxidized graphite achieve a better intercalation effect. If a conjugated structure exists in intercalants, π–π interactions may also assist the intercalation process. The oxidation and modification of graphite can significantly reduce the resistance of intercalation and exfoliation, and the introduced active sites facilitate the functional applications of composites. However, traditional methods prefer to attack the carbon atoms in the peripheral region, leading to inhomogeneous distribution of functional groups. In recent years, controllable and spatially uniform oxidation has been achieved using K_2FeO_4 or H_2O_2 [43,44]. These results inspire us to comprehensive consider when graphite oxide or modified graphite are used in intercalation polymerization. For example, the slightly but uniformly oxidized graphite achieve fully intercalation, spontaneous exfoliation and homogeneously dispersed graphene, thus leading to highly conductive and mechanically strong polymer composites [119]. The graphite oxide with a high degree of oxidation also improves the EM absorption of PPy [120].

Recently, graphene/polymer composites synthesized by intercalation polymerization have exhibited a significant improvement of performance in various fields. However, some related fundamental scientific issues should be studied. For instance, it is important to understand the structural evolution of polymers during polymerizing in the 2D space of the graphite interlayers. Thereafter, we can reveal the interaction mechanism between graphene and polymer molecules in the process of intercalation polymerization, which may aid in the further molecular regulation and functional design of polymer materials. It is believed that intercalation polymerization will offer a bright future in the field of the synthesis and application of graphene/polymer composites.

Acknowledgments: The authors gratefully acknowledge the financial support of the National Natural Science Foundation of China (No. 51573149), the Science and Technology Planning Project of Sichuan Province (Nos. 2016CZYZF0003, 2016GZYZF0008 and 2016GZ0229) and the Fundamental Research Funds for the Central Universities of China (No. 2682016CX069).

Author Contributions: Yifan Guo performed the literature review and wrote the paper. Yifan Guo and Fuxi Peng reproduced the figures used in the context. Zuowan Zhou and Fanbin Meng revised the article and directed the project. Huagao Wang, Fei Huang and David Hui polished this article.

Conflicts of Interest: The authors declare no conflict of interest.

Abbreviations

CVD	Chemical vapor deposition
NG	Natural graphite
GO	Graphene oxide
GtO	Graphite oxide
GIC	Graphite intercalated compounds
Pyr$_{14}$TFSI	N-butyl-N-methylpyrrolidinium bis-(trifluoromethanesulfonyl)imide
PC	Propylene carbonate
DMSO	Dimethyl sulfoxide
DMF	Dimethylformamide
Pyr$_{14}^+$	N-butyl-N-methylpyrrolidinium cation
LiTFSI	Lithium bis(trifluoromethanesulfonyl)imide
TFSI$^-$	Bis(trifluoromethanesulfonyl)imide anion
TBA$^+$	Tetrabutylammonium cations
1,2-DAP	1,2-diaminopropane
XRD	X-ray diffraction
ANi$^+$	Aniline cation
CL$^+$	Caprolactam onium ion
EOG	Edge-selectively oxidized graphite
C$_{14}$N$^+$	Tetradecyl-ammonium cation
TAA$^+$	Tetraalkylammonium ions
PVA	Poly(vinyl alcohol)
PDDA	Poly(diallyldimethylammonium chloride)
PVAc	Poly(vinyl acetate)

EG	Expanded graphite
ANi	Aniline
KC$_{24}$	Stage 2 potassium intercalated graphite
KC$_8$	Stage I potassium intercalated graphite
GNRs	Graphene nanoribbons
MWCNTs	Multiwalled carbon nanotubes
PANi	Polyaniline
PPy	Polypyrrole
PA-6	Polyamide-6
CEG	Chemical expanded graphite
PMMA	Polymethyl methacrylate
SEM	Scanning electron microscope
TEM	Transmission electron microscope
AFM	Atomic force microscope
FTIR	Fourier transform infrared
XPS	X-ray photoelectron spectroscopy
DSC	Differential scanning calorimetry
POM	Polarized optical microscopy
rGO	Reduced graphene oxide
PEO	Polyethylene oxide
PBT	Poly(butylene terephthalate)
PET	Poly(ethylene terephthalate)
PI	Polyimide
EM	Electromagnetic waves
RL	Refection loss
NBR	Nitrile butadiene rubber
TPU	Thermoplastic polyurethane
TC	Thermal conductivity
PS	Polystyrene
PP	Polypropylene
PLA	Poly(lactic acid)

References

1. Novoselov, K.S.; Geim, A.K.; Morozov, S.V.; Jiang, D.; Zhang, Y.; Dubonos, S.V.; Grigorieva, I.V.; Firsov, A.A. Electric field effect in atomically thin carbon films. *Science* **2004**, *306*, 666–669. [CrossRef] [PubMed]
2. Xu, Z.; Gao, C. Graphene chiral liquid crystals and macroscopic assembled fibres. *Nat. Commun.* **2011**, *2*, 571. [CrossRef] [PubMed]
3. Raccichini, R.; Varzi, A.; Passerini, S.; Scrosati, B. The role of graphene for electrochemical energy storage. *Nat. Mater.* **2015**, *14*, 271–279. [CrossRef] [PubMed]
4. Peng, L.; Xu, Z.; Liu, Z.; Guo, Y.; Li, P.; Gao, C. Ultrahigh thermal conductive yet superflexible graphene films. *Adv. Mater.* **2017**, *29*, 1700589. [CrossRef] [PubMed]
5. Xin, G.; Yao, T.; Sun, H.; Scott, S.M.; Shao, D.; Wang, G.; Lian, J. Highly thermally conductive and mechanically strong graphene fibers. *Science* **2015**, *349*, 1083–1087. [CrossRef] [PubMed]
6. Cairns, D.R.; Witte, R.P.; Sparacin, D.K.; Sachsman, S.M.; Paine, D.C.; Crawford, G.P.; Newton, R.R. Strain-dependent electrical resistance of tin-doped indium oxide on polymer substrates. *Appl. Phys. Lett.* **2000**, *76*, 1425–1427. [CrossRef]
7. Jang, H.; Park, Y.J.; Chen, X.; Das, T.; Kim, M.S.; Ahn, J.H. Graphene-based flexible and stretchable electronics. *Adv. Mater.* **2016**, *28*, 4184–4202. [CrossRef] [PubMed]
8. Cao, G. Atomistic studies of mechanical properties of graphene. *Polymers* **2014**, *6*, 2404–2432. [CrossRef]
9. Nan, L.; Chortos, A.; Lei, T.; Jin, L.; Kim, T.R.; Bae, W.G.; Zhu, C.; Wang, S.; Pfattner, R.; Chen, X. Ultratransparent and stretchable graphene electrodes. *Sci. Adv.* **2017**, *3*, e1700159.
10. Feng, C.; Wang, Y.; Kitipornchai, S.; Yang, J. Effects of reorientation of graphene platelets (GPLs) on Young's modulus of polymer nanocomposites under uni-axial stretching. *Polymers* **2017**, *9*, 532. [CrossRef]

11. Kim, D.S.; Dhand, V.; Rhee, K.Y.; Park, S.J. Study on the effect of silanization and improvement in the tensile behavior of graphene-chitosan-composite. *Polymers* **2015**, *7*, 527–551. [CrossRef]

12. Skountzos, E.N.; Anastassiou, A.; Mavrantzas, V.G.; Theodorou, D.N. Determination of the mechanical properties of a poly(methyl methacrylate) nanocomposite with functionalized graphene sheets through detailed atomistic simulations. *Macromolecules* **2014**, *47*, 8072–8088. [CrossRef]

13. Tang, Y.; Hu, X.; Liu, D.; Guo, D.; Zhang, J. Effect of microwave treatment of graphite on the electrical conductivity and electrochemical properties of polyaniline/graphene oxide composites. *Polymers* **2016**, *8*, 399. [CrossRef]

14. Wang, Z.; Shen, X.; Han, N.M.; Liu, X.; Wu, Y.; Ye, W.; Kim, J.-K. Ultralow electrical percolation in graphene aerogel/epoxy composites. *Chem. Mater.* **2016**, *28*, 6731–6741. [CrossRef]

15. Beckert, F.; Held, A.; Meier, J.; Mülhaupt, R.; Friedrich, C. Shear- and temperature-induced graphene network evolution in graphene/polystyrene nanocomposites and its influence on rheological, electrical, and morphological properties. *Macromolecules* **2014**, *47*, 8784–8794. [CrossRef]

16. Li, A.; Zhang, C.; Zhang, Y.F.; Li, A.; Zhang, C.; Zhang, Y.F. Thermal conductivity of graphene-polymer composites: Mechanisms, properties, and applications. *Polymers* **2017**, *9*, 437.

17. Hong, H.; Kim, J.; Kim, T.I. Effective assembly of nano-ceramic materials for high and anisotropic thermal conductivity in a polymer composite. *Polymers* **2017**, *9*, 413. [CrossRef]

18. Meng, F.; Huang, F.; Guo, Y.; Chen, J.; Chen, X.; Hui, D.; He, P.; Zhou, X.; Zhou, Z. In situ intercalation polymerization approach to polyamide-6/graphite nanoflakes for enhanced thermal conductivity. *Compos. Part B Eng.* **2017**, *117*, 165–173. [CrossRef]

19. Yoonessi, M.; Shi, Y.; Scheiman, D.A.; Lebron-Colon, M.; Tigelaar, D.M.; Weiss, R.A.; Meador, M.A. Graphene polyimide nanocomposites; thermal, mechanical, and high-temperature shape memory effects. *ACS Nano* **2012**, *6*, 7644–7655. [CrossRef] [PubMed]

20. Li, C.; Qiu, L.; Zhang, B.; Li, D.; Liu, C.Y. Robust vacuum-/air-dried graphene aerogels and fast recoverable shape-memory hybrid foams. *Adv. Mater.* **2016**, *28*, 1510–1516. [CrossRef] [PubMed]

21. Wang, J.; Sun, L.; Zou, M.; Gao, W.; Liu, C.; Shang, L.; Gu, Z.; Zhao, Y. Bioinspired shape-memory graphene film with tunable wettability. *Sci. Adv.* **2017**, *3*, e1700004. [CrossRef] [PubMed]

22. Chen, H.; Gao, Q.; Li, J.; Lin, J.M. Graphene materials-based chemiluminescence for sensing. *J. Photochem. Photobiol. C* **2016**, *27*, 54–71. [CrossRef]

23. Luo, J.; Xu, Y.; Yao, W.; Jiang, C.; Xu, J. Synthesis and microwave absorption properties of reduced graphene oxide-magnetic porous nanospheres-polyaniline composites. *Compos. Sci. Technol.* **2015**, *117*, 315–321. [CrossRef]

24. Wang, L.; Huang, Y.; Li, C.; Chen, J.; Sun, X. Hierarchical composites of polyaniline nanorod arrays covalently-grafted on the surfaces of graphene@Fe$_3$O$_4$@C with high microwave absorption performance. *Compos. Sci. Technol.* **2015**, *108*, 1–8. [CrossRef]

25. Yan, D.X.; Pang, H.; Li, B.; Vajtai, R.; Xu, L.; Ren, P.G.; Wang, J.H.; Li, Z.M. Structured reduced graphene oxide/polymer composites for ultra-efficient electromagnetic interference shielding. *Adv. Funct. Mater.* **2015**, *25*, 559–566. [CrossRef]

26. Marra, F.; D'Aloia, A.G.; Tamburrano, A.; Ochando, I.M.; Bellis, G.D.; Ellis, G.; Sarto, M.S. Electromagnetic and dynamic mechanical properties of epoxy and vinylester-based composites filled with graphene nanoplatelets. *Polymers* **2016**, *8*, 272. [CrossRef]

27. Shen, B.; Zhai, W.; Cao, C.; Lu, D.; Jing, W.; Zheng, W. Melt blending in situ enhances the interaction between polystyrene and graphene through π–π stacking. *ACS Appl. Mater. Interfaces* **2011**, *3*, 3103–3109. [CrossRef] [PubMed]

28. Liu, C.; Wong, H.; Yeung, K.; Tjong, S. Novel electrospun polylactic acid nanocomposite fiber mats with hybrid graphene oxide and nanohydroxyapatite reinforcements having enhanced biocompatibility. *Polymers* **2016**, *8*, 287. [CrossRef]

29. Kim, H.; Kobayashi, S.; AbdurRahim, M.A.; Zhang, M.J.; Khusainova, A.; Hillmyer, M.A.; Abdala, A.A.; Macosko, C.W. Graphene/polyethylene nanocomposites: Effect of polyethylene functionalization and blending methods. *Polymer* **2011**, *52*, 1837–1846. [CrossRef]

30. Araby, S.; Meng, Q.; Zhang, L.; Kang, H.; Majewski, P.; Tang, Y.; Ma, J. Electrically and thermally conductive elastomer/graphene nanocomposites by solution mixing. *Polymer* **2014**, *55*, 201–210. [CrossRef]

31. Gaska, K.; Xu, X.; Gubanski, S.; Kádár, R. Electrical, mechanical, and thermal properties of LDPE graphene nanoplatelets composites produced by means of melt extrusion process. *Polymers* **2017**, *9*, 11. [CrossRef]

32. Wang, X.; Hu, Y.; Song, L.; Yang, H.; Xing, W.; Lu, H. In situ polymerization of graphene nanosheets and polyurethane with enhanced mechanical and thermal properties. *J. Mater. Chem.* **2011**, *21*, 4222–4227. [CrossRef]

33. Xu, Z.; Gao, C. In situ polymerization approach to graphene-reinforced nylon-6 composites. *Macromolecules* **2010**, *43*, 6716–6723. [CrossRef]

34. Liu, S.; Wang, L.; Tian, J.; Luo, Y.; Zhang, X.; Sun, X. Aniline as a dispersing and stabilizing agent for reduced graphene oxide and its subsequent decoration with Ag nanoparticles for enzymeless hydrogen peroxide detection. *J. Colloid Interface Sci.* **2011**, *363*, 615–619. [CrossRef] [PubMed]

35. Das, S.; Wajid, A.S.; Shelburne, J.L.; Liao, Y.-C.; Green, M.J. Localized in situ polymerization on graphene surfaces for stabilized graphene dispersions. *ACS Appl. Mater. Interfaces* **2011**, *3*, 1844–1851. [CrossRef] [PubMed]

36. Wang, J.-Y.; Yang, S.-Y.; Huang, Y.-L.; Tien, H.-W.; Chin, W.-K.; Ma, C.-C.M. Preparation and properties of graphene oxide/polyimide composite films with low dielectric constant and ultrahigh strength via in situ polymerization. *J. Mater. Chem.* **2011**, *21*, 13569–13575. [CrossRef]

37. Chen, Y.; Sun, J.; Gao, J.; Du, F.; Han, Q.; Nie, Y.; Chen, Z.; Bachmatiuk, A.; Priydarshi, M.K.; Ma, D. Growing uniform graphene disks and films on molten glass for heating devices and cell culture. *Adv. Mater.* **2015**, *27*, 7839–7846. [CrossRef] [PubMed]

38. Guo, W.; Jing, F.; Xiao, J.; Zhou, C.; Lin, Y.; Wang, S. Oxidative-etching-assisted synthesis of centimeter-sized single-crystalline graphene. *Adv. Mater.* **2016**, *28*, 3152–3158. [CrossRef] [PubMed]

39. Yang, W.; Chen, G.; Shi, Z.; Liu, C.-C.; Zhang, L.; Xie, G.; Cheng, M.; Wang, D.; Yang, R.; Shi, D. Epitaxial growth of single-domain graphene on hexagonal boron nitride. *Nat. Mater.* **2013**, *12*, 792–797. [CrossRef] [PubMed]

40. Gu, W.; Zhang, W.; Li, X.; Zhu, H.; Wei, J.; Li, Z.; Shu, Q.; Wang, C.; Wang, K.; Shen, W. Graphene sheets from worm-like exfoliated graphite. *J. Mater. Chem.* **2009**, *19*, 3367–3369. [CrossRef]

41. Qian, W.; Hao, R.; Hou, Y.; Tian, Y.; Shen, C.; Gao, H.; Liang, X. Solvothermal-assisted exfoliation process to produce graphene with high yield and high quality. *Nano Res.* **2009**, *2*, 706–712. [CrossRef]

42. Ciesielski, A.; Samorì, P. Graphene via sonication assisted liquid-phase exfoliation. *Chem. Soc. Rev.* **2014**, *43*, 381–398. [CrossRef] [PubMed]

43. Geng, X.; Guo, Y.; Li, D.; Li, W.; Zhu, C.; Wei, X.; Chen, M.; Gao, S.; Qiu, S.; Gong, Y. Interlayer catalytic exfoliation realizing scalable production of large-size pristine few-layer graphene. *Sci. Rep.* **2013**, *3*, 1134. [CrossRef] [PubMed]

44. Peng, L.; Xu, Z.; Liu, Z.; Wei, Y.; Sun, H.; Li, Z.; Zhao, X.; Gao, C. An iron-based green approach to 1-h production of single-layer graphene oxide. *Nat. Commun.* **2015**, *6*, 5716. [CrossRef] [PubMed]

45. Georgakilas, V.; Tiwari, J.N.; Kemp, K.C.; Perman, J.A.; Bourlinos, A.B.; Kim, K.S.; Zboril, R. Noncovalent functionalization of graphene and graphene oxide for energy materials, biosensing, catalytic, and biomedical applications. *Chem. Rev.* **2016**, *116*, 5464–5519. [CrossRef] [PubMed]

46. Renteria, J.D.; Ramirez, S.; Malekpour, H.; Alonso, B.; Centeno, A.; Zurutuza, A.; Cocemasov, A.I.; Nika, D.L.; Balandin, A.A. Strongly anisotropic thermal conductivity of free-standing reduced graphene oxide films annealed at high temperature. *Adv. Funct. Mater.* **2015**, *25*, 4664–4672. [CrossRef]

47. Matsumoto, R.; Arakawa, M.; Yoshida, H.; Akuzawa, N. Alkali-metal-graphite intercalation compounds prepared from flexible graphite sheets exhibiting high air stability and electrical conductivity. *Synth. Met.* **2012**, *162*, 2149–2154. [CrossRef]

48. Jones, J.E.; Cheshire, M.C.; Casadonte, D.J.J.; Phifer, C.C. Facile sonochemical synthesis of graphite intercalation compounds. *Org. Lett.* **2004**, *6*, 1915–1917. [CrossRef] [PubMed]

49. Dimiev, A.M.; Ceriotti, G.; Metzger, A.; Kim, N.D.; Tour, J.M. Chemical mass production of graphene nanoplatelets in ~100% yield. *ACS Nano* **2015**, *10*, 274–279. [CrossRef] [PubMed]

50. Zhao, W.; Tan, P.H.; Liu, J.; Ferrari, A.C. Intercalation of few-layer graphite flakes with FeCl₃: Raman determination of fermi level, layer by layer decoupling, and stability. *J. Am. Chem. Soc.* **2011**, *133*, 5941–5946. [CrossRef] [PubMed]

51. Khrapach, I.; Withers, F.; Bointon, T.H.; Polyushkin, D.K.; Barnes, W.L.; Russo, S.; Craciun, M.F. Novel highly conductive and transparent graphene-based conductors. *Adv. Mater.* **2012**, *24*, 2844–2849. [CrossRef] [PubMed]

52. Vallés, C.; Drummond, C.; Saadaoui, H.; Furtado, C.A.; He, M.; Roubeau, O.; Ortolani, L.; Monthioux, M.; Pénicaud, A. Solutions of negatively charged graphene sheets and ribbons. *J. Am. Chem. Soc.* **2008**, *130*, 15802–15804. [CrossRef] [PubMed]

53. Chen, X.; Meng, F.; Zhou, Z.; Tian, X.; Shan, L.; Zhu, S.; Xu, X.; Jiang, M.; Wang, L.; Hui, D. One-step synthesis of graphene/polyaniline hybrids by in situ intercalation polymerization and their electromagnetic properties. *Nanoscale* **2014**, *6*, 8140–8148. [CrossRef] [PubMed]

54. Bissessur, R.; Scully, S.F. Intercalation of solid polymer electrolytes into graphite oxide. *Solid State Ion.* **2007**, *178*, 877–882. [CrossRef]

55. Dresselhaus, M.S.; Dresselhaus, G. Intercalation compounds of graphite. *Adv. Phys.* **2002**, *51*, 1–186. [CrossRef]

56. Srinivas, G.; Lovell, A.; Howard, C.A.; Skipper, N.T.; Ellerby, M.; Bennington, S.M. Structure and phase stability of hydrogenated first-stage alkali- and alkaline-earth metal–graphite intercalation compounds. *Synth. Met.* **2010**, *160*, 1631–1635. [CrossRef]

57. Höhne, M.; Wang, Y.; Stumpp, E.; Hummel, H.-J. New ternary alkaline earth metal-ammonia compounds. *Synth. Met.* **1989**, *34*, 41–46. [CrossRef]

58. Cahen, S.; Vangelisti, R. Synthesis, structure and magnetic properties of lanthanide trichlorides-GIC: Stage2 DyCl₃–GIC. *J. Phys. Chem. Solids* **2006**, *67*, 1223–1227. [CrossRef]

59. Hagiwara, R.; Ito, M.; Ito, Y. Graphite intercalation compounds of lanthanide metals prepared in molten chlorides. *Carbon* **1996**, *34*, 1591–1593. [CrossRef]

60. Emery, N.; Hérold, C.; Lagrange, P. The synthesis of binary metal-graphite intercalation compounds using molten lithium alloys. *Carbon* **2008**, *46*, 72–75. [CrossRef]

61. Rousseau, B.; Estrade-Szwarckopf, H. X-ray and UV photoelectron spectroscopy study of Na–halogen–graphite intercalation compounds. Comparison between donor-and acceptor-graphite compounds. *Solid State Commun.* **2003**, *126*, 583–587. [CrossRef]

62. Wang, F.; Yi, J.; Wang, Y.; Wang, C.; Wang, J.; Xia, Y. Graphite intercalation compounds (GICs): A new type of promising anode material for lithium-ion batteries. *Adv. Energy Mater.* **2014**, *4*, 5866–5874. [CrossRef]

63. Stumpp, E.; Hummel, H.-J.; Ehrhardt, C. Thermogravimetric study of gold (III) bromide graphite. *Synth. Met.* **1988**, *23*, 441–446. [CrossRef]

64. Noel, M.; Santhanam, R. Electrochemistry of graphite intercalation compounds. *J. Power Sources* **1998**, *72*, 53–65. [CrossRef]

65. Sorokina, N.; Shornikova, O.; Avdeev, V. Stability limits of graphite intercalation compounds in the systems graphite-HNO₃ (H₂SO₄)-H₂O-KMnO₄. *Inorg. Mater.* **2007**, *43*, 822–826. [CrossRef]

66. Savoskin, M.V.; Yaroshenko, A.P.; Whyman, G.; Mestechkin, M.M.; Mysyk, R.D.; Mochalin, V. Theoretical study of stability of graphite intercalation compounds with brønsted acids. *Carbon* **2003**, *41*, 2757–2760. [CrossRef]

67. Liu, Y.; Xu, Z.; Zhan, J.; Li, P.; Gao, C. Superb electrically conductive graphene fibers via doping strategy. *Adv. Mater.* **2016**, *28*, 7941–7947. [CrossRef] [PubMed]

68. Han, W.-P.; Li, Q.-Q.; Lu, Y.; Yan, X.; Zhao, H.; Long, Y.-Z. Optical contrast spectra studies for determining thickness of stage-1 graphene-FeCl₃ intercalation compounds. *AIP Adv.* **2016**, *6*, 075219. [CrossRef]

69. Bointon, T.H.; Khrapach, I.; Yakimova, R.; Shytov, A.V.; Craciun, M.F.; Russo, S. Approaching magnetic ordering in graphene materials by FeCl₃ intercalation. *Nano Lett.* **2014**, *14*, 1751–1755. [CrossRef] [PubMed]

70. Bepete, G.; Hof, F.; Huang, K.; Kampioti, K.; Anglaret, E.; Drummond, C.; Pénicaud, A. "Eau de graphene" from a KC₈ graphite intercalation compound prepared by a simple mixing of graphite and molten potassium. *Phys. Status Solidi* **2016**, *10*, 895–899.

71. Maluangnont, T.; Gotoh, K.; Fujiwara, K.; Lerner, M.M. Cation-directed orientation of amines in ternary graphite intercalation compounds. *Carbon* **2011**, *49*, 1040–1042. [CrossRef]

72. Wen, Y.; He, K.; Zhu, Y.; Han, F.; Xu, Y.; Matsuda, I.; Ishii, Y.; Cumings, J.; Wang, C. Expanded graphite as superior anode for sodium-ion batteries. *Nat. Commun.* **2014**, *5*, 4033. [CrossRef] [PubMed]

73. Yang, S.; Brüller, S.; Wu, Z.S.; Liu, Z.; Parvez, K.; Dong, R.; Richard, F.; Samori, P.; Feng, X.; Müllen, K. Organic radical-assisted electrochemical exfoliation for the scalable production of high-quality graphene. *J. Am. Chem. Soc.* **2015**, *137*, 13927–13932. [CrossRef] [PubMed]

74. Parvez, K.; Wu, Z.S.; Li, R.; Liu, X.; Graf, R.; Feng, X.; Müllen, K. Exfoliation of graphite into graphene in aqueous solutions of inorganic salts. *J. Am. Chem. Soc.* **2014**, *136*, 6083–6091. [CrossRef] [PubMed]

75. Wang, D.-Y.; Wei, C.-Y.; Lin, M.-C.; Pan, C.-J.; Chou, H.-L.; Chen, H.-A.; Gong, M.; Wu, Y.; Yuan, C.; Angell, M. Advanced rechargeable aluminium ion battery with a high-quality natural graphite cathode. *Nat. Commun.* **2017**, *8*, 14283. [CrossRef] [PubMed]

76. Rodríguez-Pérez, I.A.; Ji, X. Anion hosting cathodes in dual-ion batteries. *ACS Energy Lett.* **2017**, *2*, 1762–1770. [CrossRef]

77. Xu, J.; Dou, Y.; Wei, Z.; Ma, J.; Deng, Y.; Li, Y.; Liu, H.; Dou, S. Recent progress in graphite intercalation compounds for rechargeable metal (Li, Na, K, Al)-ion batteries. *Adv. Sci.* **2017**, *4*, 1700146. [CrossRef] [PubMed]

78. Wang, Y.; Liu, B.; Li, Q.; Cartmell, S.; Ferrara, S.; Deng, Z.D.; Xiao, J. Lithium and lithium ion batteries for applications in microelectronic devices: A review. *J. Power Sources* **2015**, *286*, 330–345. [CrossRef]

79. Li, Z.; Huang, J.; Liaw, B.Y.; Metzler, V.; Zhang, J. A review of lithium deposition in lithium-ion and lithium metal secondary batteries. *J. Power Sources* **2014**, *254*, 168–182. [CrossRef]

80. Lin, M.-C.; Gong, M.; Lu, B.; Wu, Y.; Wang, D.-Y.; Guan, M.; Angell, M.; Chen, C.; Yang, J.; Hwang, B.-J. An ultrafast rechargeable aluminium-ion battery. *Nature* **2015**, *520*, 324–328. [CrossRef] [PubMed]

81. Bonaccorso, F.; Lombardo, A.; Hasan, T.; Sun, Z.; Colombo, L.; Ferrari, A.C. Production and processing of graphene and 2d crystals. *Mater. Today* **2012**, *15*, 564–589. [CrossRef]

82. Rothermel, S.; Meister, P.; Schmuelling, G.; Fromm, O.; Meyer, H.; Nowak, S.; Winter, M.; Placke, T. Dual-graphite cells based on the reversible intercalation of bis(trifluoromethanesulfonyl)imide anions from an ionic liquid electrolyte. *Energy Environ. Sci.* **2014**, *7*, 3412–3423. [CrossRef]

83. Placke, T.; Fromm, O.; Lux, S.F.; Bieker, P.; Rothermel, S.; Meyer, H.W.; Passerini, S.; Winter, M. Reversible intercalation of bis(trifluoromethanesulfonyl)imide anions from an ionic liquid electrolyte into graphite for high performance dual-ion cells. *J. Electrochem. Soc.* **2012**, *159*, A1755–A1765. [CrossRef]

84. Li, B.; Xu, M.; Li, T.; Li, W.; Hu, S. Prop-1-ene-1,3-sultone as SEI formation additive in propylene carbonate-based electrolyte for lithium ion batteries. *Electrochem. Commun.* **2012**, *17*, 92–95. [CrossRef]

85. Meister, P.; Siozios, V.; Reiter, J.; Klamor, S.; Rothermel, S.; Fromm, O.; Meyer, H.W.; Winter, M.; Placke, T. Dual-ion cells based on the electrochemical intercalation of asymmetric fluorosulfonyl-(trifluoromethanesulfonyl) imide anions into graphite. *Electrochim. Acta* **2014**, *130*, 625–633. [CrossRef]

86. Moon, H.; Tatara, R.; Mandai, T.; Ueno, K.; Yoshida, K.; Tachikawa, N.; Yasuda, T.; Dokko, K.; Watanabe, M. Mechanism of Li ion desolvation at the interface of graphite electrode and glyme–Li salt solvate ionic liquids. *J. Phys. Chem. C* **2014**, *118*, 20246–20256. [CrossRef]

87. Takeuchi, S.; Fukutsuka, T.; Miyazaki, K.; Abe, T. Electrochemical preparation of a lithium–graphite-intercalation compound in a dimethyl sulfoxide-based electrolyte containing calcium ions. *Carbon* **2013**, *57*, 232–238. [CrossRef]

88. Tasaki, K. Density functional theory study on structural and energetic characteristics of graphite intercalation compounds. *J. Phys. Chem. C* **2015**, *118*, 1443–1450. [CrossRef]

89. Billaud, D.; Pron, A.; Vogel, F.L.; Hérold, A. Intercalation of pyrographite by NO_2^+ and NO^+ salts. *Mater. Res. Bull.* **1980**, *15*, 1627–1634. [CrossRef]

90. Jobert, A.; Touzain, P.; Bonnetain, L. Insertion des ions PF_6^-, AsF_6^- et SbF_6^- dans le graphite par methode electrochimique. Caracterisation des produits obtenus. *Carbon* **1981**, *19*, 193–198. [CrossRef]

91. Xia, Z.Y.; Giambastiani, G.; Christodoulou, C.; Nardi, M.V.; Koch, N.; Treossi, E.; Bellani, V.; Pezzini, S.; Corticelli, F.; Morandi, V. Synergic exfoliation of graphene with organic molecules and inorganic ions for the electrochemical production of flexible electrodes. *Chempluschem* **2014**, *79*, 439–446. [CrossRef]

92. Sirisaksoontorn, W.; Lerner, M.M. Preparation of a homologous series of tetraalkylammonium graphite intercalation compounds. *Inorg. Chem.* **2013**, *52*, 7139–7144. [CrossRef] [PubMed]

93. Zhang, H.; Lerner, M.M. Preparation of graphite intercalation compounds containing oligo and polyethers. *Nanoscale* **2016**, *8*, 4608–4612. [CrossRef] [PubMed]

94. Do Nascimento, G.M.; Constantino, V.R.L.; Landers, R.; Temperini, M.L.A. Aniline polymerization into montmorillonite clay: A spectroscopic investigation of the intercalated conducting polymer. *Mol. Microbiol.* **2004**, *52*, 751–761. [CrossRef]

95. Sirisaksoontorn, W.; Adenuga, A.A.; Remcho, V.T.; Lerner, M.M. Preparation and characterization of a tetrabutylammonium graphite intercalation compound. *J. Am. Chem. Soc.* **2011**, *133*, 12436–12438. [CrossRef] [PubMed]

96. Zhong, Y.L.; Swager, T.M. Enhanced electrochemical expansion of graphite for in situ electrochemical functionalization. *J. Am. Chem. Soc.* **2012**, *134*, 17896. [CrossRef] [PubMed]

97. Zhang, H.; Lerner, M.M. Preparation of graphite intercalation compounds containing crown ethers. *Inorg. Chem.* **2016**, *55*, 8281–8284. [CrossRef] [PubMed]

98. Xu, J.; Dang, D.K.; Tran, V.T.; Liu, X.; Chung, J.S.; Hur, S.H.; Choi, W.M.; Kim, E.J.; Kohl, P.A. Liquid-phase exfoliation of graphene in organic solvents with addition of naphthalene. *J. Colloid Interface Sci.* **2014**, *418*, 37–42. [CrossRef] [PubMed]

99. Wei, L.; Wu, F.; Shi, D.; Hu, C.; Li, X.; Yuan, W.; Wang, J.; Zhao, J.; Geng, H.; Wei, H. Spontaneous intercalation of long-chain alkyl ammonium into edge-selectively oxidized graphite to efficiently produce high-quality graphene. *Sci. Rep.* **2013**, *3*, 2636. [CrossRef] [PubMed]

100. Liu, Z.H.; Wang, Z.M.; Yang, X.; Ooi, K. Intercalation of organic ammonium ions into layered graphite oxide. *Langmuir* **2002**, *18*, 4926–4932. [CrossRef]

101. Matsuo, Y.; Tahara, K.; Sugie, Y. Structure and thermal properties of poly(ethylene oxide)-intercalated graphite oxide. *Carbon* **1997**, *35*, 113–120. [CrossRef]

102. Matsuo, Y.; Hatase, K.; Sugie, Y. Preparation and characterization of poly(vinyl alcohol)- and $Cu(OH)_2$-poly(vinyl alcohol)-intercalated graphite oxides. *Chem. Mater.* **1998**, *10*, 2266–2269. [CrossRef]

103. Kotov, N.A.; Dékány, I.; Fendler, J.H. Ultrathin graphite oxide-polyelectrolyte composites prepared by self-assembly: Transition between conductive and non-conductive states. *Adv. Mater.* **1996**, *8*, 637–641. [CrossRef]

104. Liu, P.; Gong, K.; Xiao, P.; Xiao, M. Preparation and characterization of poly(vinyl acetate)-intercalated graphite oxide nanocomposite. *J. Mater. Chem.* **2000**, *10*, 933–935. [CrossRef]

105. Kim, H.; Hahn, H.T.; Viculis, L.M.; Gilje, S.; Kaner, R.B. Electrical conductivity of graphite/polystyrene composites made from potassium intercalated graphite. *Carbon* **2007**, *45*, 1578–1582. [CrossRef]

106. Nadiv, R.; Shtein, M.; Buzaglo, M.; Peretz-Damari, S.; Kovalchuk, A.; Wang, T.; Tour, J.M.; Regev, O. Graphene nanoribbon—Polymer composites: The critical role of edge functionalization. *Carbon* **2016**, *99*, 444–450. [CrossRef]

107. Ma, J.; Li, Y.; Yin, X.; Xu, Y.; Yue, J.; Bao, J.; Zhou, T. Poly(vinyl alcohol)/graphene oxide nanocomposites prepared by in situ polymerization with enhanced mechanical properties and water vapor barrier properties. *RSC Adv.* **2016**, *6*, 49448–49458. [CrossRef]

108. Shioyama, H.; Tatsumi, K.; Iwashita, N.; Fujita, K.; Sawada, Y. On the interaction between the potassium—GIC and unsaturated hydrocarbons. *Synth. Met.* **1998**, *96*, 229–233. [CrossRef]

109. Shioyama, H. On the interaction between alkali metal-GICs and unsaturated hydrocarbons. *Mol. Cryst. Liq. Cryst.* **2000**, *340*, 101–106. [CrossRef]

110. Lu, W.; Ruan, G.; Genorio, B.; Zhu, Y.; Novosel, B.; Peng, Z.; Tour, J.M. Functionalized graphene nanoribbons via anionic polymerization initiated by alkali metal-intercalated carbon nanotubes. *ACS Nano* **2013**, *7*, 2669–2675. [CrossRef] [PubMed]

111. Kosynkin, D.V.; Lu, W.; Sinitskii, A.; Pera, G.; Sun, Z.; Tour, J.M. Highly conductive graphene nanoribbons by longitudinal splitting of carbon nanotubes using potassium vapor. *ACS Nano* **2011**, *5*, 968–974. [CrossRef] [PubMed]

112. Genorio, B.; Wei, L.; Dimiev, A.M.; Zhu, Y.; Raji, A.R.O.; Novosel, B.; Alemany, L.B.; Tour, J.M. In situ intercalation replacement and selective functionalization of graphene nanoribbon stacks. *ACS Nano* **2012**, *6*, 4231–4240. [CrossRef] [PubMed]

113. Zhang, C.; Peng, Z.; Lin, J.; Zhu, Y.; Ruan, G.; Hwang, C.C.; Lu, W.; Hauge, R.H.; Tour, J.M. Splitting of a vertical multiwalled carbon nanotube carpet to a graphene nanoribbon carpet and its use in supercapacitors. *ACS Nano* **2013**, *7*, 5151–5159. [CrossRef] [PubMed]

114. Liang, F.; Beach, J.M.; Kobashi, K.; Sadana, A.K.; Vegacantu, Y.I.; Tour, J.M.; Billups, W.E. In situ polymerization initiated by single-walled carbon nanotube salts. *Chem. Mater.* **2013**, *18*, 4764–4767. [CrossRef]

115. Xiang, C.; Cox, P.J.; Kukovecz, A.; Genorio, B.; Hashim, D.P.; Yan, Z.; Peng, Z.; Hwang, C.C.; Ruan, G.; Samuel, E.L. Functionalized low defect graphene nanoribbons and polyurethane composite film for improved gas barrier and mechanical performances. *ACS Nano* **2013**, *7*, 10380. [CrossRef] [PubMed]

116. Genorio, B.; Znidarsic, A. Functionalization of graphene nanoribbons. *J. Phys. D* **2014**, *10*, 229A. [CrossRef]

117. Shan, L.; Chen, X.; Tian, X.; Chen, J.; Zhou, Z.; Jiang, M.; Xu, X.; Hui, D. Fabrication of polypyrrole/nano-exfoliated graphite composites by in situ intercalation polymerization and their microwave absorption properties. *Compos. Part B Eng.* **2014**, *73*, 181–187. [CrossRef]

118. Wang, R.X.; Huang, L.F.; Tian, X.Y. Understanding the protonation of polyaniline and polyaniline–graphene interaction. *J. Phys. Chem. C* **2012**, *116*, 13120–13126. [CrossRef]

119. Wang, P.; Zhang, J.; Dong, L.; Sun, C.; Zhao, X.; Ruan, Y.; Lu, H. Interlayer polymerization in chemically expanded graphite for preparation of highly conductive, mechanically strong polymer composites. *Chem. Mater.* **2017**, *29*, 3412–3422. [CrossRef]

120. Chen, X.; Chen, J.; Meng, F.; Shan, L.; Jiang, M.; Xu, X.; Lu, J.; Wang, Y.; Zhou, Z. Hierarchical composites of polypyrrole/graphene oxide synthesized by in situ intercalation polymerization for high efficiency and broadband responses of electromagnetic absorption. *Compos. Sci. Technol.* **2016**, *127*, 71–78. [CrossRef]

121. Lin, S.; Dong, L.; Zhang, J.; Lu, H. Room-temperature intercalation and ~1000-fold chemical expansion for scalable preparation of high-quality graphene. *Chem. Mater.* **2016**, *28*, 2138–2146. [CrossRef]

122. Georgakilas, V.; Otyepka, M.; Bourlinos, A.B.; Chandra, V.; Kim, N.; Kemp, K.C.; Hobza, P.; Zboril, R.; Kim, K.S. Functionalization of graphene: Covalent and non-covalent approaches, derivatives and applications. *Chem. Rev.* **2012**, *112*, 6156–6214. [CrossRef] [PubMed]

123. Dong, L.; Chen, Z.; Lin, S.; Wang, K.; Ma, C.; Lu, H. Reactivity-controlled preparation of ultralarge graphene oxide by chemical expansion of graphite. *Chem. Mater.* **2017**, *29*, 564–572. [CrossRef]

124. Yang, Q.; Su, Y.; Chi, C.; Cherian, C.; Huang, K.; Kravets, V.; Wang, F.; Zhang, J.; Pratt, A.; Grigorenko, A. Ultrathin graphene-based membrane with precise molecular sieving and ultrafast solvent permeation. *Nat. Mater.* **2017**, *16*, 1198–1203. [CrossRef] [PubMed]

125. Kuila, T.; Bose, S.; Khanra, P.; Kim, N.H.; Rhee, K.Y.; Lee, J.H. Characterization and properties of in situ emulsion polymerized poly(methyl methacrylate)/graphene nanocomposites. *Compos. Part A Appl. Sci. Manuf.* **2011**, *42*, 1856–1861. [CrossRef]

126. Zeng, X.; Yang, J.; Yuan, W. Preparation of a poly(methyl methacrylate)-reduced graphene oxide composite with enhanced properties by a solution blending method. *Eur. Polym. J.* **2012**, *48*, 1674–1682. [CrossRef]

127. Malas, A.; Bharati, A.; Verkinderen, O.; Goderis, B.; Moldenaers, P.; Cardinaels, R. Effect of the GO reduction method on the dielectric properties, electrical conductivity and crystalline behavior of PEO/rGO nanocomposites. *Polymers* **2017**, *9*, 613. [CrossRef]

128. Colonna, S.; Monticelli, O.; Gomez, J.; Novara, C.; Saracco, G.; Fina, A. Effect of morphology and defectiveness of graphene-related materials on the electrical and thermal conductivity of their polymer nanocomposites. *Polymer* **2016**, *102*, 292–300. [CrossRef]

129. Li, M.; Jeong, Y.G. Poly(ethylene terephthalate)/exfoliated graphite nanocomposites with improved thermal stability, mechanical and electrical properties. *Compos. Part A Appl. Sci. Manuf.* **2011**, *42*, 560–566. [CrossRef]

130. Zhu, J.; Lim, J.; Lee, C.H.; Joh, H.I.; Kim, H.C.; Park, B.; You, N.H.; Lee, S. Multifunctional polyimide/graphene oxide composites via in situ polymerization. *J. Appl. Polym. Sci.* **2014**, *131*, 40177. [CrossRef]

131. Ming, P.; Zhang, Y.; Bao, J.; Liu, G.; Li, Z.; Jiang, L.; Cheng, Q.D. Bioinspired highly electrically conductive graphene-epoxy layered composites. *RSC Adv.* **2015**, *5*, 22283–22288. [CrossRef]

132. Wang, L.; Huang, Y.; Huang, H. N-doped graphene@polyaniline nanorod arrays hierarchical structures: Synthesis and enhanced electromagnetic absorption properties. *Mater. Lett.* **2014**, *124*, 89–92. [CrossRef]

133. Bai, X.; Zhai, Y.; Zhang, Y. Green approach to prepare graphene-based composites with high microwave absorption capacity. *J. Phys. Chem. C* **2011**, *115*, 11673–11677. [CrossRef]

134. Ashwin, S.; Isupova, O. Microwave absorbing properties of a thermally reduced graphene oxide/nitrile butadiene rubber composite. *Carbon* **2012**, *50*, 2202–2208.

135. Wang, Y.; Wu, X.; Zhang, W. Synthesis and high-performance microwave absorption of graphene foam/polyaniline nanorods. *Mater. Lett.* **2016**, *165*, 71–74. [CrossRef]

136. Liu, P.; Huang, Y. Decoration of reduced graphene oxide with polyaniline film and their enhanced microwave absorption properties. *J. Polym. Res.* **2014**, *21*, 1–5. [CrossRef]

137. Tripathi, S.N.; Saini, P.; Gupta, D.; Choudhary, V. Electrical and mechanical properties of PMMA/reduced graphene oxide nanocomposites prepared via in situ polymerization. *J. Mater. Sci.* **2013**, *48*, 6223–6232. [CrossRef]

138. Vallés, C.; Abdelkader, A.M.; Young, R.J.; Kinloch, I.A. The effect of flake diameter on the reinforcement of few-layer graphene–PMMA composites. *Compos. Sci. Technol.* **2015**, *111*, 17–22. [CrossRef]

139. Tang, L.C.; Wan, Y.J.; Yan, D.; Pei, Y.B.; Zhao, L.; Li, Y.B.; Wu, L.B.; Jiang, J.X.; Lai, G.Q. The effect of graphene dispersion on the mechanical properties of graphene/epoxy composites. *Carbon* **2013**, *60*, 16–27. [CrossRef]

140. Wan, Y.J.; Tang, L.C.; Gong, L.X.; Yan, D.; Li, Y.B.; Wu, L.B.; Jiang, J.X.; Lai, G.Q. Grafting of epoxy chains onto graphene oxide for epoxy composites with improved mechanical and thermal properties. *Carbon* **2014**, *69*, 467–480. [CrossRef]

141. Nguyen, D.A.; Yu, R.L.; Raghu, A.V.; Han, M.J.; Shin, C.M.; Kim, B.K. Morphological and physical properties of a thermoplastic polyurethane reinforced with functionalized graphene sheet. *Polym. Int.* **2009**, *58*, 412–417. [CrossRef]

142. Chen, J.; Chen, X.; Meng, F.; Li, D.; Tian, X.; Wang, Z.; Zhou, Z. Super-high thermal conductivity of polyamide-6/graphene-graphene oxide composites through in situ polymerization. *High Perform. Polym.* **2016**, *29*, 585–594. [CrossRef]

143. Ding, P.; Su, S.; Song, N.; Tang, S.; Liu, Y.; Shi, L. Highly thermal conductive composites with polyamide-6 covalently-grafted graphene by an in situ polymerization and thermal reduction process. *Carbon* **2014**, *66*, 576–584. [CrossRef]

144. Li, X.; Shao, L.; Song, N.; Shi, L.; Ding, P. Enhanced thermal-conductive and anti-dripping properties of polyamide composites by 3d graphene structures at low filler content. *Compos. Part A Appl. Sci. Manuf.* **2016**, *88*, 305–314. [CrossRef]

145. Zhou, S.; Chen, Y.; Zou, H.; Liang, M. Thermally conductive composites obtained by flake graphite filling immiscible polyamide 6/polycarbonate blends. *Thermochim. Acta* **2013**, *566*, 84–91. [CrossRef]

146. Wu, K.; Lei, C.; Huang, R.; Yang, W.; Chai, S.; Geng, C.; Chen, F.; Fu, Q. Design and preparation of a unique segregated double network with excellent thermal conductive property. *ACS Appl. Mater. Interfaces* **2017**, *9*, 7637. [CrossRef] [PubMed]

147. Alam, F.E.; Dai, W.; Yang, M.; Du, S.; Li, X.; Yu, J.; Jiang, N.; Lin, C.T. In-situ formation of cellular graphene framework in thermoplastic composites leading to superior thermal conductivity. *J. Mater. Chem. A* **2017**, *5*. [CrossRef]

148. Kim, S.Y.; Ye, J.N.; Yu, J. Thermal conductivity of graphene nanoplatelets filled composites fabricated by solvent-free processing for the excellent filler dispersion and a theoretical approach for the composites containing the geometrized fillers. *Compos. Part A Appl. Sci. Manuf.* **2015**, *69*, 219–225. [CrossRef]

149. Han, H.S.; You, J.M.; Jeong, H.; Jeon, S. Synthesis of graphene oxide grafted poly(lactic acid) with palladium nanoparticles and its application to serotonin sensing. *Appl. Surf. Sci.* **2013**, *284*, 438–445. [CrossRef]

150. Konwer, S.; Guha, A.K.; Dolui, S.K. Graphene oxide-filled conducting polyaniline composites as methanol-sensing materials. *J. Mater. Sci.* **2013**, *48*, 1729–1739. [CrossRef]

151. Konwer, S. Graphene oxide-polyaniline nanocomposites for high performance supercapacitor and their optical, electrical and electrochemical properties. *J. Mater. Sci.* **2016**, *27*, 4139–4146. [CrossRef]

polymers

MDPI

Review

Recent Achievements of Self-Healing Graphene/Polymer Composites

Yongxu Du [†], Dong Li [†], Libin Liu *and Guangjie Gai

School of Chemistry and Pharmaceutical Engineering, Qilu University of Technology
(Shandong Academy of Sciences), Jinan 250353, China; Duyongxu2009@163.com (Y.D.);
LD18396813772@163.com (D.L.); 18396814715@163.com (G.G.)
* Correspondence: lbliu@qlu.edu.cn; Tel.: +86-187-6398-8232
† The two authors contribute to the paper equally.

Received: 21 December 2017; Accepted: 22 January 2018; Published: 25 January 2018

Abstract: Self-healing materials have attracted much attention because that they possess the ability to increase the lifetime of materials and reduce the total cost of systems during the process of long-term use; incorporation of functional material enlarges their applications. Graphene, as a promising additive, has received great attention due to its large specific surface area, ultrahigh conductivity, strong antioxidant characteristics, thermal stability, high thermal conductivity, and good mechanical properties. In this brief review, graphene-containing polymer composites with self-healing properties are summarized including their preparations, self-healing conditions, properties, and applications. In addition, future perspectives of graphene/polymer composites are briefly discussed.

Keywords: self-healing; graphene; polymer; composite

1. Introduction

Regenerative abilities allow creatures to repair damaged functions to prolong their life span. Researchers are inspired to design and prepare self-healing materials to increase the lifetime of materials and reduce the total cost of systems during the process of long-term use. Recently, great progress has been made in self-healing composite materials that possess the ability to restore their structure and functionality after damage. Early self-healing materials were focused on microcapsule or microtubule by release of healing agents to achieve repairing. However, the self-healing times of these methods are dependent on the amounts of healing agents in the microcapsule or microtubules [1]. To address these limitations, dynamic chemistry involving dynamic covalent chemistry (e.g., imine bonds [2,3], disulfide bonds [4–6], acylhydrazone bonds [7–9], and boronate ester bonds [7]) and non-covalent interactions, such as hydrogen bonds [10,11], $\pi-\pi$ stacking [12], hydrophobic interactions [13,14], host-guest interactions [15], ionic interactions [16], electrostatic interactions [17], and metal-coordination interactions [18–20], has been recently introduced to construct self-healing materials with multiple reversible healing ability.

Graphene, as a new type of two-dimensional planar monolayer of sp^2 carbon atoms, has attracted widespread attention in all kinds of research areas due to its large specific surface area, excellent electrical conductivity, thermal conductivity, and unique mechanical properties [21–25]. Recently, graphene or graphene derivatives have been widely introduced into polymer matrices. The excellent performance of graphene or graphene derivatives, combined with the advantages of the polymer matrix, makes graphene/polymer composites suitable for application in conductive devices, coating, and biological and pharmaceutical field [26–30]. Although graphene-based composites have been well established [31–34], graphene-containing composites with self-healing capacity have not been summarized up to now. Introduction of self-healing capability into graphene/polymer composites will endow them with the ability of repairing themselves after damage and enlarge their service life.

A lot of studies have been reported on the self-healing of the graphene/polymer composites due to their wide applications (Figure 1). Therefore, it is necessary to review self-healing graphene/polymer composites, which combine the outstanding properties of graphene with advantages of the polymeric matrix and can be used in the field of mechanics, thermology, photology and electricity.

In this review, the current advances in self-healing graphene/polymer composites have been summarized, including their preparation methods, self-healing conditions, properties and applications. Finally, the future prospects of the self-healing graphene/polymer composites are discussed.

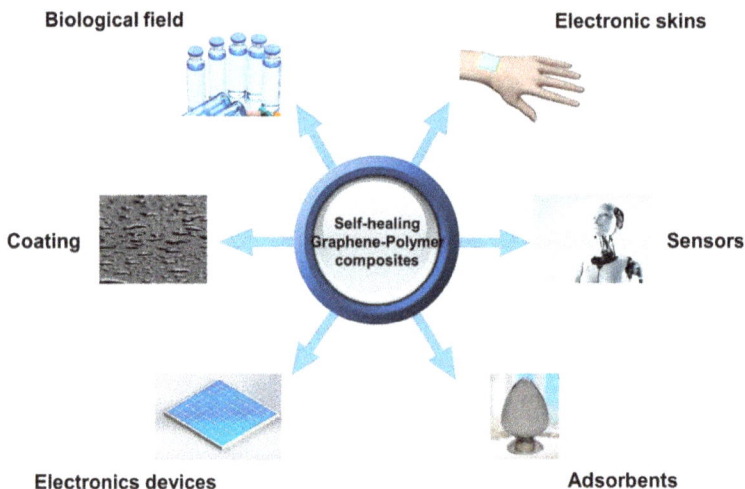

Figure 1. Applications of self-healing graphene/polymer composites.

2. Fabrication Methods

2.1. Simple Mixing

The simple mixing is the simplest and commonest method for preparing graphene or graphene derivative/polymer composites [35–43]. Usually, graphene or graphene derivatives are blended with polymers by mechanical mixing or ultrasonic dispersion. For example, Sabzi et al. [44] prepared poly(vinyl alcohol) (PVA)/Agar/graphene self-healing hydrogels by simply mechanical stirring and ultrasonication.

Considering its easy agglomerate due to the strong π-π interactions, usually, graphene has been oxidized into graphene oxide (GO). Yan et al. fabricated chitosan/GO supramolecular hydrogels with self-healing properties [45]. It was found that at high GO concentration, a hydrogel can be obtained by simple mixing chitosan and GO at room temperature. However, at low GO concentration, the supramolecular hydrogel formed only at high temperature (95 °C). Walther et al. fabricated rapid self-healing supramolecular elastomers by simple mixing graphene and supramolecular pseudo-copolymer [37]. The copolymer system was formed by co-assembly of diaminotriazine (DAT) functionalized polyglycidols (PG) and cyanuric acid (CA) functionalized PG. Thermally reduced graphene oxide (TRGO) was added from a freshly sonicated dispersion to a heated supramolecular pseudo-copolymer to reach 0.1 wt % in the final nanocomposite. The excellent photothermal effect was enhanced by TRGO, which made the hydrogen bonds break and bond in co-assembled elastomers (Figure 2).

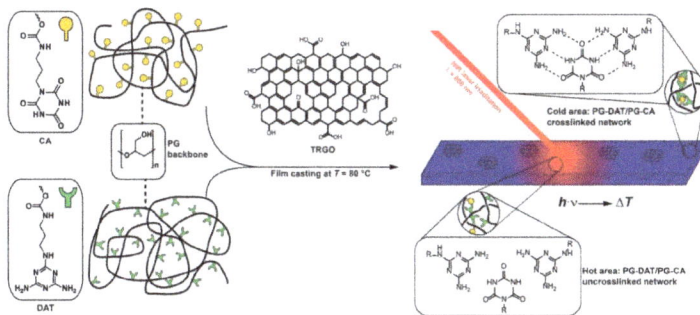

Figure 2. Remote, spatiotemporal, light-fueled modulation of the mechanical properties of co-assembled PG-DAT/PG-CA films hybridized with TRGO. The incorporation of TRGO in low amounts (0.1 wt %) allows localized heating via a NIR laser (808 nm) to break the hydrogen bonds, thus allowing molecular motion and relaxation. Reproduced with permission from [37]. Copyright (2017) Advanced Functional Materials.

2.2. In Situ Polymerization

In situ polymerization can be interpreted to mean that the monomer and graphene or graphene derivative were mixed firstly and subsequently polymerized by the addition of initiators [46–58]. For example, Green et al. demonstrated physically cross-linked graphene-polyacrylamide (PAM) self-healing hydrogels with increased thermal stability and electrical conductivity [54]. All the reactants, acrylamide (AM), N,N-methylenebisacrylamide (MBA), and potassium persulfate, were added to the graphene dispersion in water and polymerized.

Taking advantage of this method, our group fabricated cationic PAM/GO self-healing hydrogels with tough, stretchable, compressive property [46]. The GO aqueous dispersion was adjusted to pH 10 by dropping ammonium hydroxide. Successively, the monomers 2-(dimethylamino)ethylacrylatemethochloride (DAC) and AM were added into the GO suspension under stirring followed by the addition of MBA. After adding initiators, polymerization was carried out in an oven at 35 °C for 12 h. Ran et al. fabricated self-healing GO/hydrophobically associated polyacrylamide (HAPAM) composite hydrogels [49]. During the synthesis, GO, the hydrophilic monomer AM and the hydrophobic monomer stearyl methacrylate were mixed to make a uniform solution. After that, potassium persulfate was added to the solution to initiate the polymerization. As shown in Figure 3, a dual cross-linked network was formed after introducing GO into HAPAM through a facile one-pot in situ polymerization.

During the in situ polymerization process, graphene can also be modified and participate in the polymerization, forming the covalent bond between graphene and polymers. For example, Karak et al. fabricated a tough self-healing elastomeric nanocomposite containing a castor oil-based hyperbranched polyurethane (PU) and an iron oxide nanoparticle decorated reduced graphene oxide (IORGO) nanohybrid [59]. The IORGO was prepared by the co-precipitation of ferrous and ferric ions on the GO sheets, followed by the reduction of GO. The reaction was carried out by in situ polymerization of poly(ε-caprolactone) diol, 1,4-butanediol, 2,4/2,6-toluene diisocyanate and IORGO. After formation of pre-polymer, a monoglyceride of castor oil as a chain extender was added to form the resulting PU/IORGO nanocomposites. The reaction process is a conventional condensation reaction and is shown in Figure 4a. The same group also prepared sulfur nanoparticle decorated reduced graphene oxide (SRGO) [60] and fabricated self-healing hyperbranched PU/SRGO nanocomposites [61].

Similarly, Kim et al. synthesized a phenyl isocyanate modified GO and obtained the self-healing composites by condensation reaction of poly(tetramethylene glycol) and 4,4′-methylene diphenyl diisocyanate and phenyl isocyanate modified GO in the presence of phenylhydrazine [62] (Figure 4b,c).

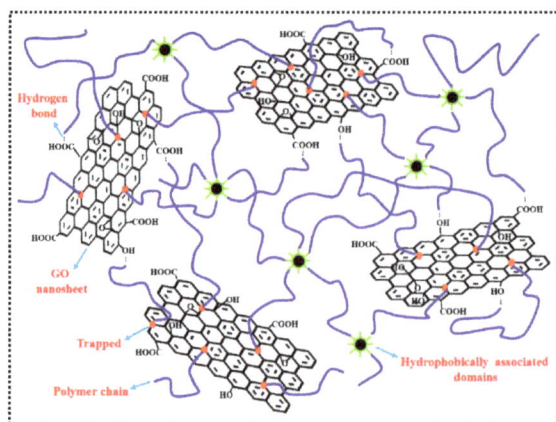

Figure 3. A proposed structure illustration of GO/hydrophobically associated polyacrylamide (HAPAM) composite hydrogels. Reproduced with permission from [49]. Copyright (2015) Journal of Materials Chemistry A.

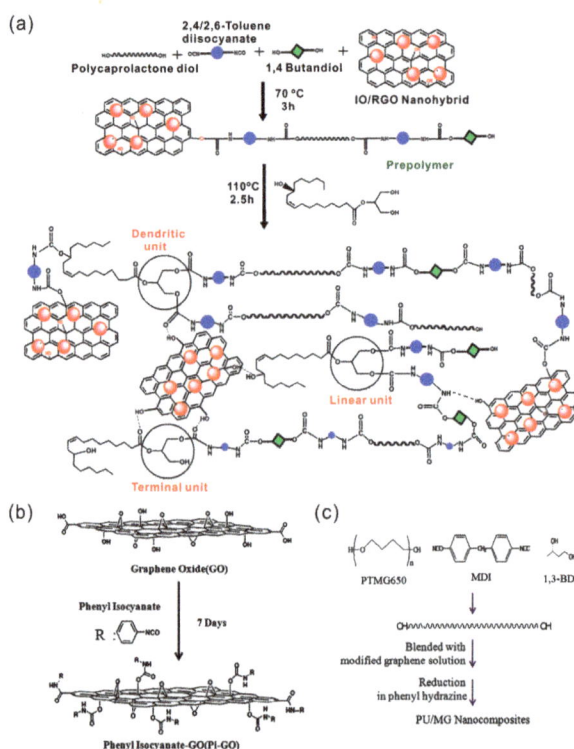

Figure 4. (**a**) Synthesis of a hyperbranched PU/IORGO nanocomposite. Reproduced with permission from [59]. Copyright (2015) New Journal of Chemistry. (**b**) Modification of graphene oxide (GO) by phenyl isocyanate. (**c**) Overall reaction scheme to prepare a PU/MG nanocomposite. Reproduced with permission from [62]. Copyright (2013) European Polymer Journal.

2.3. Diels-Alder (DA) Reactions

The DA reaction and its retro-Diels-Alder (rDA) analogue is a promising route to introduce self-healing properties to polymeric systems, which can be performed under mild conditions without any catalyst or healing agent [63–66]. To realize the DA reaction, GO is usually functionalized to react with polymers. For example, Liu and coworkers synthesized the maleimide functionalized GO [67], which can produce a DA cross-linked bond with furan groups of the polyurethane chains (Figure 5). Zhang et al. synthesized an ultrafast self-healing composite material based on DA reactions. The surface modification of graphene involved hydramine-functionalized graphene oxide (FGO) and reduction of FGO to afford hydramine-functionalized graphene nanosheets. The resulted composite was formed by introduction of functionalized graphene nanosheets into the pre-PU which was prepared from the condensation of NCO-terminated PU and a DA resultant of furfury alcohol and bismaleimide [68,69].

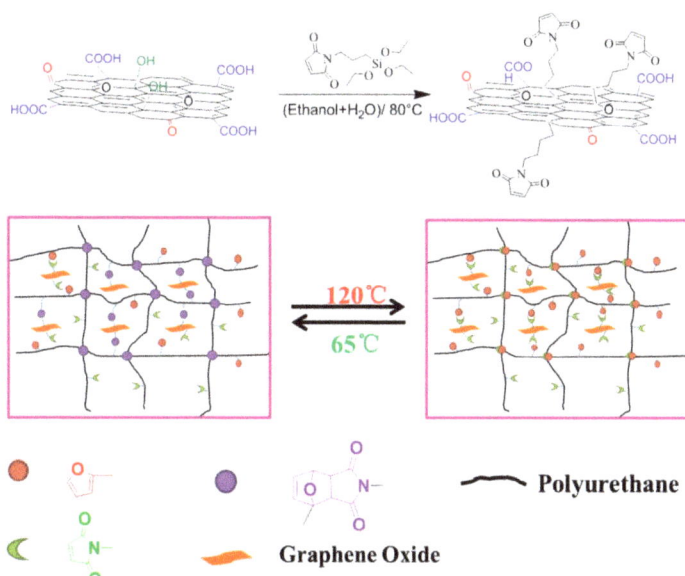

Figure 5. Synthetic routes of nanocomposites. Reproduced with permission from [67]. Copyright (2017) Polymer.

2.4. Layer-by-Layer Self-Assembly Technique

The layer-by-layer (LBL) self-assembly technique is a versatile approach to fabricate multilayered nanostructural composites [70–73]. The first implementation of this technique is attributed to J. J. Kirkland and R. K. Iler, who carried it out using microparticles in 1966 [74]. The LBL self-assembly technique now can be accomplished by using various methods such as immersion, spin, spray, electromagnetism, or fluidics. For example, Fan et al. introduced a self-healing anticorrosion coating on a magnesium alloy (AZ31) substrate [75]. Firstly, cerium nitrate hexahydrate was coated on AZ31 substrate and then the sample was heated at 80 °C for 30 min to partially convert the oxide Ce(III) to Ce(IV). Subsequently, poly(ethyleneimine) (PEI) and GO was coated on the sample to form the PEI/GO layer. Finally, the PEI/GO coated sample was immersed in PEI, deionized water, poly(acrylic acid) (PAA), and deionized water, alternatively (Figure 6a). Graphene oxide was incorporated as corrosion inhibitor and the self-healing ability was attributed to the PEI/PAA multilayers. Ge et al. also prepared a self-healing multilayer polyelectrolyte film based on branched PEI, PAA and graphene by a LBL self-assembly technique [76].

Sun and coworkers reported an intrinsically healable, reduced graphene oxide (RGO)-reinforced polymer composite film via LBL assembly [77]. RGO modified with β-cyclodextrin (β-CD) (denoted as RGO-CD) can complex with branched PEI grafted with ferrocene groups (Fc) (denoted as bPEI-Fc) based on host-guest interactions to form bPEI-Fc&RGO-CD complexes. The bPEI-Fc&RGO-CD complexes are LBL assembled with PAA to fabricate PAA/bPEI-Fc&RGO-CD composite films. The reversible host-guest interactions between nanofillers and LBL-assembled polyelectrolyte films make the composites possess excellent mechanical robustness and highly efficient self-healing properties simultaneously (Figure 6b,c).

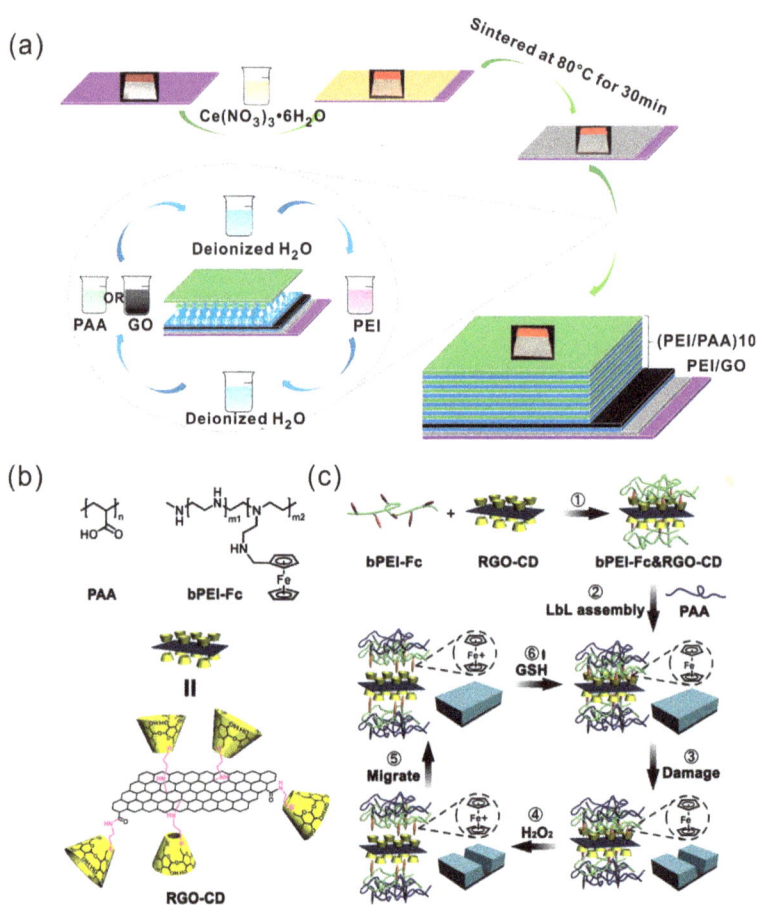

Figure 6. (a) Preparation process of the self-healing anticorrosion coating. Reproduced with permission from [75]. Copyright (2015) ACS Applied Materials and Interfaces. (b) The chemical structures of bPEI-Fc, PAA and RGO-CD. (c) Schematic illustration of the fabrication and healing process of the PAA/bPEI-Fc&RGO-CD composite films. Reproduced with permission from [77]. Copyright (2017) ACS Nano.

2.5. Hydrothermal Methods

Apart from the aforementioned fabrication methods, there are other ways to fabricate self-healing graphene/polymer composites. Tang et al. reported conductive and self-healing nanocomposite hydrogels though a simple hydrothermal method [78]. All the reagents were poured into a Teflon-lined stainless steel autoclave and heated at 100 °C for 10 h, forming nanocomposite hydrogels.

3. Self-Healing Condition

To restore their original properties, graphene/polymer composites need to repair themselves autonomously or require external energy/stimuli such as mechanical force, light, heat, pH changes. In this section, different self-healing conditions of graphene/polymer composites were summarized.

3.1. Heating

Heating is a common self-healing condition due to its easy operation. By heating, the mobility of polymer chains increases, thus facilitating the self-healing of graphene/polymer composites [40,41,43,48,51,53,54,56,67,79–82]. Pugno and Valentini's group fabricated a negative temperature coefficient (exhibiting electrical resistance decrease with temperature increase) silicone rubber (SR)-graphene nanoplatelets (GNPs) composite that can be healed by simple thermal annealing in an oven up to 250 °C for 2 h [43]. After heating treatment, the composite showed a healing efficiency of ~87% by tensile strength. The reversible crosslinking among the damaged network of SR/GNP composite can be thermally activated due to free silanol groups. In conductive composites, the electrical conductivity commonly decreased due to the destroyed conductive network during tensile cycles. However, Zhan et al. reported a conductive graphene/natural rubber composite [40], in which the electrical conductivity rose to nearly two times higher than that of the original one after four tensile cycles and subsequently thermal treatment. The increased conductivity indicates that the destroyed networks, which occur during the tensile process, can be healed during the post thermal treatment.

3.2. Light Radiation

Although heating is an easy way to repair damaged graphene/polymer composites, during heating other parts of devices are susceptive to heat, thus heating will cause interference of other parts of the devices and result in the deterioration of devices. Therefore, light radiation is an alternative way to repair the damaged composites. Graphene has good ability of photothermal energy transformation that makes the self-healing graphene/polymer composites usually show the same capacity [83]. Zhang et al. synthesized an ultrafast infrared (IR) laser-triggered self-healing composite material [68]. Due to the IR absorbing capacity of functionalized graphene nanosheets (FGNS), the temperature of the composites increased from 30 °C to 150 °C over within 20 s under IR laser irradiation, which reaches the healing temperature of rDA chemistry. The healing efficiency of the snipped specimen reached more than 96% in terms of Young's modulus, fracture strength, fracture elongation only by IR laser irradiation in 1 min (Figure 7).

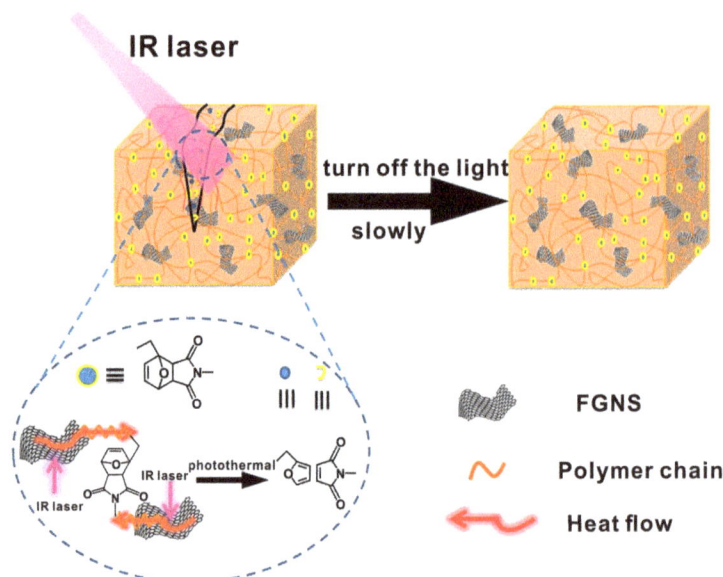

Figure 7. Illustration of the self-healing process of FGNS-PU-DA nanocomposite. Reproduced with permission from [68]. Copyright (2017) ACS Applied Materials and Interfaces.

The near-infrared (NIR) irradiation has been suggested as a non-invasive, harmless and highly efficient skin-penetrating biomedical technique [84,85]. Tong and coworkers demonstrated a fast self-healing GO-hectorite clay-poly(*N,N*-dimethylacrylamide) hybrid hydrogel realized by NIR irradiation for only 2~3 min up to the strength recovery of ~96% [52]. GO acted not only as a collaborative cross-linking agent, but also as a NIR irradiation energy absorber to transform it to thermal energy rapidly and efficiently to promote the mutual diffusion of the polymer chains across the cut interface. Liang and coworkers reported a self-healing bilayer hydrogel system [86]. When the fracture surfaces of the cut hydrogel were contacted, the healing is achieved by irradiating with a NIR illuminant with a wavelength of 808 nm and a power of 1.25 W. The self-healing behavior was ascribed to the photothermal energy transformation property of GO. With the increase of GO content, the heating rate of the hydrogels increased. Kim et al. synthesized self-healing PU/graphene nanocomposites with mechanical, thermal, optical properties [62]. The self-healing ability of PU/graphene nanocomposites was achieved because of intermolecular diffusion of polymer chains, which can be accelerated by NIR absorptions.

Considering NIR as a non-invasive, harmless and highly efficient biomedical technique, it can be used in photothermal therapy. Wang et al. reported a hydrogel made by cross-linking poly(*N,N*-dimethylacrylamide) chains on 3D graphene networks that exhibits good neural compatibility, high conductivity, low impedance and efficient NIR-triggered photothermal self-healing [57,87]. The self-healing hydrogel can act effectively as a promising artificial tissue material.

In addition to IR or NIR, other wavelengths of light radiation can also achieve self-healing capability. For example, Fei et al. fabricated a tri-layered, light-triggered healable and highly electrically conductive fibrous membrane by depositing RGO and silver nanowires onto gold nanoparticles incorporated poly(ε-caprolactone). The polymer chains interdiffuse across the crack surface of the damaged fibrous membrane under 532 nm light irradiation, and recrystallize upon cooling after turning off the continuous-wave diode laser. The surface conductivity recovery by 91% and tensile strength of the membrane are still well maintained after multiple cutting-healing cycles [88].

3.3. Microwave

Graphene has an excellent microwaves absorption capability due to its large area conjugated π-structure [89,90]. Under microwaves absorption, the π-structure of graphene will make it a giant electric dipole and transform microwaves into heat in the form of dipoles distortion [91], thus the microwaves absorption can also be used for self-healing of the graphene/polymer composites. For example, Zhang et al. utilized DA chemistry to prepare covalently crosslinked reduced functionalized GO/PU composites with self-healing ability using microwaves [63]. The two broken surfaces were immediately reunited when subjected to a gentle pressure and exposed to a 800 W domestic microwave oven operating at 2.45 GHz for 5 min followed by 2 h at 70 °C without any continuous pressure. The microwaves absorbed by reduced functionalized GO turned into heat and then promoted the healing process of the composites based on DA chemistry. The healed samples still possess good Young's modulus, fracture strain and fracture stress. Iron oxide (IO) nanoparticles have intrinsic microwave absorbing capacity and high thermal conductivity, which can also be incorporated in self-healing composites. Karak et al. reported a tough self-healing elastomeric nanocomposite containing IORGO, which exhibits the healing efficiency of 99% or more by microwave treatment [59].

3.4. Solvent-Assistant Self-Healing

Deionized water or organic solvent is a practicable external stimulus to heal damage [71,92–94]. Solvents can be helpful to the recombination of chemical bonds. In our group the hydrogels fabricated by copolymerization of AM and DAC in the presence of GO can be healed by dropping of water [46]. The cut pieces were slightly put together with the fracture surfaces contacting each other. Because the fresh fracture surfaces are relatively adhesive when the hydrogel was cut, no additional external force is required for connecting the broken parts. After a drop of water was dropped on the fracture surfaces, the healed sample can be stretched to a large strain by hand (Figure 8a–c). Without water assistance, the healed hydrogel has a fracture stress of 248.9 kPa and the healing efficiency is about 45.6%. After self-healing with water assistance, the hydrogel reaches a stress of 503.4 kPa and the healing efficiency on the base of fracture stress is 92.3%. These results indicate that ionic bonds and hydrogen bonds can be reformed via water assistance.

In addition, salt solution can also facilitate the self-healing process. For example, Wang and Tong et al. prepared a multiple shape memory, self-healable, and super tough PAA-GO-Fe^{3+} hydrogel [47]. The self-healing process is due to the strong ionic binding of Fe^{3+} ions to the carboxyl groups on the PAA chains. Keeping the cut surfaces in contact and immersing them in $FeCl_3$/HCl solution for a certain time gave rise to the self-healing process. After 15 h immersion, the healed hydrogel can bear a dumbbell of 5.5 kg. Importantly, the break position of the healed hydrogel after tensile test is not the healing position, proving the perfect connection of cut surfaces by Fe^{3+} ions (Figure 8d–i).

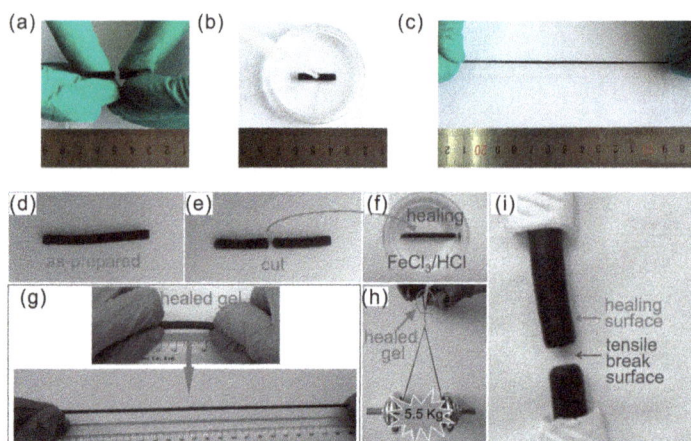

Figure 8. Self-healing properties of hydrogels. A pristine cylinder of the sample was cut in half (**a**). The two-halves were simply brought into contact, and a drop of water was dropped on the cut surface (**b**). After standing for 24 h, the sample can be stretched to a large strain by hand (**c**). Reproduced with permission from [46]. Copyright (2017) ACS Applied Materials and Interfaces. Photos of the self-healing hydrogel: (**d**) as-prepared; (**e**) cut sample; (**f**) contacted and immersed in $FeCl_3/HCl$; (**g**) stretched to ≈600% after immersing in $FeCl_3/HCl$ for 5 h; (**h**) loaded with ≈5.5 kg after healing for 15 h; (**i**) the healing surface and broken surface after tensile test of the hydrogel healed for 15 h. Reproduced with permission from [47]. Copyright (2016) Macromolecular Materials and Engineering.

3.5. Simple Contacting without Any External Stimuli

This self-healing process is realized at room temperature without any external stimuli, which is more practical considering the economic and operation aspects [78,95–99]. For instance, Peng and Turng et al. fabricated a mussel-inspired electroactive chitosan/GO composite hydrogel [95]. The self-healing property was due to dynamic covalent bonds, hydrogen bonding and π-π stacking. The two pieces of the hydrogel can be reconnected when the fractured surface contacts. The stress–strain curves of the recovered hydrogel almost coincided with that of the original one. Polyborosiloxane (PBS) is a well-known "solid–liquid" material whose viscoelastic properties (it flows as a highly viscous liquid at low strain rates but behaves as a solid at high strain rates) promote fast and complete healing due to its dynamic dative bonds. However Saiz et al. developed self-healing graphene-based PBS composites [98]. The mechanical and electrical properties of the composites can be autonomously and fully healed no matter scratched or complete ruptured just by placing the fracture surfaces in contact at ambient conditions for 10 min without external stimulus. Since the healing is driven by polymer flow and the dynamic dative bond interactions between polymer chains, the self-healing can be repeated multiple times.

Self-healing speed is another key factor for practical application. Designing materials with high self-healing speed without any external stimuli is highly desired. Bao et al. synthesized a rapid and efficient self-healing thermo-reversible elastomer (HBN-GO) based on GO and amine-terminated randomly branched oligomers [99]. Both the amorphous structure of the elastomer and its low glass transition temperature (T_g) allow the polymer chains to diffuse and mix at room temperature, and be able to self-heal at room temperature without the need of any plasticizer, solvent, healing agents or external stimuli. Two cut pieces of samples were gently brought into contact and healed in less than 60 s. The obtained stress–strain curves were similar to that of the original uncut samples (Figure 9). The HBN-1% GO sample can be healed to 60% of its original tensile strength prior to cutting by bringing the two cut pieces back in contact for 1 min. In addition, complete healing of the mechanical properties can be obtained in 1 h. Increasing the amount of GO allows for more covalent cross-linking and more restricted movement of polymer chains, which may explain the observed trend for increased time for the self-healing process. Both the samples in HBN-2% GO and HBN-4% GO also possess fast self-healing capabilities, in which they are able to recover 36% and 20% of their original extensibilities, respectively, in 1 min. the self-healing capability only dropped to 90% efficiency after the two fractured surfaces were left apart for 24 h, and dropped further to 50% after 96 h (Figure 9).

Kim et al. developed a mussel-mimetic nanocomposite hydrogel based on catechol-containing polyaspartamide and GO [97]. Two pieces of damaged gels can be strongly healed by contacting with each other due to the dynamic complexation between B^{3+} and catechol. No obvious cut line was seen and the healed sample can be stretched without fracturing by the tweezers.

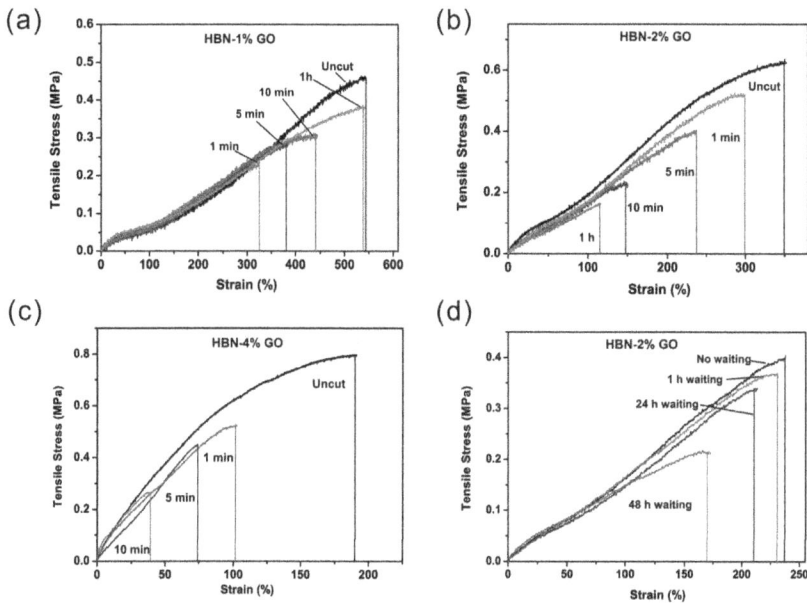

Figure 9. Strain-stress curves of the self-healing composites with GO at (**a**) 1 wt %, (**b**) 2 wt % and (**c**) 4 wt % (termed HBN-1% GO, HBN-2% GO, HBN-4% GO, respectively) upon different healing time at room temperature; (**d**) Strain-stress curves of the HBN-2% GO samples of 10 min healing at different waiting time. The waiting was performed at approx. 0% relative humidity from [99]. Copyright (2013) Advanced Materials.

3.6. Microcapsules and Others Healing Methods

Apart from the aforementioned self-healing conditions, microcapsules are also utilized to heal the damaged composites, although this method cannot realize multiple healing in one damaged area. Gao et al. fabricated GO microcapsules (GOMCs) in Pickering emulsions containing linseed oil as the healing agent [36]. The GO microcapsules were embedded into a waterborne PU matrix to prepare self-healing and anticorrosive coatings. The self-healing of graphene/polymer composites can be achieved through multiple methods. For example, Huang and Chen et al. synthesized mechanical enhanced graphene-thermoplastic PU composites [91], which can be self-healed by IR light, electricity and electromagnetic waves with high healing efficiencies (Figure 10). Zhang et al. also reported a recyclable epoxy resin (ER)/graphene nanocomposite, where a graphene crosslinked ER matrix via DA reaction can be rapidly and efficiently healed via multiple approaches, including heat, IR light, and microwave [35].

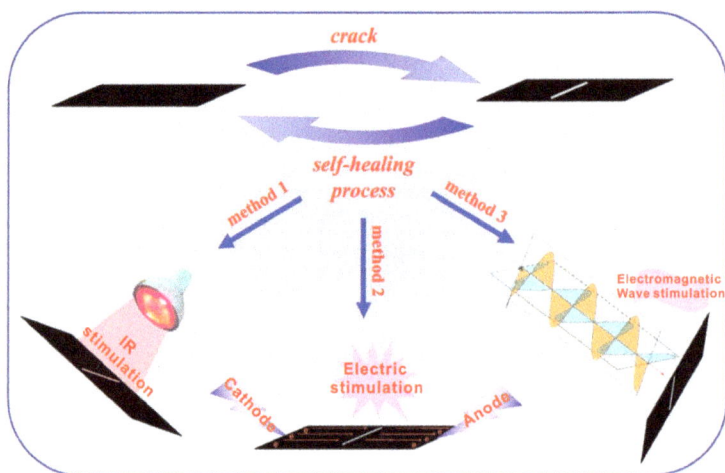

Figure 10. The graphene-thermoplastic PU composites were healed by IR light, electricity and electromagnetic waves with high healing efficiencies. Reproduced with permission from [91]. Copyright (2013) Advanced Materials.

4. Properties and Applications

4.1. Mechanical Property

Self-healing materials usually have poor mechanical properties, which limit their applications. In order to improve the mechanical behaviors, graphene and its derivates were usually added into the polymer matrix [36,52,63,67,81,100]. For instance, Deng and Wang et al. synthesized a gar-PAM/GO nanocomposite double network (DN) hydrogel with good fatigue resistance and self-healing ability [48]. The hydrogels exhibit excellent mechanical properties with a fracture strain of 4600%, fracture strength of 332 kPa and fracture dissipated energies of 11.5 MJ m^{-3}. The healed hydrogels also have favorable mechanical properties with a fracture strain of 2000% and a fracture strength of 153 kPa. GO played a role as the cross-link agent in the polyacrylamide network, which improved the tensile property (Figure 11).

The distribution of GO sheets in the composites affects the properties of the polymer composites especially the mechanical property. GO sheets were slightly agglomerated at high contents (0.5 wt %, 1.0 wt %). Primary GO sheets are well dispersed in the polymer matrix at low content (0.1 wt %). Sun et al. fabricated a covalent bonding GO/PU composite with significant mechanical reinforcement and thermal healable property. The Young's modulus (21.95 ± 2.56 MPa) and fracture strain (449 ± 16%) increased twice and the fracture stress (8.01 ± 0.71 MPa) even increased nearly 4 times at a GO loading of only 0.1 wt %. The enhanced mechanical properties of the composites are ascribed to the good dispersion of GO and covalent linking based on DA chemistry [79]. To improve the mechanical properties, functional groups are usually introduced on the graphene surface. Liu et al. synthesized maleimide functionalized GO and then added it into a PU matrix, forming the self-healing composite [67]. The composites with maleimide functionalized GO exhibit higher self-healing efficiency and mechanical properties than the composites with unmodified GO because the maleimide functionalized GO serves as cross-linking points, improved the dispersion and interfacial interaction with the polymer matrix, which effectively dissipates energy and obviously increases the mechanical properties.

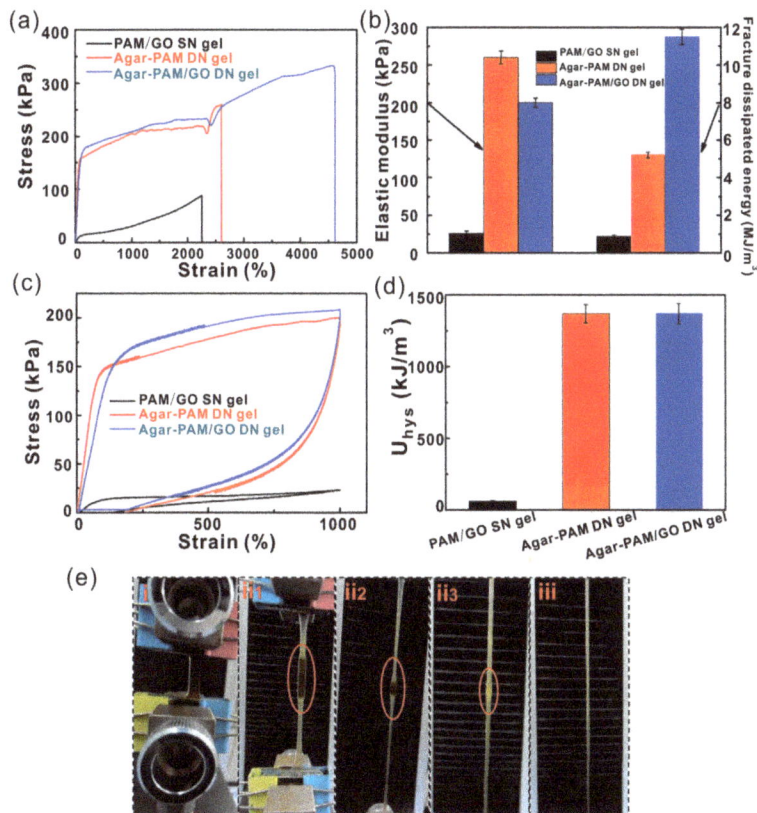

Figure 11. (**a**) Tensile stress–strain curves, and (**b**) elastic modulus and fracture dissipated energy of a PAM/GO single network (SN) gel, agar-PAM DN gel, and agar-PAM/GO DN gel. (**c**) Loading–unloading cyclic tensile stress–stain curves at a maximal strain of 1000%, and (**d**) corresponding dissipated energy of three gels. (**e**) Photos of the stretching process of agar-PAM/GO DN gel (from i to iii). Reproduced with permission from [48]. Copyright (2016) Advanced Engineering Materials.

4.2. Shape Memory Property

Besides high mechanical properties, shape memory behavior is an attractive property for self-healing graphene/polymer composites due to their wide applications in machinery, electronic device, chemical industry, biology and other fields [47,50,59,61,80,82]. For instance, Weng and Dai et al. prepared a series of graphene–poly(acrylamide-*co*-acrylic acid) hybrid materials with shape memory behavior and self-healing ability [80]. An unfolded cube box was designed as the original shape and heated at 35 °C. The compressed box can recover its original shape after heated at 37 °C for 30 s and can be repeated 10 times (Figure 12a). The hyperbranched PU–TiO$_2$/RGO nanocomposite fabricated by Karak et al. also possessed excellent shape memory behavior [50]. As RGO is a conducting material, it absorbs the energy from sunlight and efficiently transfers the absorbed energy to the PU matrix. Therefore, the nanocomposite reaches its transition temperature easily to recover its shape. The surface temperature was measured to be 38.1–38.4 °C at the time of shape recovery.

For conductive graphene/polymer composites, the shape memory can be realized by electrical conductivity. When the content of conductive filler in a polymer matrix exceeds the percolation threshold, the conductive network can generate Joule heating. Therefore, the speed of the shape memory depends on the electrical conductivity of the filler. For instance, Mohammad and coworkers synthesized fast electroactive shape memory and self-healing PVA/graphene nanocomposites. The electrically triggered shape recovery experiments of the composite containing 3 and 4.5 wt % of graphene (designated as PVA/Gr3 and PVA/Gr4.5, respectively) were conducted under four different DC voltages (40, 50, 60 and 70 V) and the resulting recovery ratio as a function of recovery time is presented in Figure 12b,c, respectively. It can be clearly seen that increasing the graphene content from 3 to 4.5 wt % leads to a dramatically faster recovery response [41].

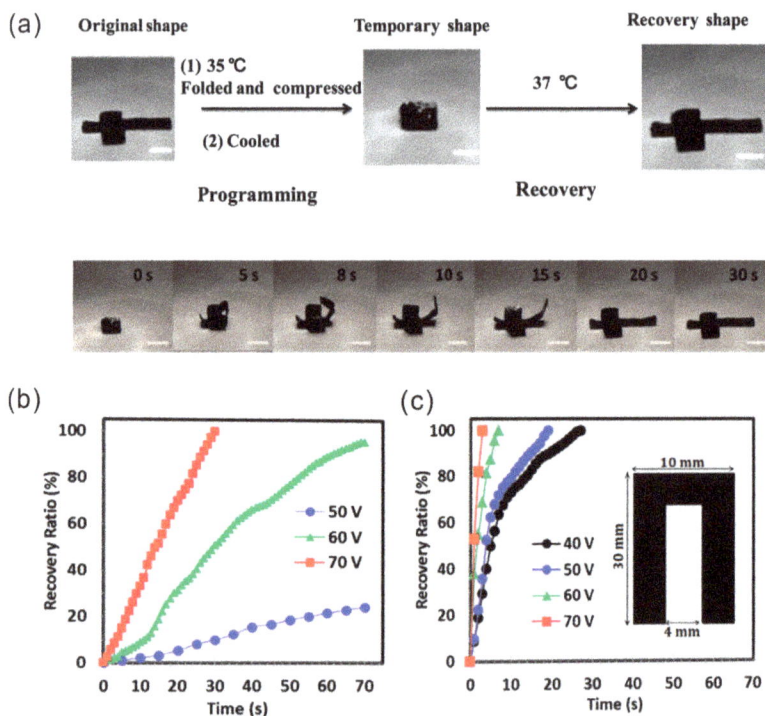

Figure 12. (**a**) The shape memory behavior of an unfolded cube box (10 mm × 10 mm × 10 mm) (top). Photographs demonstrating the shape memory behavior of the unfolded cube box in different times (bottom). Scale = 10 mm. Reproduced with permission from [80]. Copyright (2013) Macromolecular Rapid Communications. Electrically activated shape recovery ratio as a function of time under various triggering voltages for (**b**) PVAc/Gr3 and (**c**) PVAc/Gr4.5. The inset shows the sample geometry. Reproduced with permission from [41]. Copyright (2016) Society of Chemical Industry.

4.3. Conductivity and Electrical Devices

Generally, a polymer is partially an insulator because the covalent bond of the polymer chain does not have free carries (electrons), on the other hand, as the polymer molecules pile together by van der Waals forces, the distances between the molecules are large, electrons overlap between the molecules are poor, so the free carries are very difficult to mobile in polymer [101]. Therefore, graphene can be used as an ideal filler to prepare conductive self-healing composite due to its inherent good conductivity [57,68,88,99,102]. The content of graphene highly affects the conductivity of self-healing composites and the conductivity of composites increases with the increase of graphene content. The charge-transfer resistance of a bPEI/(PAA-graphene) multilayer polyelectrolyte film electrode prepared by Ge et al. is 750 Ω, which is better than the pure bPEI/PAA electrode without graphene [76]. In conductive hydrogels, the water content of hydrogels is also a factor affecting the conductivity [103]. For example, Tang et al. fabricated a versatile hydrogel composite based on a commercial superabsorbent polymer and a hyperbranched polymer with RGO though a simple hydrothermal method [78]. The hydrogel possesses a super water-absorption ability and a fast electrical self-healing ability. The main factor influencing conductivity is the water content, and the RGO nanosheets improve the sensitivity of samples of water content because RGO can contribute to water dispersion and free ion (e.g., Na^+) transportation.

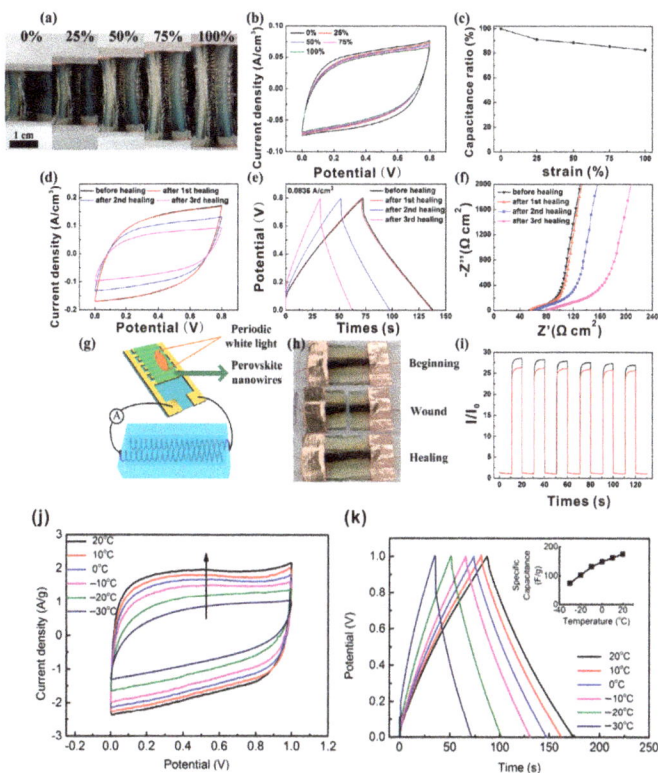

Figure 13. Electrochemical measurements and application for as-prepared stretchable and self-healing supercapacitors. (**a**) Photographs, (**b**) cyclic voltammogram curves, and (**c**) evolutions of specific capacitance of the supercapacitor before and after stretching to 100%. (**d**) Cyclic voltammogram curves, (**e**) galvonostatic charge–discharge measurements, and (**f**) Nyquist plots of the supercapacitor before healing and after self-healing cycles. (**g**) Illustration of the supercapacitor driving a photodetector of perovskite nanowires. (**h**) Photographs of the supercapacitor before and after self-healing. (**i**) Photocurrent dependence on time of the photodetector under illumination of on/off states driven by the original and self-healing supercapacitor after a healing cycle; red corresponds to the self-healing supercapacitor and black to the original. Reproduced with permission from [39]. Copyright (2017) ACS Nano. (**j**) Fabricated supercapacitor temperature dependence of cyclic voltammetry profiles at a scan rate of 20 mV s^{-1}. The arrow indicates the direction of increasing temperatures. (**k**) Galvanostatic charging–discharging profiles at a current density of 1 A g^{-1}. The inset indicates calculated specific capacitance with respect to temperature change. Reproduced with permission from [104]. Copyright (2017) Scientific Reports.

Supercapacitors are very common applications of conductive self-healing composite. For instance, Liu and Gao et al. assembled a stretchable and self-healable supercapacitor [39]. The supercapacitor was fabricated on two parallel RGO-based fiber springs wrapped with gel electrolyte and PU. After being cut and healed, the electrochemical performances of the device were still maintained at a high level (Figure 13a–i). Low temperature is a big challenge for applications of conductive self-healing composite. Chung and Ok et al. reported a supercapacitor with a high energy density that can work at low temperatures (even dropped to $-30\ ^{\circ}$C) [104]. The supercapacitor was fabricated with a combination of biochar-RGO electrodes and a polyampholyte hydrogel. The reason why the supercapacitor performance is improved at low temperature is that water molecules strongly adsorbed on hydrophilic polymer chains cannot participate in ice formation (Figure 13j,k).

4.4. Anticorrosive Coating

Fabrication of self-healing anticorrosive coatings has attracted attentions as it has the ability to extend the service life and prevent the substrate from corrosive attack. The self-healing coating on the surface of the material can repair the damaged surface automatically to protect the material [35,36,38,75,105–110]. For example, Lu and coworker reported a novel nanocomposite coating that consisted of lignin-modified graphene and waterborne PU [38]. The self-healing, electrically conductive coatings with UV resistant can be used as corrosion preventive or antistatic coatings. Gao et al. prepared a self-healing and anticorrosive coating [36]. The anticorrosion properties of neat PU coatings and GOMCs/PU composite coatings were characterized by the salt spray test on hot-dip galvanized steel (HDG) substrates with a 5% NaCl solution. The HDG plate coated with GOMCs/PU coatings showed no visual evidence of corrosion even after 116 h of salt spray test, while some white corrosion products were observed on the neat PU coated HDG plate (Figure 14).

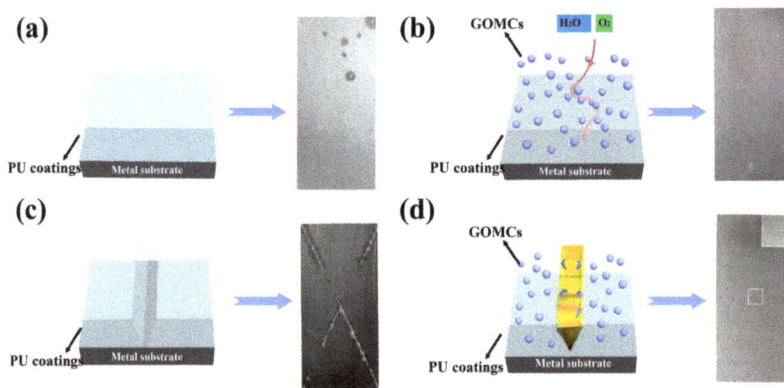

Figure 14. Schematic and images of (**a**) neat PU coating and (**b**) GOMCs/PU coatings subjected to the salt spray test for 116 h. Schematic and images of (**c**) neat PU coating and (**d**) GOMCs/PU coatings after scratching and 15 days of healing, subjected to the salt spray test for 43 h. Inset: the enlarged view of the white block. Reproduced with permission from [36]. Copyright (2016) Composites Science and Technology.

4.5. Biological and Pharmaceutical Applications

One of the threats to human health is microbial contaminations or infections. Microbial fouling is the critical factor to degradation of polymeric self-healing materials. Therefore, antimicrobial activity is a fantastic property of self-healing composites. The nanocomposites prepared by Karak et al. exhibited good antimicrobial activity against Staphylococcus aureus, Escherichia coli and Candida albicans because sulfur-containing compounds and polysulfanes are generally considered to be antimicrobial agents [61].

The medical field also has great interest in self-healing materials. For example, Hu et al. fabricated a double network hydrogel based on β-CD functionalized graphene and N,N-dimethylacrylamide that can achieve a self-healing ability at 37 °C [53]. Camptothecin (CPT) as a model anticancer drug was loaded into the hydrogel before the second hydrogel network was introduced. The content of loaded CPT and the cumulated CPT release in a β-CD functionalized graphene hydrogel are both better than that of a pristine graphene hydrogel (Figure 15a). Therefore, this system had potential capacity as anticancer drug carrier.

Due to the biocompatibility of DNA, the hydrogels can possess a variety of biological and environmental applications. Shi et al. reported GO/DNA self-healing hydrogels with high mechanical strength, excellent environmental stability, high dye-adsorption capacity and can be applied in tissue engineering, drug delivery, and removing organic pollutant [81].

Self-healing hydrogels were proposed to be used as biomaterials because of the capability of spontaneously healing any injury. Lu et al. prepared a mussel-inspired conductive, self-adhesive, and self-healing tough hydrogel [102]. The hydrogel could be implanted because of its good long-term biocompatibility and detection in vivo, which showed more accuracy in detecting/stimulating specific muscles in deep tissue. The intramuscular hydrogel electrodes yielded excellent signals from the dorsal muscle after implantation. The magnitude of the signals was in the range of 0.1–40 mV (Figure 15b–d), which was much higher than the signals detected by the surface electrodes.

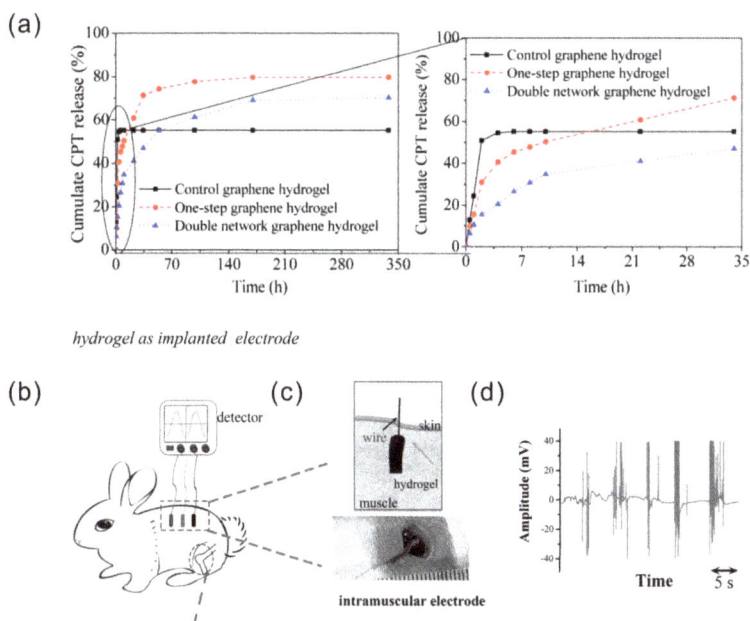

Figure 15. (**a**) Cumulate CPT release of hydrogels in PBS. Reproduced with permission from [53]. Copyright (2014) Materials Technology Advanced Performance Materials. The hydrogel as intramuscular electrodes. (**b**) Three hydrogel electrodes implanted into the dorsal muscle and the wires from the electrodes were transcutaneously connected to the signal detector. (**c**) Photos of the hydrogel implantation. (**d**) Example of the electromyographic signal recorded by the implanted hydrogel electrodes from the muscle when the rabbit was interfered by external stimulation. Reproduced with permission from [102]. Copyright (2016) Small.

5. Future Perspectives

Self-healing graphene/polymer composites are one of the most promising intelligent materials. The latest research progress in the preparation, properties and applications of self-healing graphene/polymer composites is reviewed and some of the composites are summarized in Table 1. Despite the rapid development of this research area, there are still many challenges that should be solved and considered in view of the practical applications.

(1) Graphene, as a component of the self-healing composites, can play an active role in the self-healing process because of the photothermal energy transformation ability. In this case, graphene acts as an energy absorber to transform irradiation or sunlight to thermal energy rapidly and efficiently to promote the mutual diffusion of the polymer chains across the damaged interface, thus facilitating the self-healing process. On the other hand, graphene in the composites is only used as a reinforcing agent to enhance the mechanical strength. In this case, the self-healing process may be impeded because some interactions between graphene and polymers will limit the movement and diffusion of the polymer chains. In addition, the glass transition temperature of the composites may be changed after the addition of graphene, thus influencing the self-healing efficiency. Therefore, graphene as a reinforcing agent can improve the mechanical performance. However, the self-healing efficiency of the composite will be reduced. The mechanical properties and self-healing efficiency are two contradictory aspects. How to balance the two aspects to achieve graphene/polymer composites with both high mechanical strength and high self-healing efficiency is a challenge.

(2) For intrinsically self-healing graphene-based materials, simple contacting without any external stimuli is considered to be most potential considering practical application. Therefore, exploring this kind of self-healing graphene-based composites is desired.

(3) For electrical conductive graphene/polymer composites, the mechanical strength and the conductivity are two contradictory aspects. Increasing the graphene content will increase the conductivity of the composite, but excess graphene will reduce the mechanical strength. How to obtain high mechanical strength and high conductivity of the graphene/composites with self-healing properties is a challenge.

(4) A lot of polymeric materials should be developed which exhibit self-healing properties. Also, it is highly challenging to endow the combined properties in a single material. The structural incompatibility among such types of polymers is one of the crucial reasons for this difficulty.

(5) To improve the compatibility of polymer and graphene, graphene usually needs to be modified. However, modification of graphene will lose some inherent properties of graphene. Therefore, how to improve the compatibility of polymer and graphene and retain the inherent properties of graphene as much as possible in the composite is a challenge.

Table 1. The properties of various self-healing graphene/polymers composites.

Materials	Self-Healing Mechanisms	Self-Healing Condition	Self-Healing Efficiency	Original Mechanical Property	Applications	Reference
FG/TPU material	PU chains diffuse	IR light, Electricity, Microwave (2.45 GHz)	98% of electrical conductivity	Tensile strength 40 MPa	Transport industries, construction industries, electronics	[91]
P(AM-co-DAC)/GO Hydrogels	Hydrogen bonds, electrostatic interaction	Drop water	>92% of tensile strength, >99% of tensile strain and >93% of toughness	Young's modulus 1 MPa Tensile strength 2 MPa	-	[46]
Chitosan/GO Hydrogel	π-π stacking, hydrogen bonds	Contact (room temperature)	-	Adhesive strength 1 MPa compressive stress 14 KPa	Electroactive tissue engineering applications	[95]
PU-DA-mGO	Dynamic covalent bonds	Heating (120 °C after 10 min)	90% of tensile strength	Stress 38 MPa	Aerospace, automobile, coating, electronics, energy, etc.	[67]
Graphene/PU	Diels–Alder chemistry	IR	96% of tensile strength	Breaking strength 36 MPa Young's modulus 127 MPa	Flexible electronics	[68]
PVAc/graphene nanocomposites	Diffusion of the polymer chains	60 °C for 1 h	89% (mechanical properties)	-	Sensors and fast deployable and actuating devices	[41]
RfGO/PU composites	Diels–Alder chemistry	Microwaves	93% (mechanical properties)	Stress 24 MPa Young's module 52 MPa	Flexible conductors, strain sensors	[63]
PAA-GO-Fe^{3+} Hydrogel	Ionic binding	Contact and immersed in FeCl$_3$/HCl	Nearly 100% tensile	Tensile strength 2.5 MPa Elongation 700%	Soft actuators, robots	[47]
PDA-pGO-PAM hydrogel	Non-covalent bonds	Contact	60% of tensile strength 95% of electrical conductivity	Tensile strength 75 kPa	Bioelectronics	[102]
PAM/GO DN gel	Hydrogen bond	Heating 80 °C for 3 h	48% of tensile strength	Elongation 4600% Fracture strength 332 kPa	Engineering fields	[48]
Au@PCLx/rGO/Ag	Fibers soften and flow	Light irradiation (532 nm)	90% of tensile strength 91% of conductivity	Tensile strength 4.85 MPa	Optoelectronic devices	[88]
SR/GNP composite	Reversible bonds	Thermal annealing	87% of tensile strength	Stress 1.3 MPa	Seals, hoses and automotive sector	[43]
HPU-IO-RGO	Diffusion of the polymer chains	Microwave sunlight	99%	Tensile strength 24.15 MPa Tensile modulus 28.55 MPa Toughness 110.8 MJ m^3	Transport, construction, electronics	[59]
GO-Clay-PDMAA Hybrid Hydrogels	Diffusion of the polymer chains hydrogen bonds	NIR	96% (mechanical strength) Strength 184 kPa Elongation 1890%		Surgical dressing	[52]
GO/PU composites	Covalent bonding	Heating	78% of tensile stress	Tensile modulus 21.95 MPa Fracture stress 8 MPa Young's modulus 22 MPa	Smart materials and structural material	[79]
SHPU/graphene composites	Interchain diffusion	NIR	39%	Stress 4 MPa	Functional polymer	[62]
GO/PAA composite hydrogels	Diffusion of polymer chains hydrogen bond	Contact at different temperatures	88% (mechanical properties)	Tensile strength 0.35 MPa Elongation 4900% (healed)	Biomedical and engineering fields	[56]

6. Conclusions

In this review, we have summarized the recent progress of graphene-containing polymer composites with self-healing capability. The preparation methods, self-healing conditions, and properties and applications of the graphene/polymer composites have been briefly discussed. Finally, the further perspectives of the composites have been proposed. Intelligent materials and self-healing materials, specifically for graphene-containing composites, are still in the initial stage. There will be a great research space in the field. The progress will guide further development of the self-healing graphene/polymer composites.

Acknowledgments: We acknowledge the support by the National Natural Science Foundation of China (51702178) and the Natural Science Foundation of Shandong Province (ZR2017MB009).

Author Contributions: Yongxu Du and Dong Li collected all the references, Yongxu Du, Dong Li and Libin Liu summarized and wrote the paper, Guangjie Gai discussed the data.

Conflicts of Interest: The authors declare no conflict of interest.

References

1. White, S.R.; Sottos, N.R.; Geubelle, P.H.; Moore, J.S.; Kessler, M.R.; Sriram, S.R.; Brown, E.N.; Viswanathan, S. Autonomic healing of polymer composites. *Nature* **2001**, *409*, 794–797. [CrossRef] [PubMed]
2. Zhang, B.; Zhang, P.; Zhang, H.; Yan, C.; Zheng, Z.; Wu, B.; Yu, Y. A Transparent, Highly stretchable, autonomous self-healing poly(dimethyl siloxane) elastomer. *Macromol. Rapid Commun.* **2017**, *38*. [CrossRef] [PubMed]
3. Chao, A.; Negulescu, I.; Zhang, D. Dynamic covalent polymer networks based on degenerative imine bond exchange: Tuning the malleability and self-healing properties by solvent. *Macromolecules* **2016**, *49*, 6277–6284. [CrossRef]
4. Jian, X.; Hu, Y.; Zhou, W.; Xiao, L. Self-healing polyurethane based on disulfide bond and hydrogen bond. *Polym. Adv. Technol.* **2017**, *29*, 463–469. [CrossRef]
5. Wang, Z.; Tian, H.; He, Q.; Cai, S. Reprogrammable, reprocessible, and self-healable liquid crystal elastomer with exchangeable disulfide bonds. *ACS Appl. Mater. Interfaces* **2017**, *9*, 33119–33128. [CrossRef] [PubMed]
6. Xu, Y.; Chen, D. Self-healing Polyurethane/attapulgite nanocomposites based on disulfide bonds and shape memory effect. *Mater. Chem. Phys.* **2017**, *195*, 40–48. [CrossRef]
7. Guo, Z.; Ma, W.; Gu, H.; Feng, Y.; He, Z.; Chen, Q.; Mao, X.; Zhang, J.; Zheng, L. pH-Switchable and self-healable hydrogels based on ketone type acylhydrazone dynamic covalent bonds. *Soft Matter* **2017**, *13*, 7371–7380. [CrossRef] [PubMed]
8. Zhang, D.D.; Ruan, Y.B.; Zhang, B.Q.; Qiao, X.; Deng, G.; Chen, Y.; Liu, C.Y. A self-healing PDMS elastomer based on acylhydrazone groups and the role of hydrogen bonds. *Polymer* **2017**, *120*, 189–196. [CrossRef]
9. Yu, F.; Cao, X.; Du, J.; Wang, G.; Chen, X. Multifunctional hydrogel with good structure integrity, self-healing, and tissue-adhesive property formed by combining diels-alder click reaction and acylhydrazone bond. *ACS Appl. Mater. Interfaces* **2015**, *7*, 24023–24031. [CrossRef] [PubMed]
10. Cordier, P.; Tournilhac, F.; Soulié-Ziakovic, C.; Leibler, L. Self-healing and thermoreversible rubber from supramolecular assembly. *Nature* **2008**, *451*, 977–980. [CrossRef] [PubMed]
11. Chen, Y.; Kushner, A.M.; Williams, G.A.; Guan, Z. Multiphase design of autonomic self-healing thermoplastic elastomers. *Nat. Chem.* **2012**, *4*, 467–472. [CrossRef] [PubMed]
12. Hart, L.R.; Nguyen, N.A.; Harries, J.L.; Mackay, M.E.; Colquhoun, H.M.; Hayes, W. Perylene as an electron-rich moiety in healable, complementary π–π stacked, supramolecular polymer systems. *Polymer* **2015**, *69*, 293–300. [CrossRef]
13. Tuncaboylu, D.C.; Sari, M.; Oppermann, W.; Okay, O. Tough and self-healing hydrogels formed via hydrophobic interactions. *Macromolecules* **2011**, *44*, 4997–5005. [CrossRef]
14. Tuncaboylu, D.C.; Argun, A.; Sahin, M.; Sari, M.; Okay, O. Structure optimization of self-healing hydrogels formed via hydrophobic interactions. *Polymer* **2012**, *53*, 5513–5522. [CrossRef]

15. Yan, X.; Xu, D.; Chi, X.; Chen, J.; Dong, S.; Ding, X.; Yu, Y.; Huang, F. A multiresponsive, shape-persistent, and elastic supramolecular polymer network gel constructed by orthogonal self-assembly. *Adv. Mater.* **2012**, *24*, 362–369. [CrossRef] [PubMed]

16. Khamrai, M.; Banerjee, S.L.; Kundu, P.P. Modified bacterial cellulose based self-healable polyeloctrolyte film for wound dressing application. *Carbohydr. Polym.* **2017**, *174*, 580–590. [CrossRef] [PubMed]

17. Wang, Q.; Mynar, J.L.; Yoshida, M.; Lee, E.; Lee, M.; Okuro, K.; Kinbara, K.; Aida, T. High-water-content mouldable hydrogels by mixing clay and a dendritic molecular binder. *Nature* **2010**, *463*, 339–343. [CrossRef] [PubMed]

18. Mozhdehi, D.; Ayala, S.; Cromwell, O.R.; Guan, Z. Self-healing multiphase polymers via dynamic metal-ligand interactions. *J. Am. Chem. Soc.* **2014**, *136*, 16128–16131. [CrossRef] [PubMed]

19. Rao, Y.L.; Chortos, A.; Pfattner, R.; Lissel, F.; Chiu, Y.C.; Feig, V.; Xu, J.; Kurosawa, T.; Gu, X.; Wang, C. Stretchable self-healing polymeric dielectrics cross-linked through metal-ligand coordination. *J. Am. Chem. Soc.* **2016**, *138*, 6020–6027. [CrossRef] [PubMed]

20. Shi, Y.; Wang, M.; Ma, C.; Wang, Y.; Li, X.; Yu, G. A Conductive self-healing hybrid gel enabled by metal-ligand supramolecule and nanostructured conductive polymer. *Nano Lett.* **2015**, *15*, 6276–6281. [CrossRef] [PubMed]

21. Novoselov, K.S.; Geim, A.K.; Morozov, S.V.; Jiang, D.; Zhang, Y.; Dubonos, S.V.; Grigorieva, I.V.; Firsov, A.A. Electric field effect in atomically thin carbon films. *Science* **2004**, *306*, 666–669. [CrossRef] [PubMed]

22. Zhu, J.; Yang, D.; Yin, Z.; Yan, Q.; Zhang, H. Graphene and graphene-based materials for energy storage applications. *Small* **2014**, *10*, 3480–3498. [CrossRef] [PubMed]

23. Kim, K.S.; Zhao, Y.; Jang, H.; Lee, S.Y.; Kim, J.M.; Kim, K.S.; Ahn, J.H.; Kim, P.; Choi, J.Y.; Hong, B.H. Large-scale pattern growth of graphene films for stretchable transparent electrodes. *Nature* **2009**, *457*, 706–710. [CrossRef] [PubMed]

24. Bonaccorso, F.; Sun, Z.; Hasan, T.; Ferrari, A.C. Graphene photonics and optoelectronics. *Nat. Photonics* **2010**, *4*, 611–622. [CrossRef]

25. Geim, A.K.; Macdonald, A.H. Graphene: Exploring carbon flatland. *Phys. Today* **2007**, *60*, 35–41.

26. Chen, G.; Wu, D.; Weng, W.; Wu, C. Exfoliation of graphite flake and its nanocomposites. *Carbon* **2003**, *41*, 619–621. [CrossRef]

27. Silva, M.; Alves, N.M.; Paiva, M.C. Graphene-polymer nanocomposites for biomedical applications. *Polym. Adv. Technol.* **2017**, *29*, 687–700. [CrossRef]

28. Tee, B.C.; Wang, C.; Allen, R.; Bao, Z. An electrically and mechanically self-healing composite with pressure-and flexion-sensitive properties for electronic skin applications. *Nat. Nanotechnol.* **2012**, *7*, 825–832. [CrossRef] [PubMed]

29. Nistor, R.A.; Newns, D.M.; Martyna, G.J. The role of chemistry in graphene doping for carbon-based electronics. *ACS Nano* **2011**, *5*, 3096–3103. [CrossRef] [PubMed]

30. Lou, Z.; Chen, S.; Wang, L.; Jiang, K.; Shen, G. An ultra-sensitive and rapid response speed graphene pressure sensors for electronic skin and health monitoring. *Nano Energy* **2016**, *23*, 7–14. [CrossRef]

31. Zhang, Q.; Liu, L.; Pan, C.; Li, D. Review of recent achievements in self-healing conductive materials and their applications. *J. Mater. Sci.* **2017**, *53*, 27–46. [CrossRef]

32. Cui, Y.; Kundalwal, S.I.; Kumar, S. Gas barrier performance of graphene/polymer nanocomposites. *Carbon* **2016**, *98*, 313–333. [CrossRef]

33. Wang, M.; Duan, X.; Xu, Y.; Duan, X. Functional three-dimensional graphene/polymer composites. *ACS Nano* **2016**, *10*, 7231–7247. [CrossRef] [PubMed]

34. Tjong, S.C. Polymer composites with graphene nanofillers: Electrical properties and applications. *J. Nanosci. Nanotechnol.* **2014**, *14*, 1154–1168. [CrossRef] [PubMed]

35. Cai, C.; Zhang, Y.; Zou, X.; Zhang, R.; Wang, X.; Wu, Q.; Sun, P. Rapid self-healing and recycling of multiple-responsive mechanically enhanced epoxy resin/graphene nanocomposites. *RSC Adv.* **2017**, *7*, 46336–46343. [CrossRef]

36. Li, J.; Feng, Q.; Cui, J.; Yuan, Q.; Qiu, H.; Gao, S.; Yang, J. Self-assembled graphene oxide microcapsules in Pickering emulsions for self-healing waterborne polyurethane coatings. *Compos. Sci. Technol.* **2017**, *151*, 282–290. [CrossRef]

37. Noack, M.; Merindol, R.; Zhu, B.; Benitez, A.; Hackelbusch, S.; Beckert, F.; Seiffert, S.; Mülhaupt, R.; Walther, A. Light-fueled, spatiotemporal modulation of mechanical properties and rapid self-healing of graphene-doped supramolecular elastomers. *Adv. Funct. Mater.* **2017**, *27*. [CrossRef]

38. Shahabadi, S.I.S.; Kong, J.; Lu, X. Aqueous-only, green route to self-healable, UV-resistant, and electrically conductive polyurethane/graphene/lignin nanocomposite coatings. *ACS Sustain. Chem. Eng.* **2017**, *5*, 3148–3157. [CrossRef]

39. Wang, S.; Liu, N.; Su, J.; Li, L.; Long, F.; Zou, Z.; Jiang, X.; Gao, Y. Highly stretchable and self-healable supercapacitor with reduced graphene oxide based fiber springs. *ACS Nano* **2017**, *11*, 2066–2074. [CrossRef] [PubMed]

40. Zhan, Y.; Meng, Y.; Li, Y. Electric heating behavior of flexible graphene/natural rubber conductor with self-healing conductive network. *Mater. Lett.* **2016**, *192*, 115–118. [CrossRef]

41. Sabzi, M.; Babaahmadi, M.; Samadi, N.; Mahdavinia, G.R.; Keramati, M.; Nikfarjam, N. Graphene network enabled high speed electrical actuation of shape memory nanocomposite based on poly(vinyl acetate). *Polym. Int.* **2016**, *66*, 665–671. [CrossRef]

42. Dhar, P.; Katiyar, A.; Maganti, L.S. Smart viscoelastic and self-healing characteristics of graphene nano-gels. *J. Appl. Phys.* **2016**, *120*. [CrossRef]

43. Valentini, L.; Bon, S.B.; Pugno, N.M. Severe graphene nanoplatelets aggregation as building block for the preparation of negative temperature coefficient and healable silicone rubber composites. *Compos. Sci. Technol.* **2016**, *134*, 125–131. [CrossRef]

44. Samadi, N.; Sabzi, M.; Babaahmadi, M. Self-healing and tough hydrogels with physically cross-linked triple networks based on Agar/PVA/Graphene. *Int. J. Biol. Macromol.* **2017**, *107*, 2291–2297. [CrossRef] [PubMed]

45. Han, D.; Yan, L. Supramolecular hydrogel of chitosan in the presence of graphene oxide nanosheets as 2D cross-linkers. *ACS Sustain. Chem. Eng.* **2013**, *2*, 296–300. [CrossRef]

46. Pan, C.; Liu, L.; Chen, Q.; Zhang, Q.; Guo, G. Tough, stretchable, compressive novel polymer/graphene oxide nanocomposite hydrogels with excellent self-healing performance. *ACS Appl. Mater. Interfaces* **2017**, *9*, 38052–38061. [CrossRef] [PubMed]

47. Zhao, L.; Huang, J.; Wang, T.; Sun, W.; Tong, Z. Multiple shape memory, self-healable, and supertough PAA-GO-Fe^{3+} hydrogel. *Macromol. Mater. Eng.* **2016**, *302*. [CrossRef]

48. Zhu, P.; Hu, M.; Deng, Y.; Wang, C. One-Pot Fabrication of a Novel agar-polyacrylamide/graphene oxide nanocomposite double network hydrogel with high mechanical properties. *Adv. Eng. Mater.* **2016**, *18*, 1799–1807. [CrossRef]

49. Cui, W.; Ji, J.; Cai, F.Y.; Li, H.; Rong, R. Robust, anti-fatigue, and self-healing graphene oxide/hydrophobic association composite hydrogels and their use as recyclable adsorbents for dye wastewater treatment. *J. Mater. Chem. A* **2015**, *3*, 17445–17458. [CrossRef]

50. Thakur, S.; Karak, N. Tuning of sunlight-induced self-cleaning and self-healing attributes of an elastomeric nanocomposite by judicious compositional variation of the TiO$_2$–reduced graphene oxide nanohybrid. *J. Mater. Chem. A* **2015**, *3*, 12334–12342. [CrossRef]

51. Liu, Y.T.; Zhong, M.; Xie, X.M. Self-healable, super tough graphene oxide/poly(acrylic acid) nanocomposite hydrogels facilitated by dual cross-linking effects through dynamic ionic interactions. *J. Mater. Chem. B* **2015**, *3*, 4001–4008.

52. Zhang, E.; Wang, T.; Zhao, L.; Sun, W.; Liu, X.; Tong, Z. Fast self-healing of graphene oxide-hectorite clay-poly(*N*,*N*-dimethylacrylamide) hybrid hydrogels realized by near-infrared irradiation. *ACS Appl. Mater. Interfaces* **2014**, *6*, 22855–22861. [CrossRef] [PubMed]

53. Chen, P.; Wang, X.; Wang, G.Y.; Duo, Y.R.; Zhang, X.Y.; Hu, X.H.; Zhang, X.J. Double network self-healing graphene hydrogel by two step method for anticancer drug delivery. *Mater. Technol.* **2014**, *29*, 210–213. [CrossRef]

54. Das, S.; Irin, F.; Ma, L.; Bhattacharia, S.K.; Hedden, R.C.; Green, M.J. Rheology and morphology of pristine graphene/polyacrylamide gels. *ACS Appl. Mater. Interfaces* **2013**, *5*, 8633–8640. [CrossRef] [PubMed]

55. Cong, H.P.; Wang, P.; Yu, S.H. Stretchable and self-healing graphene oxide–polymer composite hydrogels: A dual-network design. *Chem. Mater.* **2013**, *25*, 3357–3362. [CrossRef]

56. Liu, J.; Song, G.; He, C.; Wang, H. Self-healing in tough graphene oxide composite hydrogels. *Macromol. Rapid Commun.* **2013**, *34*, 1002–1007. [CrossRef] [PubMed]

57. Hou, C.; Duan, Y.; Zhang, Q.; Wang, H.; Li, Y. Bio-applicable and electroactive near-infrared laser-triggered self-healing hydrogels based on graphene networks. *J. Mater. Chem.* **2012**, *22*, 14991–14996. [CrossRef]

58. Hou, C.; Huang, T.; Wang, H.; Yu, H.; Zhang, Q.; Li, Y. A strong and stretchable self-healing film with self-activated pressure sensitivity for potential artificial skin applications. *Sci. Rep.* **2013**, *3*, 3138. [CrossRef] [PubMed]

59. Thakur, S.; Karak, N. A tough smart elastomeric bio-based hyperbranched polyurethane nanocomposite. *New J. Chem.* **2015**, *39*, 2146–2154. [CrossRef]

60. Thakur, S.; Das, G.; Raul, P.K.; Karak, N. Green one-step approach to prepare sulfur/reduced graphene oxide nanohybrid for effective mercury ions removal. *J. Phys. Chem. C* **2013**, *117*, 7636–7642. [CrossRef]

61. Thakur, S.; Barua, S.; Karak, N. Self-healable castor oil based tough smart hyperbranched polyurethane nanocomposite with antimicrobial attributes. *RSC Adv.* **2014**, *5*, 2167–2176. [CrossRef]

62. Jin, T.K.; Kim, B.K.; Kim, E.Y.; Sun, H.K.; Han, M.J. Synthesis and properties of near IR induced self-healable polyurethane/graphene nanocomposites. *Eur. Polym. J.* **2013**, *49*, 3889–3896.

63. Li, J.; Zhang, G.; Sun, R.; Wong, C.P. A covalently cross-linked reduced functionalized graphene oxide/polyurethane composite based on Diels–Alder chemistry and its potential application in healable flexible electronics. *J. Mater. Chem. C* **2016**, *5*, 220–228. [CrossRef]

64. Wool, R.P. Self-healing materials: A review. *Soft Matter* **2008**, *4*, 400–418. [CrossRef]

65. Zhang, Y.; Broekhuis, A.A.; Picchioni, F. Thermally self-healing polymeric materials: The next step to recycling thermoset polymers? *Macromolecules* **2009**, *42*, 1906–1912. [CrossRef]

66. Kuang, X.; Liu, G.; Dong, X.; Liu, X.; Xu, J.; Wang, D. Facile fabrication of fast recyclable and multiple self-healing epoxy materials through diels-alder adduct cross-linker. *J. Polym. Sci. Part A Polym. Chem.* **2015**, *53*, 2094–2103. [CrossRef]

67. Lin, C.; Sheng, D.; Liu, X.; Xu, S.; Ji, F.; Dong, L.; Zhou, Y.; Yang, Y. A self-healable nanocomposite based on dual-crosslinked Graphene oxide/polyurethane. *Polymer* **2017**, *127*, 241–250. [CrossRef]

68. Wu, S.; Li, J.; Zhang, G.; Yao, Y.; Li, G.; Sun, R.; Wong, C.P. Ultrafastly self-healing nanocomposites via infrared laser and its application in flexible electronics. *ACS Appl. Mater. Interfaces* **2017**, *9*, 3040–3049. [CrossRef] [PubMed]

69. Sun, X.; Liu, Z.; Welsher, K.; Robinson, J.T.; Goodwin, A.; Zaric, S.; Dai, H. Nano-graphene oxide for cellular imaging and drug delivery. *Nano Res.* **2008**, *1*, 203–212. [CrossRef] [PubMed]

70. Tugba, C.B.; Oytun, F.; Hasan, A.M.; Jeong, S.H.; Choi, H.; Basarir, F. Fabrication of a transparent conducting electrode based on graphene/silver nanowires via layer-by-layer method for organic photovoltaic devices. *J. Colloid Interface Sci.* **2017**, *505*, 79–86. [CrossRef] [PubMed]

71. Li, Y.; Chen, S.; Wu, M.; Sun, J. Polyelectrolyte multilayers impart healability to highly electrically conductive films. *Adv. Mater.* **2012**, *24*, 4578–4582. [CrossRef] [PubMed]

72. Zhang, J.; Wang, T.; Zhang, L.; Tong, Y.; Zhang, Z.; Shi, X.; Guo, L.; Huang, H.; Yang, Q.; Huang, W. Tracking deep crust by zircon xenocrysts within igneous rocks from the northern Alxa, China: Constraints on the southern boundary of the Central Asian Orogenic Belt. *J. Asian Earth Sci.* **2015**, *108*, 150–169. [CrossRef]

73. Kovtyukhova, N.I.; Ollivier, P.J.; Martin, B.R.; Mallouk, T.E.; Chizhik, S.A.; Buzaneva, E.V.; Gorchinskiy, A.D. Layer-by-layer assembly of ultrathin composite films from micron-sized graphite oxide sheets and polycations. *Chem. Mater.* **1999**, *11*, 771–778. [CrossRef]

74. Kirkland, J.J. Porous Thin-layer modified glass bead supports for gas liquid chromatography. *Anal. Chem.* **1965**, *37*, 1458–1461. [CrossRef]

75. Fan, F.; Zhou, C.; Wang, X.; Szpunar, J. Layer-by-layer assembly of a self-healing anticorrosion coating on magnesium alloys. *ACS Appl. Mater. Interfaces* **2015**, *7*, 27271–27278. [CrossRef] [PubMed]

76. Zhu, Y.; Yao, C.; Ren, J.; Liu, C.; Ge, L. Graphene improved electrochemical property in self-healing multilayer polyelectrolyte film. *Colloids Surf. A Physicochem. Eng. Asp.* **2015**, *465*, 26–31. [CrossRef]

77. Xiang, Z.; Zhang, L.; Li, Y.; Yuan, T.; Zhang, W.; Sun, J. Reduced graphene oxide-reinforced polymeric films with excellent mechanical robustness and rapid and highly efficient healing properties. *ACS Nano* **2017**, *11*, 7134–7141. [CrossRef] [PubMed]

78. Peng, R.; Yu, Y.; Chen, S.; Yang, Y.; Tang, Y. Conductive nanocomposite hydrogels with self-healing property. *RSC Adv.* **2014**, *4*, 35149–35155. [CrossRef]

79. Li, J.; Zhang, G.; Deng, L.; Zhao, S.; Gao, Y.; Jiang, K.; Sun, R.; Wong, C. In situ polymerization of mechanically reinforced, thermally healable graphene oxide/polyurethane composites based on Diels-Alder chemistry. *J. Mater. Chem. A* **2014**, *2*, 20642–20649. [CrossRef]

80. Dong, J.; Ding, J.; Weng, J.; Dai, L. Graphene enhances the shape memory of poly (acrylamide-*co*-acrylic acid) grafted on graphene. *Macromol. Rapid Commun.* **2013**, *34*, 659–664. [CrossRef] [PubMed]

81. Xu, Y.; Wu, Q.; Sun, Y.; Bai, H.; Shi, G. Three-dimensional self-assembly of graphene oxide and DNA into multifunctional hydrogels. *ACS Nano* **2010**, *4*, 7358–7362. [CrossRef] [PubMed]

82. Xiao, X.; Xie, T.; Cheng, Y.T. Self-healable graphene polymer composites. *J. Mater. Chem.* **2010**, *20*, 3508–3514. [CrossRef]

83. Li, Y.; Chen, S.; Wu, M.; Sun, J. Rapid and efficient multiple healing of flexible conductive films by near-infrared light irradiation. *ACS Appl. Mater. Interfaces* **2014**, *6*, 16409–16415. [CrossRef] [PubMed]

84. Huang, X.; El-Sayed, I.H.; Qian, W.; El-Sayed, M.A. Cancer cell imaging and photothermal therapy in the near-infrared region by using gold nanorods. *J. Am. Chem. Soc.* **2006**, *128*, 2115–2120. [CrossRef] [PubMed]

85. Robinson, J.T.; Tabakman, S.M.; Liang, Y.; Wang, H.; Casalongue, H.S.; Vinh, D.; Dai, H. Ultrasmall reduced graphene oxide with high near-infrared absorbance for photothermal therapy. *J. Am. Chem. Soc.* **2011**, *133*, 6825–6831. [CrossRef] [PubMed]

86. Zhao, Q.; Hou, W.; Liang, Y.; Zhang, Z.; Ren, L.; Zhao, Q.; Hou, W.; Liang, Y.; Zhang, Z.; Ren, L. Design and fabrication of bilayer hydrogel system with self-healing and detachment properties achieved by near-infrared irradiation. *Polymers* **2017**, *9*, 237. [CrossRef]

87. Markovic, Z.M.; Harhaji-Trajkovic, L.M.; Todorovic-Markovic, B.M.; Kepić, D.P.; Arsikin, K.M.; Jovanović, S.P.; Pantovic, A.C.; Dramićanin, M.D.; Trajkovic, V.S. In vitro comparison of the photothermal anticancer activity of graphene nanoparticles and carbon nanotubes. *Biomaterials* **2011**, *32*, 1121–1129. [CrossRef] [PubMed]

88. Chen, L.; Si, L.; Wu, F.; Chan, S.Y.; Yu, P.; Fei, B. Electrical and mechanical self-healing membrane using gold nanoparticles as localized "nano-heaters". *J. Mater. Chem. C* **2016**, *4*, 10018–10025. [CrossRef]

89. Hu, H.; Zhao, Z.; Wan, W.; Gogotsi, Y.; Qiu, J. Ultralight and highly compressible graphene aerogels. *Adv. Mater.* **2013**, *25*, 2219–2223. [CrossRef] [PubMed]

90. Zhu, Y.; Murali, S.; Stoller, M.D.; Velamakanni, A.; Piner, R.D.; Ruoff, R.S. Microwave assisted exfoliation and reduction of graphite oxide for ultracapacitors. *Carbon* **2010**, *48*, 2118–2122. [CrossRef]

91. Huang, L.; Yi, N.; Wu, Y.; Zhang, Y.; Zhang, Q.; Huang, Y.; Ma, Y.; Chen, Y. Multichannel and repeatable self-healing of mechanical enhanced graphene-thermoplastic polyurethane composites. *Adv. Mater.* **2013**, *25*, 2224–2228. [CrossRef] [PubMed]

92. South, A.B.; Lyon, L.A. Autonomic self-healing of hydrogel thin films. *Angew. Chem.* **2010**, *49*, 767–771. [CrossRef] [PubMed]

93. Wang, X.; Liu, F.; Zheng, X.; Sun, J. Water-enabled self-healing of polyelectrolyte multilayer coatings. *Angew. Chem.* **2011**, *123*, 11580–11583. [CrossRef]

94. Yang, Y.; Urban, M.W. Self-repairable polyurethane networks by atmospheric carbon dioxide and water. *Angew. Chem.* **2014**, *53*, 12142–12147. [CrossRef] [PubMed]

95. Jing, X.; Mi, H.Y.; Napiwocki, B.N.; Peng, X.F.; Turng, L.S. Mussel-inspired electroactive chitosan/graphene oxide composite hydrogel with rapid self-healing and recovery behavior for tissue engineering. *Carbon* **2017**, *125*, 557–570. [CrossRef]

96. Konwar, A.; Kalita, S.; Kotoky, J.; Chowdhury, D. Chitosan-iron oxide coated graphene oxide nanocomposite hydrogel: A robust and soft antimicrobial bio-film. *ACS Appl. Mater. Interfaces* **2016**, *8*, 20625–20634. [CrossRef] [PubMed]

97. Wang, B.; Jeon, Y.S.; Park, H.S.; Kim, J.H. Self-healable mussel-mimetic nanocomposite hydrogel based on catechol-containing polyaspartamide and graphene oxide. *Mater. Sci. Eng. C* **2016**, *69*, 160–170. [CrossRef] [PubMed]

98. D'Elia, E.; Barg, S.; Ni, N.; Rocha, V.G.; Saiz, E. Self-healing graphene-based composites with sensing capabilities. *Adv. Mater.* **2015**, *27*, 4788–4794. [CrossRef] [PubMed]

99. Wang, C.; Liu, N.; Allen, R.; Tok, J.B.; Wu, Y.; Zhang, F.; Chen, Y.; Bao, Z. A rapid and efficient self-healing thermo-reversible elastomer crosslinked with graphene oxide. *Adv. Mater.* **2013**, *25*, 5785–5790. [CrossRef] [PubMed]

100. Qi, K.; He, J.; Wang, H.; Zhou, Y.; You, X.; Nan, N.; Shao, W.; Wang, L.; Ding, B.; Cui, S. A highly stretchable nanofiber-based electronic skin with pressure-, strain-, and flexion-sensitive properties for health and motion monitoring. *ACS Appl. Mater. Interfaces* **2017**, *9*, 42951–42960. [CrossRef] [PubMed]

101. Palleau, E.; Reece, S.; Desai, S.C.; Smith, M.E.; Dickey, M.D. Self-healing stretchable wires for reconfigurable circuit wiring and 3D microfluidics. *Adv. Mater.* **2013**, *25*, 1589–1592. [CrossRef] [PubMed]

102. Han, L.; Lu, X.; Wang, M.; Gan, D.; Deng, W.; Wang, K.; Fang, L.; Liu, K.; Chan, C.W.; Tang, Y. A Mussel-inspired conductive, self-adhesive, and self-healable tough hydrogel as cell stimulators and implantable bioelectronics. *Small* **2017**, *13*. [CrossRef] [PubMed]

103. Wu, T.; Chen, B. A mechanically and electrically self-healing graphite composite dough for stencil-printable stretchable conductors. *J. Mater. Chem. C* **2016**, *4*, 4150–4154. [CrossRef]

104. Li, X.; Liu, L.; Wang, X.; Ok, Y.S.; Jaw, E.; Chang, S.X.; Chung, H.J. Flexible and self-healing aqueous supercapacitors for low temperature applications: Polyampholyte gel electrolytes with biochar electrodes. *Sci. Rep.* **2017**, *7*, 1685. [CrossRef] [PubMed]

105. Das, A.; Deka, J.; Raidongia, K.; Manna, U. Robust and self-healable bulk-superhydrophobic polymeric coating. *Chem. Mater.* **2017**, *29*, 8720–8728. [CrossRef]

106. Qiu, S.; Li, W.; Zheng, W.; Zhao, H.; Wang, L. Synergistic effect of polypyrrole-intercalated graphene for enhanced corrosion protection of aqueous coating in 3.5% NaCl solution. *ACS Appl. Mater. Interfaces* **2017**, *9*, 34294–34304. [CrossRef] [PubMed]

107. Cheng, H.; Huang, Y.; Cheng, Q.; Shi, G.; Jiang, L.; Qu, L. Self-healing graphene oxide based functional architectures triggered by moisture. *Adv. Funct. Mater.* **2017**, *27*. [CrossRef]

108. Sazou, D.; Deshpande, P.P. Conducting polyaniline nanocomposite-based paints for corrosion protection of steel. *Chem. Pap.* **2017**, *71*, 459–487. [CrossRef]

109. Wang, H.; Liu, Z.; Zhang, X.; Lv, C.; Yuan, R.; Zhu, Y.; Mu, L.; Zhu, J. Durable Self-healing superhydrophobic coating with biomimic "Chloroplast" analogous structure. *Adv. Mater. Interfaces* **2016**, *3*. [CrossRef]

110. Niratiwongkorn, T.; Luckachan, G.E.; Mittal, V. Self-healing protective coatings of polyvinyl butyral/polypyrrole-carbon black composite on carbon steel. *RSC Adv.* **2016**, *6*, 43237–43249. [CrossRef]

polymers

MDPI

Article

Graphene Oxide-Graft-Poly(L-lactide)/Poly(L-lactide) Nanocomposites: Mechanical and Thermal Properties

Li-Na Wang [1,2], Pei-Yao Guo Wang [1] and Jun-Chao Wei [1,*]

[1] College of Chemistry, Nanchang University, Nanchang 330031, China; linawang@nit.edu.cn (L.-N.W.); 18942337364@163.com (P.-Y.G.W.)

[2] College of Science, Nanchang Institute of Technology, Nanchang 330029, China

* Correspondence: weijunchao@ncu.edu.cn; Tel.: +86-0791-83969514

Received: 31 July 2017; Accepted: 3 September 2017; Published: 7 September 2017

Abstract: The surface modification of graphene sheets with polymer chains may greatly hinder its aggregation and improve its phase compatibility with a polymer matrix. In this work, poly(L-lactic acid)-grafted graphene oxide (GO-g-PLLA) was prepared via a simple condensation polymerization method, realizing its dispersion well in organic solvents, which demonstrated that the surface of GO changed from hydrophilic to hydrophobic. GO-g-PLLA can disperse homogeneously in the PLLA matrix, and the tensile test showed that the mechanical properties of GO-g-PLLA/PLLA were much better than that of GO/PLLA; compared with GO, only 3% GO-g-PLLA content can realize a 37.8% increase in the tensile strength for their PLLA composites. Furthermore, the differential scanning calorimetry (DSC) and polarized optical microscopy (POM) results demonstrated that GO-g-PLLA shows a nucleating agent effect and can promote the crystallization of PLLA.

Keywords: graphene; poly(L-lactide); surface modification; nanocomposite

1. Introduction

Poly(L-lactic acid) (PLLA), as an important biocompatible and biodegradable polymer, has been widely used in package materials [1], tissue regeneration [2], drug delivery, and many biomedical fields [3]. However, its poor mechanical properties and slow crystallization rate have greatly restricted its applications [4]. In order to overcome this issue, many inorganic fillers, such as bioactive glass nanoparticles [5], hydroxyapatite [6], carbon nanotubes [7], and graphene oxide [8], have been used to reinforce the PLLA matrix and enhance its crystallization rate. However, a vital problem for nanofiller-reinforced composites is that the nanoparticles always aggregate easily in the polymer matrix, and thus the properties of the composites are far from their theoretical values [9,10]. So many works have focused on the surface modification of nanofillers to enhance their phase compatibility and homogeneous dispersion state [11]. For example, grafting polymer chains on the surface of hydroxyapatite [12], as well as the treatment of carbon nanotubes and many other nanofillers have been widely reported [13–15].

Graphene, as a kind of novel nanofiller, can significantly improves the mechanical and thermal properties of a polymer matrix [16,17]. However, the strong π-π interaction leads to the intrinsic aggregation of graphene sheets and makes it difficult to disperse homogeneously in a polymer matrix, and thus many methods including covalent and noncovalent surface modification of graphene sheets have been designed to tune its surface properties and phase compatibility with polymers [16–18]. Up to now, several reports on PLLA-grafted graphene hybrids have been reported. He et al. used the in situ ring opening polymerization of D-lactide on the surface of GO sheets and prepared poly(D-lactide)-grafted graphene (GO-g-PDLA) [19]. GO-g-PDLA can form stereo-complex with PLLA and thus enhance the phase interaction between GO sheets and PLLA chains, in addition to exhibiting an effective nucleating agent effect. Lee's group also used the in situ ring opening polymerization method to prepare graphene oxide/PLLA composites with PLLA chains grafted on the surface of GO sheets [20]. However, the in situ grafting process happens in a rigid anhydrous and inert environment, and is furthermore a time-consuming process. Xu's group prepared PLLA-grafted graphene oxide via a melting polycondensation method of GO and L-lactic acid monomer; in this method, a catalyst was used, and the reaction was carried out at a high temperature (180 °C) [8].

Generally, to prepare PLLA chains, three methods are always used; in situ ring opening polymerization, melting polymerization, and polycondensation in the solvent. Each method has its advantages and drawbacks. Polycondensation of L-lactic acid in the solvent is a convenient method to prepare PLLA chains, as this method is simple, does not require a rigid anhydrous condition, and is a catalyst-free and relatively low-temperature process, but this method cannot form high molecular weight PLLA chains. However, it is a very convenient method to graft oligo PLLA chains on the surface of nanoparticles. For example, Chen's group used this method to prepare PLLA-grafted hyroxyapatite nanoparticles, and the results demonstrated that the oligo PLLA is enough to tune the surface properties of nanofillers [21,22].

In this manuscript, PLLA-grafted graphene oxide (GO-g-PLLA) was prepared via the simple condensation polymerization method, and the reaction was conducted at 120 °C, much lower than the reported 180 °C [8]. Furthermore, no catalyst was used in this method. The mechanical and thermal properties of GO-g-PLLA/PLLA composites were investigated, and the results showed that the mechanical properties of Go-g-PLLA/PLLA composites were much better than those of GO/PLLA composites. Moreover, the GO-g-PLLA hybrids can work as nucleating agents and promote the crystallization of PLLA.

2. Experimental

2.1. Materials

GO was prepared by the oxidation of graphite according to the modified Hummers method [23]. PLLA (revode 190) and L-lactic acid was donated from Hisun Biomaterials Co., Ltd. (Taizhou, Zhejiang, China).

2.2. Preparation of GO-g-PLLA

Firstly, two hundred milligrams of GO were dispersed homogeneously in 10 mL of H_2O, and then 600 mg L-lactic acid and 350 mL toluene were mixed with the GO solution. The mixture was heated at 120 °C for 12 h under vigorous stirring; in the meantime, the water was removed with the boiling of toluene. In the end, the solid product was collected by centrifugation, and washed with chloroform several times. The final product (GO-g-PLLA) was dried under vacuum at 40 °C.

2.3. Preparation of GO-g-PLLA/PLLA Nanocomposites

The GO-g-PLLA/PLLA nanocomposites were prepared via a solvent mixing method. Briefly, PLLA and GO-g-PLLA were dissolved in chloroform, respectively, and then PLLA and the GO-g-PLLA solution were mixed together under vigorous stirring. The mixed solutions were poured into a glass dish, when the solvent evaporated completely, the GO-g-PLLA/PLLA nanocomposite film was obtained.

Composites with different GO-g-PLLA content, such as 0.3, 0.7, 1.0, 3.0, 4.0, and 5.0 wt % were prepared, and the corresponding composites are denoted as GO-g-PLLA0.3/PLLA, GO-g-PLLA0.7/PLLA, GO-g-PLLA1/PLLA, GO-g-PLLA3/PLLA, and GO-g-PLLA5/PLLA, respectively.

2.4. Characterizations

The Fourier transform infrared (FTIR) spectra of GO and GO-g-PLLA were recorded at room temperature in a Shimadzu IRPrestige-21 Fourier transform infrared spectra-photometer.

The X-ray photo-electron spectroscopy (XPS) of GO and GO-g-PLLA was obtained with a thermo-VG Scientific ESCALAB 250 photo-electron spectrometer using a monochromatic Al Ka (1486.6 eV) X-ray source.

The atomic force microscopy (AFM) of GO and GO-g-PLLA was measured on a nanoscope III A (Digital Instruments) scanning probe microscope via the tapping mode.

Raman spectra of GO and GO-g-PLLA were acquired using a Raman spectrometer (Horiba Jobin Yvon) operating with an Xplora microraman system.

The thermogravimetric analysis (TGA) measurements were performed by PyrisDiamond TG/DTA. The samples were heated from room temperature to 800 °C at a rate of 10 °C/min under nitrogen flow.

The dispersion test of GO and GO-g-PLLA in chloroform and water were measured at 2 mg/mL concentration. The samples were dispersed in solvent by sonication and then the optical photos were taken with a general digital camera. The tensile tests of GO/PLLA and GO-g-PLLA/PLLA nanocomposites were measured on an electronic universal testing machine (CMT8502, Shiji Tianyuan Instrument Co., Ltd., Shenzhen, China) at a tensile speed of 10 mm/min at room temperature. The test of each sample type was repeated three times. The morphologies of tensile fractured surfaces were observed by scanning electron microscopy (SEM, QuanTA-200F, FEI Company, Hillsboro, OR, USA).

Thermal analysis of neat PLLA and GO-g-PLLA/PLLA was conducted using a TA Instruments differential scanning calorimeter (Q2000, TA instruments company, New Castle, DE, USA) under nitrogen purge. For non-isothermal melt crystallization, the samples were heated from room temperature to 200 °C at a heating rate of 10 °C/min, and kept for 3 min, cooled to 20 °C at 10 °C/min, and then heated to 200 °C at 10 °C/min.

The crystalline morphology of neat PLLA and its nanocomposites were observed via a Nikon E600 polarized microscope (POM) equipped with a temperature controller (Hs400, Instec Company, Boulder, CO, USA). The samples were first heated at 190 °C for 3 min and cooled to 130 °C at a cooling rate of 100 °C/min^{-1}, then isothermally crystallized for 50 min.

Wide angle X-ray diffraction (WAXD) patterns of GO, PLLA, and PLLA/GO-g-PLLA nanocomposites were acquired with a D8 focus X-ray Diffractometer operating at 30 kV and 20 mA with a copper target ($\lambda = 1.54$ Å). The scanning range was from 5° to 40°, and the scanning speed was 2° min^{-1} at room temperature.

3. Results and Discussion

3.1. Synthesis and Characterization of GO-g-PLLA

As shown in Figure 1A, curve (a), the characteristic peaks at 3417 and 1750 cm^{-1} are attributed to the stretching vibration of the hydroxyl groups and carboxyl groups of GO, which are reacted with the carboxyl or hydroxyl groups of lactic acid, respectively. When GO was mixed with L-lactic acid at a higher temperature, the polycondensation of L-lactic acid occurred; at the same time, the L-lactic acid monomer or oligo-polymers also reacted with the hydroxyl or carboxyl groups of GO, and thus poly(L-lactic acid) chains were grafted on the surface of the GO sheets. Compared with the IR spectra of GO, the intensity of absorption peaks at 3400 and 1750 of GO-g-PLLA (shown in Figure 1A, curve (b)) was much weaker, while a new peak appeared at 1712 cm^{-1}. This new peak is attributed to the stretch vibration of ester groups in PLLA chains, demonstrating that PLLA was successfully grafted on the surface of GO.

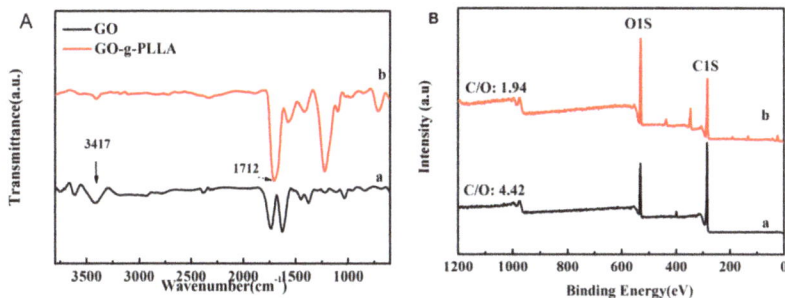

Figure 1. Fourier transform infrared spectra (**A**) and X-ray photo-electron spectroscopy; (**B**) spectra of GO (a) and GO-g-PLLA (b).

When PLLA chains were grafted on the surface of GO, the surface composition of GO changes considerably. As shown in Figure 1B, two typical peaks at 288 eV and 532 eV were recorded, which are the characteristic peaks of C1s and O1s. The C/O ratio of GO was 4.42, while the C/O ratio of GO-g-PLLA was 1.94, which is much lower than that of GO, indicating the increase of oxygen content, and this change was due to the higher content of oxygen in PLLA chains. On the other hand, the existence of PLLA chains on the GO sheet may also change its thickness, as shown in Figure 2; the average height of a GO sheet is about 1.0 nm, while the average height of GO-g-PLLA is about 3.8 nm, so it can be judged that the thickness of the PLLA layers may be 2.8 nm.

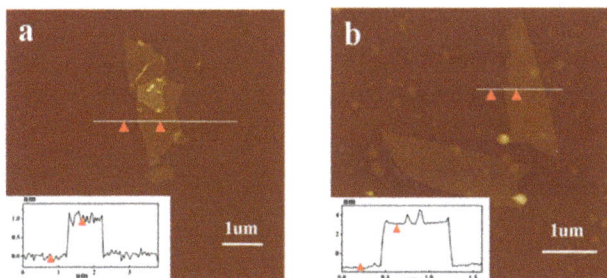

Figure 2. Atomic force microscopy images of GO (**a**) and GO-g-PLLA (**b**).

The TGA curves of GO and GO-g-PLLA are shown in Figure 3. The weight loss below 150 °C was ascribed to the evaporation of adsorbed solvent; because of the strong hydrophilic properties of GO, it can easily adsorb water in the air, and even if it is lyophilized a long time, it is still difficult to remove the adsorbed water. On the contrary, GO-g-PLLA is hydrophobic due to the existence of PLLA chains, and thus it is not easy to absorb water. Therefore, in the TGA test, the weight loss of GO-g-PLLA is much lower than that of GO. With the increase of temperature, the surface functional groups such as hydroxyl and carboxyl groups begin to detach from the GO sheet, especially when the temperature is above 200 °C, causing the GO sample to show a quick weight loss. The total weight loss of GO in the range between 150 °C and 800 °C was about 49.6%. As for the GO-g-PLLA samples (Figure 3, curve (b)), the weight loss may derive from both the decomposition of the polymer chains and the functional groups of the GO sheet. In the range between 150 °C and 800 °C, the weight loss of GO-g-PLLA was about 59.4%, much higher than that of the GO sample, confirming that PLLA was successfully grafted on the surface of GO. However, the TGA curve of GO-g-PLLA is different from the reported results of PLLA, which shows a decomposition temperature around 330 °C [8]. Because the PLLA chains grafted on the surface of GO were oligo-chains, the molecular weight is much lower than that of the commercial PLLA samples. Thus, GO-g-PLLA decomposed from 200 °C, a relative lower temperature than PLLA, and this result is similar with the decomposition of oligo PLLA-grafted hydroxyapatite [21].

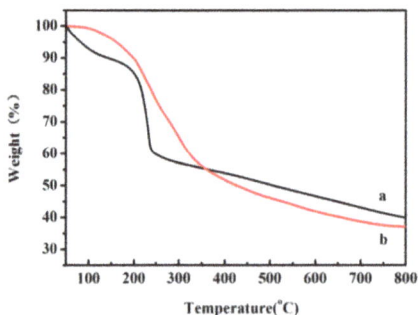

Figure 3. Thermal degradation curves of GO (**a**) and GO-g-PLLA (**b**).

The main purpose of grafting PLLA on the surface of GO is to tune its surface properties, especially wettability. GO is hydrophilic, while PLLA is hydrophobic, so GO-g-PLLA can also exhibit a hydrophobic surface, which can be verified by the sedimentation experiment. As shown in Figure 4, when GO and GO-g-PLLA were dissolved in chloroform, respectively, the GO-g-PLLA could disperse homogeneously (2.0 mg/mL) and the suspension could keep stable for a long time; in fact, 10 h later, the GO-g-PLLA was still dispersed well in the solvent. On the contrary, GO could not enter chloroform, and 5 min later, most of the GO sheets had precipitated at the bottom of the bottle. To further show the change of surface properties, GO and GO-g-PLLA were dispersed in a two-phase solvent composed of chloroform and water (v/v 1:1). As shown in Figure 4B, GO-g-PLLA dispersed well in chloroform and could not enter water phase, showing its hydrophobic properties. As a control, owing to its hydrophilic properties, GO dispersed homogeneously in the water and could not enter the chloroform phase. These results demonstrated that the surface of GO changed from hydrophilic to hydrophobic by grafting PLLA chains, which may further improve its phase compatibility with organic polymers.

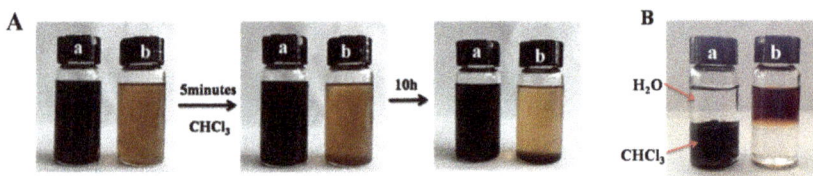

Figure 4. Photos of GO-g-PLLA (**a**) and GO (**b**) dispersed in chloroform (**A**) and a water-chloroform two-phase solvent (**B**).

3.2. Mechanical and Thermal Properties

When nanofillers are added into a polymer matrix, the mechanical properties of polymers should be reinforced; however, due to the aggregation of nanofillers, the mechanical properties of the composites are always far from their theoretical values, sometimes becoming even worse than the pure polymers. When GO was mixed with PLLA, the aggregation of GO may have reduced the mechanical properties of PLLA. As shown in Figure 5, the tensile strength of PLLA was about 51 MPa; when GO was added, the tensile strength of the GO/PLLA composites was lower than that of pure PLLA. When GO-g-PLLA was used as a nanofiller, the surface of GO was connected with PLLA chains via covalent bonds, and thus the phase compatibility of GO-g-PLLA with PLLA was much better.

When exterior forces were added to the composites, the interior stress may transfer easily from PLLA chains to GO sheets, and thus the tensile strength of GO-g-PLLA/PLLA nanocomposites was much higher than that of PLLA and GO/PLLA nanocomposites (shown in Figure 5). For example, when 3.0 wt % of GO-g-PLLA was added into the PLLA matrix, the tensile strength of the composite was 60 MPa, about 37.8% higher than that of GO/PLLA composites.

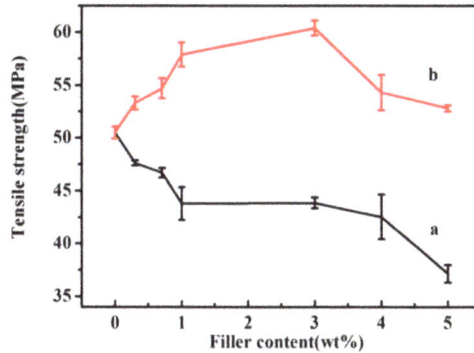

Figure 5. Tensile strength of GO/PLLA (**a**) and GO-g-PLLA/PLLA (**b**) at different filler contents.

The tensile fracture surface of PLLA, GO-g-PLLA 3, and GO-g-PLLA 5 are shown in Figure 6. The tensile fracture surface of PLLA (Figure 6a) was much smoother, which is in accordance with the brittle characteristic of PLLA. When GO-g-PLLA was mixed with PLLA, due to the interface interaction, the tensile strength increased, the tensile fracture surface was coarse, and some fibers were observed on the fracture surfaces of GO-g-PLLA3/PLLA nanocomposites (Figure 6b). This may result from the stress transfer from PLLA to GO sheets, confirming that the phase compatibility of GO-g-PLLA with the PLLA matrix was very good and the phase interaction was strong. However, with the increase of the GO-g-PLLA content, the fillers may also aggregate in the polymer matrix, and thus lead to the decrease of the tensile strength (Figure 5). As shown in Figure 6, the tensile surface of Go-g-PLLA5/PLLA was much coarser than that of the PLLA sample, but no fibril structure appeared; only some aggregates were found, implying that GO-g-PLLA did not disperse well. However, our primary test showed that the surface-grafted PLLA chains can change the surface wettability of GO from hydrophilic to hydrophobic, and thus they can enhance the phase compatibility between the PLLA matrix and GO fillers; therefore, it may be deduced that if the composites preparation method was optimized, much better mechanical properties may be obtained.

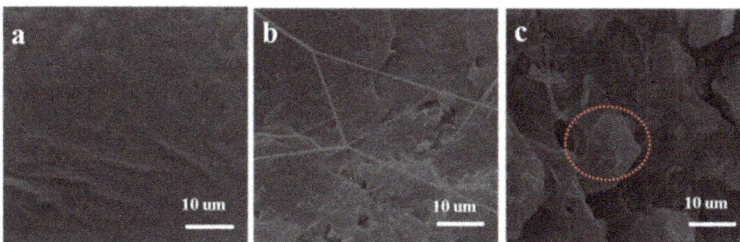

Figure 6. Scanning electron microscopy (SEM) images of the tensile failure section for PLLA (**a**), GO-g-PLLA3/PLLA (**b**) and GO-g-PLLA5/PLLA (**c**). Images in the red circle demonstrates the aggregation of nanofiller.

The thermal parameters of neat PLLA and GO-g-PLLA/PLLA nanocomposites were measured with DSC by a heating-cooling-heating process. The crystallinity (X_c) of different samples were acquired from the DSC curves of the first heating process (Figure 7A) by the following equation:

$$X_c = \frac{\Delta H_m}{f \times \Delta H^0} \times 100\%$$

where ΔH_m is the melting enthalpy of PLLA or GO-g-PLLA/PLLA composites, ΔH^0 (93.0 J/g) is the melting enthalpy of PLLA with 100% crystalline [24], and f is the content of GO-g-PLLA in the composites.

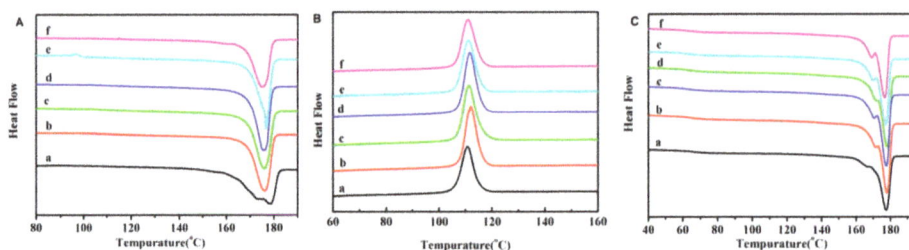

Figure 7. Differential scanning calorimetry (DSC) curves of the first heating process (**A**), cooling process (**B**), and the second heating process (**C**) for different samples. PLLA (a), GO-g-PLLA0.3/PLLA (b), GO-g-PLLA0.7/PLLA (c), GO-g-PLLA1/PLLA (d), GO-g-PLLA3/PLLA (e), and GO-g-PLLA5/PLLA (f).

As shown in Table 1, the crystallinity of PLLA was about 52.9%. With the addition of GO-g-PLLA, the GO-g-PLLA worked as a nucleating agent and enhanced the crystallization PLLA, and thus the crystallinity of the PLLA in the composites increased. However, when the filler content was higher than 3 wt %, the crystallinity showed a slight decrease; this may have resulted from the aggregation of nanofillers in the PLLA matrix, as too many aggregates will hinder the arrangement of PLLA chains and thus have a negative effect on its crystallization process. As mentioned in the mechanical properties, the GO-g-PLLA 3 has the highest tensile strength, and this can also be explained by its high crystallinity.

Table 1. Summary of the thermal parameters of PLLA and GO-g-PLLA/PLLA nanocomposites.

Samples	T_g (°C)	T_c (°C)	T_m (°C)	X_c (%)
PLLA	66.1	110.4	175.1	52.9
GO-g-PLLA0.3/PLLA	63.7	112.1	176.1	56.7
GO-g-PLLA0.7/PLLA	63.1	111.4	176.3	57.0
GO-g-PLLA1.0/PLLA	63.5	111.6	176.2	62.1
GO-g-PLLA3.0/PLLA	62.4	111.2	176.9	61.0
GO-g-PLLA5.0/PLLA	62.7	111.1	175.6	59.1

The crystalline temperature (T_c, the peak temperature of the exothermic curves obtained during the cooling process), melting temperature (T_m, the peak temperature of the second heating process), and glass transition temperature (T_g) can be measured from the DSC curves of the cooling process and the second heating process (Figure 7B,C); the corresponding results are shown in Table 1. The T_c and T_m of PLLA was 110.4 and 175.1 °C, respectively, while the T_c and T_m of GO-g-PLLA/PLLA shifted to a higher temperature, indicating the nucleating effect of GO-g-PLLA. The T_g of the GO-g-PLLA/PLLA nanocomposites was lower than that of neat PLLA, demonstrating that the polymer chain fragments in GO-g-PLLA/PLLA composites are much more easily moved than that in the pure

PLLA matrix, perhaps because the filling of GO-g-PLLA decreased the entanglements of PPLA chains. Furthermore, the melting peaks in the second heating process were a little different from the first heating process. There were weak shoulder peaks in the curves (Figure 7, curve (c)); in other words, two melting peaks were found. This finding of two melting points for PLLA samples has also been observed in many other references [25,26], and this phenomenon can be ascribed to the differences of crystal morphology. As a matter of fact, when PLLA samples crystallize at a faster speed, there will be more uncompleted crystalline phase, which will melt at a lower temperature. When nanofillers were blended with PLLA, the PLLA composites crystallized more quickly than pure PLLA samples, and thus some uncompleted crystalline formed. Hence, in the second heating process, there were minor melting peaks ascribed to the uncompleted crystals. Moreover, the first melting peaks of the composites are much clearer than that of the PLLA sample. All of these results demonstrated that GO-g-PLLA and PLLA may have strong interactions and a positive effect in enhancing the crystalline speed of PLLA.

The spherulite morphology of PLLA and its nanocomposites isothermally crystallized at 130 °C were observed with POM. The well-developed PLLA spherulites grew to about 25 μm with clear boundaries (Figure 8a). However, the diameter of GO-g-PLLA/PLLA composites spherulites were smaller than that of PLLA, while the number of spherulites was much greater than that of PLLA, which further confirmed that GO-g-PLLA may act as a nucleating agent that can largely enhances PLLA molecular chain mobility and increases the number of nucleation sites (Figure 8b–d). However, it should be noted that the size of PLLA spherulites first decreased and then increased with increasing the GO-g-PLLA loading, which indicated that the nucleation sites decreased with further increasing the content of GO-g-PLLA (Figure 8d), this may resulted from the aggregation of GO-g-PLLA filler.

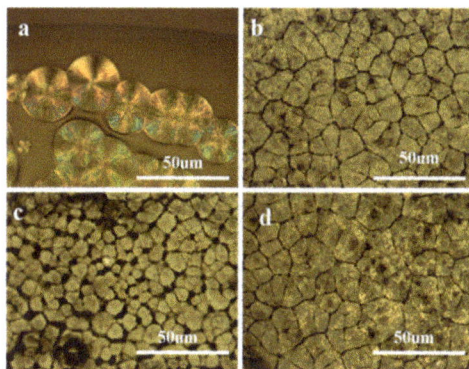

Figure 8. Polarized microscope (POM) images of PLLA and its nanocomposites isothermally crystallized at 130 °C. Neat PLLA (**a**), GO-g-PLLA1/PLLA (**b**), GO-g-PLLA3/PLLA (**c**), and GO-g-PLLA5/PLLA (**d**).

The effect of GO-g-PLLA on the crystal structure of PLLA was detected by WAXD. Figure 9 shows the patterns of neat PLLA, GO, GO-g-PLLA1/PLLA, GO-g-PLLA3/PLLA, and GO-g-PLLA5/PLLA. For neat PLLA, two sharp characteristic peaks are located at 16.6° and 18.9°, corresponding to (200)/(110) and (203) planes, which demonstrates the crystallization in α form. For GO-g-PLLA/PLLA composites, the diffraction patterns are the same as those of neat PLLA. This indicated that incorporation of GO-g-PLLA does not alter the crystalline structures of PLLA in the GO-g-PLLA/PLLA nanocomposites. In addition, the crystal peak of GO is present at about $2\theta = 10.8$, suggesting that the GO was exfoliated from graphite. On the other hand, in the composite, no GO diffraction peaks were found, and this confirms that the GO-g-PLLA dispersed well in the composites, and did not form ordered arrangements.

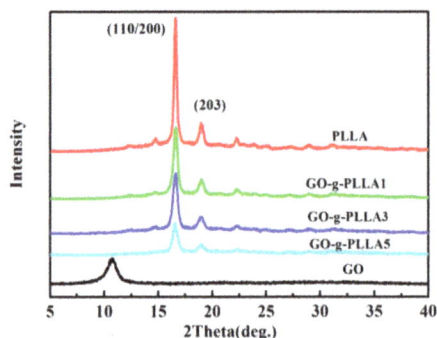

Figure 9. Wide angle X-ray diffraction (WAXD) patterns of neat PLLA and its nanocomposites.

4. Conclusions

In this work, PLLA-grafted graphene oxide (GO-g-PLLA) was fabricated via a simple condensation polymerization method with GO and L-lactic acid monomer, and used as a nanofiller to reinforce PLLA. The surface-grafted PLLA chains cause the GO surface to transform from hydrophilic to hydrophobic, and render it able to disperse well in chloroform. The SEM, WAXD, and tensile test results showed that GO is exfoliated and uniformly dispersed in the PLLA matrix, which demonstrated that surface modification is an effective method to improve the interfacial interactions between nanofillers and a polymer matrix, showing a much better reinforcing effect than GO alone. The non-isothermal melting crystallization behavior and spherulite morphology observation demonstrated that the GO-g-PLLA may act as a nucleating agent and improve the crystallization speed of PLLA.

Acknowledgments: This work was financially supported by the National Natural Science Foundation of China (Nos. 51463013 and 51663017) and the Natural Science Foundation of Jiangxi Province of China (No. 20142BAB203018).

Author Contributions: Peiyao Guowang prepared the nanocomposites, Lina Wang carried out the thermal analysis, and both of them have the same contribution to this work. Junchao Wei supervised this work and prepared the manuscript. All the authors contributed to the date interpretation and discussed the results.

Conflicts of Interest: The authors declared no conflict of interest.

References

1. Armentano, I.; Bitinis, N.; Fortunati, E.; Mattioli, S.; Rescignano, N.; Verdejo, R.; Lopez-Manchado, M.A.; Kenny, J.M. Multifunctional nanostructured PLA materials for packaging and tissue engineering. *Prog. Polym. Sci.* **2013**, *38*, 1720–1747. [CrossRef]

2. Okamoto, M.; John, B. Synthetic biopolymer nanocomposites for tissue engineering scaffolds. *Prog. Polym. Sci.* **2013**, *38*, 1487–1503. [CrossRef]

3. Tian, H.Y.; Tang, Z.H.; Zhuang, X.L.; Chen, X.S.; Jing, X.B. Biodegradable synthetic polymers: Preparation, functionalization and biomedical application. *Prog. Polym. Sci.* **2012**, *37*, 237–280. [CrossRef]

4. Wei, J.C.; Sun, J.R.; Wang, H.J.; Chen, X.S.; Jing, X.B. Isothermal crystallization behavior and unique banded spherulites of hydroxyapatite/poly(L-lactide) nanocomposites. *Chin. J. Polym. Sci.* **2010**, *28*, 499–507. [CrossRef]

5. Liu, A.X.; Hong, Z.K.; Zhuang, X.L.; Chen, X.S.; Cui, Y.; Liu, Y.; Jing, X.B. Surface modification of bioactive glass nanoparticles and the mechanical and biological properties of poly(L-lactide) composites. *Acta Biomater.* **2008**, *4*, 1005–1015. [CrossRef] [PubMed]

6. Hong, Z.K.; Qiu, X.Y.; Sun, J.R.; Deng, M.X.; Chen, X.S.; Jing, X.B. Grafting polymerization of L-lactide on the surface of hydroxyapatite nano-crystals. *Polymer* **2004**, *45*, 6699–6706. [CrossRef]

7. Gonçalves, C.; Gonçalves, I.C.; Magalhães, F.D.; Pinto, A.M. Poly(lactic acid) Composites Containing Carbon-Based Nanomaterials: A Review. *Polymers* **2017**, *9*, 269. [CrossRef]

8. Li, W.; Xu, Z.; Chen, L.; Shan, M.; Tian, X.; Yang, C.; Lv, H.; Qian, X. A facile method to produce graphene oxide-g-poly(L-lactic acid) as an promising reinforcement for PLLA nanocomposites. *Chem. Eng. J.* **2014**, *237*, 291–299. [CrossRef]

9. Podsiadlo, P.; Kaushik, A.K.; Arruda, E.M.; Waas, A.M.; Shim, B.S.; Xu, J.; Nandivada, H.; Pumplin, B.G.; Lahann, J.; Ramamoorthy, A.; et al. Ultrastrong and Stiff Layered Polymer Nanocomposites. *Science* **2007**, *318*, 80–83. [CrossRef] [PubMed]

10. Bonderer, L.J.; Studart, A.R.; Gauckler, L.J. Bioinspired Design and Assembly of Platelet Reinforced Polymer Films. *Science* **2008**, *319*, 1069–1073. [CrossRef] [PubMed]

11. Song, K.A.; Zhang, Y.Y.; Meng, J.S.; Green, E.C.; Tajaddod, N.; Li, H.; Minus, M.L. Structural Polymer-Based Carbon Nanotube Composite Fibers: Understanding the Processing-Structure-Performance Relationship. *Materials* **2013**, *6*, 2543–2577. [CrossRef] [PubMed]

12. Wei, J.C.; Liu, A.X.; Chen, L.; Zhang, P.B.; Chen, X.S.; Jing, X.B. The Surface Modification of Hydroxyapatite Nanoparticles by the Ring Opening Polymerization of gamma-Benzyl-L-glutamate N-carboxyanhydride. *Macromol. Biosci.* **2009**, *9*, 631–638. [CrossRef] [PubMed]

13. Mallakpour, S.; Soltanian, S. Surface functionalization of carbon nanotubes: Fabrication and applications. *RSC Adv.* **2016**, *6*, 109916–109935. [CrossRef]

14. Ahmad, N.N.R.; Mukhtar, H.; Mohshim, D.F.; Nasir, R.; Man, Z. Surface modification in inorganic filler of mixed matrix membrane for enhancing the gas separation performance. *Rev. Chem. Eng.* **2016**, *32*, 181–200. [CrossRef]

15. Sun, Z.; Zhang, L.; Dang, F.; Liu, Y.; Fei, Z.; Shao, Q.; Lin, H.; Guo, J.; Xiang, L.; Yerra, N.; et al. Experimental and simulation-based understanding of morphology controlled barium titanate nanoparticles under co-adsorption of surfactants. *Crystengcomm* **2017**, *19*, 3288–3298. [CrossRef]

16. Kuilla, T.; Bhadra, S.; Yao, D.; Kim, N.H.; Bose, S.; Lee, J.H. Recent advances in graphene based polymer composites. *Prog. Polym. Sci.* **2010**, *35*, 1350–1375. [CrossRef]

17. Punetha, V.D.; Rana, S.; Yoo, H.J.; Chaurasia, A.; McLeskey, J.T., Jr.; Ramasamy, M.S.; Sahoo, N.G.; Cho, J.W. Functionalization of carbon nanomaterials for advanced polymer nanocomposites: A comparison study between CNT and graphene. *Prog. Polym. Sci.* **2017**, *67*, 1–47. [CrossRef]

18. Kuila, T.; Bose, S.; Mishra, A.K.; Khanra, P.; Kim, N.H.; Lee, J.H. Chemical functionalization of graphene and its applications. *Prog. Mater. Sci.* **2012**, *57*, 1061–1105. [CrossRef]

19. Sun, Y.; He, C.B. Synthesis and Stereocomplex Crystallization of Poly(lactide)-Graphene Oxide Nanocomposites. *ACS Macro Lett.* **2012**, *1*, 709–713. [CrossRef]

20. Yang, J.H.; Lin, S.H.; Lee, Y.D. Preparation and characterization of poly(L-lactide)-graphene composites using the in situ ring-opening polymerization of PLLA with graphene as the initiator. *J. Mater. Chem.* **2012**, *22*, 10805–10815. [CrossRef]

21. Qiu, X.Y.; Chen, L.; Hu, J.L.; Sun, J.R.; Hong, Z.K.; Liu, A.X.; Chen, X.S.; Jing, X.B. Surface-modified hydroxyapatite linked by L-lactic acid oligomer in the absence of catalyst. *J. Polym. Sci. Part A Polym. Chem.* **2005**, *43*, 5177–5185. [CrossRef]

22. Qiu, X.Y.; Hong, Z.K.; Hu, J.L.; Chen, L.; Chen, X.S.; Jing, X.B. Hydroxyapatite surface modified by L-lactic acid and its subsequent grafting polymerization of L-lactide. *Biomacromolecules* **2005**, *6*, 1193–1199. [CrossRef] [PubMed]

23. Hummers, W.S.; Offeman, R.E. Preparation of Graphitic Oxide. *J. Am. Chem. Soc.* **1958**, *80*, 1339. [CrossRef]

24. Nam, J.Y.; Ray, S.S.; Okamoto, M. Crystallization behavior and morphology of biodegradable polylactide/layered silicate nanocomposite. *Macromolecules* **2003**, *36*, 7126–7131. [CrossRef]

25. Gonçalves, C.; Pinto, A.; Machado, A.V.; Moreira, J.; Gonçalves, I.C.; Magalhães, F. Biocompatible reinforcement of poly(Lactic Acid) with grapheme nanoplates. *Polym. Compos.* **2016**. [CrossRef]

26. Wei, J.C.; Dai, Y.F.; Chen, Y.W.; Chen, X.S. Mechanical and thermal properties of polypeptide modified hydroxyapatite/poly(L-lactide) nanocomposites. *Sci. China Chem.* **2011**, *54*, 431–437. [CrossRef]

![polymers logo]

polymers

MDPI

Article

Effect of Graphene Oxide on the Reaction Kinetics of Methyl Methacrylate In Situ Radical Polymerization via the Bulk or Solution Technique

Ioannis S. Tsagkalias, Triantafyllos K. Manios and Dimitris S. Achilias *

Laboratory of Polymer Chemistry and Technology, Department of Chemistry,
Aristotle University of Thessaloniki, 541 24 Thessaloniki, Greece;
itsagkal@chem.auth.gr (I.S.T.); trfgr1@gmail.com (T.K.M.)
* Correspondence: axilias@chem.auth.gr; Tel.: +30-231-099-7822

Received: 31 July 2017; Accepted: 5 September 2017; Published: 8 September 2017

Abstract: The synthesis of nanocomposite materials based on poly(methyl methacrylate) and graphene oxide (GO) is presented using the in situ polymerization technique, starting from methyl methacrylate, graphite oxide, and an initiator, and carried out either with (solution) or without (bulk) in the presence of a suitable solvent. Reaction kinetics was followed gravimetrically and the appropriate characterization of the products took place using several experimental techniques. X-ray diffraction (XRD) data showed that graphite oxide had been transformed to graphene oxide during polymerization, whereas FTIR spectra revealed no significant interactions between the polymer matrix and GO. It appears that during polymerization, the initiator efficiency was reduced by the presence of GO, resulting in a reduction of the reaction rate and a slight increase in the average molecular weight of the polymer formed, measured by gel permeation chromatography (GPC), along with an increase in the glass transition temperature obtained from differential scanning calorimetry (DSC). The presence of the solvent results in the suppression of the gel-effect in the reaction rate curves, the synthesis of polymers with lower average molecular weights and polydispersities of the Molecular Weight Distribution, and lower glass transition temperatures. Finally, from thermogravimetric analysis (TG), it was verified that the presence of GO slightly enhances the thermal stability of the nano-hybrids formed.

Keywords: PMMA; graphene oxide; polymerization kinetics; bulk; solution

---

1. Introduction

Graphene is a single-atomic, two-dimensional layer of sp^2 hybridized carbon atoms arranged in a honeycomb lattice. Because of its unique mechanical, electrical, thermal, and optical properties, it has recently attracted the research interest of the scientific community [1,2]. These properties make graphene one of the most popular candidates for the development of functional and structural graphene-reinforced polymer composites. Graphene can be obtained from the exfoliation of graphite sheets. However, it is easier to obtain graphene oxide (GO) sheets through the exfoliation of graphite oxide. The latter can be produced by the oxidation of graphite and consists of many oxygen-containing groups, such as carboxyl, epoxy, and hydroxyl groups in the basal planes and edges [2,3]. Thus, graphite oxide exhibits an increased interlayer spacing from the original 3.4 Å of graphite to 6.0–10 Å [3]. Such functional groups aim to produce graphite oxide hydrophilic and to weaken the van der Waals forces between layers. Thus, graphite oxides can be readily dispersed in aqueous media to form colloidal suspensions. This facilitates the exfoliation of layered graphite oxide into GO sheets via sonication or stirring [4].

Various techniques have been developed for the synthesis of polymers based on nano-composites, including solution mixing, melt blending, and in-situ polymerization [5–7]. The latter technique usually ensures the good dispersion of the nano-additive in the polymer matrix and has improved the final product properties. The in situ polymerization in the presence of several nano-additives was also the basis of an extensive experimental study conducted by our group [8–12].

Although nanocomposites of GO with several polymers have been the subject of studies in the past [1,7], their incorporation in methyl methacrylate (MMA) polymerization under bulk or solution conditions has not yet appeared in scientific papers and journals [5,6]. Therefore, the challenge in this work is to experimentally investigate the kinetics of the in-situ polymerization of MMA in the presence of a material having unique properties, such as the graphene oxide (GO) nano-additive. GO was formed during the reaction by the exfoliation of graphite oxide obtained from the oxidation of graphite. One of the major problems in such polymerizations is the formation of stable dispersion throughout the reaction. For this reason, solvents are usually chosen. Dimethylformamide (DMF) is an organic solvent which dissolves both the monomer MMA and the polymer PMMA and is also hydrophilic and polar with a rather high boiling point (153 °C). It has been proved to be a good solvent for the dispersion of GO [13]. Therefore, in this research, besides carrying out the polymerization in bulk, the possibility of using DMF as a reaction solvent was examined. Thus, for the first time, the effect of GO on the in situ solution polymerization of MMA was explored.

In the past, we have prepared nanocomposite materials of PMMA with several nano-clays in our laboratory and studied the influence of the nano-filler on the reaction kinetics [14–16]. As a continuation of this work, nanocomposite materials of PMMA with graphene oxide are produced in this study using an in-situ polymerization technique carried out with (solution) or without (bulk) in the presence of a solvent. The effect of GO on the reaction kinetics is investigated gravimetrically by measuring the variation of conversion with time, as well as the molecular weight distribution of the polymer formed. The properties of the PMMA/GO nanohybrids were measured via a variety of techniques such as X-ray diffraction, FTIR spectrometry, thermogravimetric analysis (TGA), gel permeation chromatography (GPC), and differential scanning calorimetry (DSC).

2. Materials and Methods

2.1. Materials

Methyl methacrylate, used as the monomer, was purchased from Alfa Aesar (Haverhill, MA, USA purity ≥ 99%). The inhibitor, hydroquinone, was removed by passing the monomer before any use, thrice, through disposable inhibitor-remover packed columns (Aldrich, Hamburg, Germany). Benzoyl peroxide (BPO) was used as a free radical initiator and was provided by Alfa Aesar (purity > 97%). The initiator was purified by fractional recrystallization twice from methanol (Chem Lab, Zedelgem, Belgium). Dimethylformamide (DMF) was used as a solvent for the solution polymerization and as a means of exfoliation of graphite oxide to graphene oxide, and was supplied from the J.T. Baker company (Radnor, PA, USA). Dichloromethane was used as the polymer solvent purchased from the company Chem Lab, while the methanol used for the precipitation of the polymer was also supplied by the same company. All other chemicals used were of analytical grade and were used as received without further purification.

2.2. Preparation of Graphite Oxide

Graphite oxide (GO) was prepared by oxidizing graphite powder which was purchased from Sigma-Aldrich (St. Louis, MO, USA), in accordance with the Hummers method. Accordingly, 10 g of commercial graphite powder was dispersed in sulfuric acid (230 mL) at 0 °C. Subsequently, 30 g of potassium permanganate ($KMnO_4$) was slowly added to the suspension by controlling the addition rate and maintaining the temperature below 20 °C. Following this, the reaction mixture was cooled to 2 °C. Then, the mixture was removed from the ice bath and stirred with a magnetic stirrer at room temperature for 30 min. Subsequently, 230 mL of deionised water was added, again controlling the addition rate, while the temperature was kept below 20 °C. Thereafter, the mixture was resuspended under mechanical agitation for 15 min, followed by the addition of 1.4 L deionized water and 100 mL of hydrogen peroxide solution (30 wt %). The mixture was allowed to stand for 24 h. The GO particles that settled at the bottom were separated from the excess liquid by decantation. The gelatinous texture material was placed in an osmotic membrane, to stop the formation of precipitate $BaSO_4$, which appeared during the addition of $BaCl_2$ aqueous solution. The material remained in the membrane for about eight days. Finally, the final product was obtained by the freeze-drying method.

2.3. Preparation of the Initial Monomer/GO Mixtures

Three different relative amounts of GO to monomer, i.e., 0.1, 0.5, and 1.0 wt % were prepared. Monomer with graphite oxide (for bulk polymerization), or monomer with the solvent (DMF) and graphite oxide (for solution polymerization), underwent ultrasonication for one hour to ensure a satisfactory colloidal dispersion of graphite oxide in the solution, while the exfoliation of graphite oxide to graphene oxide started. In the final suspension, the initiator BPO 0.03 M was added and the mixture was degassed by passing nitrogen through it, after which it was immediately used. During solution polymerization, two different relative amounts of monomer/solvent were employed, i.e., 80:20 and 50:50 *v:v*.

2.4. Synthesis of PMMA/GO Nanocomposites by the In-Situ Bulk or Solution Radical Polymerization

Bulk free-radical polymerization was carried out in small test-tubes at a constant temperature of 80 °C for a suitable time. According to this technique, 1 cm^3 of the pre-weighed mixture of monomer with the initiator and each amount of GO were placed into a series of 10 small test-tubes. They were degassed with nitrogen, sealed, and placed into a pre-heated reaction temperature bath. Each test-tube was removed from the bath at pre-specified time intervals, a few drops of hydroquinone were added in order to stop the reaction, and it was immediately frozen. The product was isolated after dissolution in dichloromethane (CH_2Cl_2) and precipitation in methanol. A different procedure for the isolation of the product was followed in the last samples of each experiment. Since polymerization had already finished and the product was a hard solid, the test-tubes were broken and nanocomposites were obtained as such. In this manner, the filler was enclosed in the polymer matrix. Subsequently, all isolated materials were dried to a constant weight in a vacuum oven at room temperature, weighed, and the degree of conversion was estimated gravimetrically.

Neat polymer was also synthesized under the above conditions and used as the reference material.

Exactly the same procedure was repeated in the solution polymerization experiments but by using the monomer MMA dissolved in DMF in all experiments.

2.5. Measurements

X-ray diffraction. The crystalline structure of the prepared PMMA/GO materials was characterized using X-ray diffraction (XRD) in a Rigaku Miniflex II instrument (Tokyo, Japan) equipped with a CuKα generator (λ = 0.1540 nm). The XRD patterns were recorded at the range $2\theta = 5$–$65°$ and a scan speed of $2°\cdot min^{-1}$.

Fourier-Transform Infra-Red (FTIR). The chemical structure of neat PMMA and PMMA/GO nanocomposites was established by recording their IR spectra. The spectrophotometer used was Spectrum 1 (Perkin Elmer, Waltham, MA, USA) equipped with an attenuated total reflectance (ATR) device. Spectra were recorded over the range 4000 to 600 cm^{-1} at a resolution of 2 cm^{-1} and 32 scans were averaged to reduce noise. Thin films were used for the measurements prepared in a hot hydraulic press and the instrument's software was employed to identify characteristic peaks.

Differential Scanning Calorimetry (DSC). The glass transition temperature, T_g, of the material prepared, was measured using the DSC-Diamond (Perkin-Elmer). Samples of approximately 5–6 mg were used and sealed in standard Perkin-Elmer sample pans. The temperature program followed included, initially heating to 180 °C at a rate of 10 °C·min^{-1} to ensure the complete polymerization of the residual monomer, cooling to 30 °C, and heating again to 130 °C at a rate of 20 °C·min^{-1}. The glass transition temperature was estimated from the second heating recordings.

Gel Permeation Chromatography (GPC). In order to estimate the average molecular weights and the full molecular weight distribution (MWD) of neat PMMA and the PMMA/GO nanocomposites, GPC was used. The chromatograph used was from Polymer Laboratories (Church Stretton, UK), model PL-GPC 50 Plus, and included an isocratic pump, three PLgel 5 μ MIXED-C columns in series, and a differential refractive index (DRI) detector. The elution solvent was tetrahydrofurane (THF) at a constant flow rate of 1 mL·min^{-1}, and the entire system was kept at a constant temperature of 30 °C. All samples were dissolved in THF at a concentration of 1 mg·mL^{-1}, filtered, and 200 μL was used for the injection into the chromatograph. Calibration of GPC was carried out with standard poly(methyl methacrylate) samples having peak molecular weights ranging from 690 to 1,944,000 (from Polymer Laboratories).

Thermogravimetric Analysis (TGA). The thermal stability of the samples was evaluated by measuring their mass loss with increasing temperature using TGA. Measurements were performed on a Pyris 1 TGA (Perkin-Elmer) thermal analyzer. Samples of approximately 8–10 mg were used and the measurements included heating from ambient temperature to 600 °C at a heating rate of 20 °C·min^{-1} under inert atmosphere (nitrogen flow).

3. Results

3.1. Characterization of the PMMA/GO Nanocomposites

In order to identify the exfoliation of graphite oxide to graphene oxide after polymerization, XRD measurements were carried for graphite, graphite oxide, neat PMMA, and the nanocomposites of PMMA/GO. From the XRD spectra shown in Figure 1, graphite shows a sharp peak at 26.5°. When it is transformed to graphite oxide, this peak is shifted to 11°. Therefore, complete oxidation is verified. Furthermore, the XRD patterns of neat PMMA and PMMA/GO materials were recorded in the angle range of 2θ (5° < 2θ < 60°) and are included in Figure 1. It is seen that in pure PMMA, three very broad peaks appear at 16°, 32°, and 43°, respectively, denoting the amorphous structure of the polymer. When GO was incorporated into the polymer matrix, the same spectrum was recorded, without any obvious peak at 11°. This is an indication that graphite oxide has been exfoliated into graphene oxide during the reaction. TEM measurements could verify these observations.

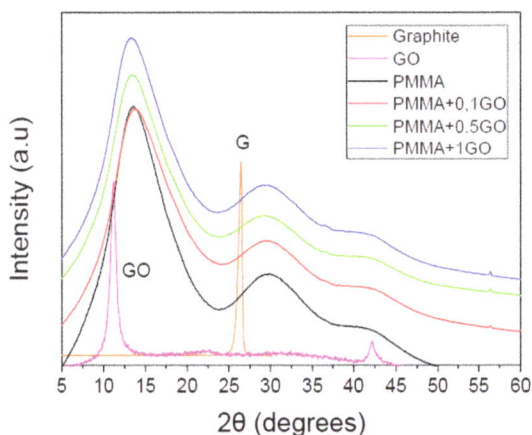

Figure 1. X-ray diffraction patterns of graphite, GO, neat PMMA, and PMMA/GO nanocomposites.

Possible physicochemical interactions between GO and the PMMA matrix were tested using FTIR-ATR measurements. The FTIR spectra of the material prepared appear in Figure 2. The spectrum of pure PMMA and all the nanocomposites show a sharp peak at 1724 cm^{-1}, which corresponds to the carbonyl group, C=O. Two small peaks at 3000/2940 cm^{-1} are attributed to methyl ester C–H stretching vibrations. An additional small peak at 2855 cm^{-1} is due to –CH$_3$ stretching vibrations. The peaks at 1436/1482 cm^{-1} correspond to C–H deformations. The peak at 1365 cm^{-1} corresponds to –CH$_3$ symmetrical deformation. Finally, the peaks at 1271/1233/1143/985 cm^{-1} are attributed to C–O stretching. Similar reflectance bands have been observed in the FTIR-ATR of pure PMMA in the literature [16]. From Figure 2, the spectra of pure PMMA and all nanocomposites appear similar, indicating that the inclusion of GO in the polymer matrix is rather physical without a strong chemical bond. An analogous observation has been reported in the literature [16].

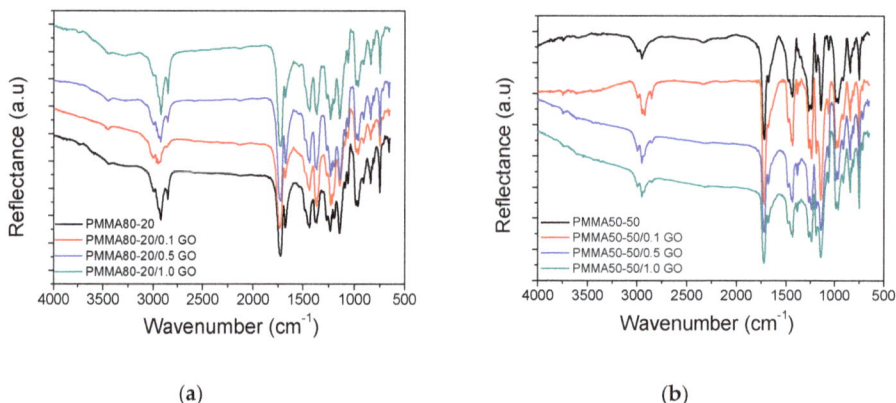

Figure 2. FTIR spectra of neat PMMA and PMMA with 0.1, 0.5, and 1.0% GO obtained under solution polymerization with 80–20 (**a**); 50–50 (**b**) ratio of monomer to solvent.

3.2. Polymerization Kinetics

The effect of carrying out the polymerization in bulk or in solution with two different solvent ratios on the variation of conversion with time is illustrated in Figure 3 for neat PMMA and its nanocomposites with 0.1, 0.5, and 1.0 wt % GO. Conversion time curves approximately follow classical radical polymerization kinetics until near 30% conversion, whereas afterwards, an increase in the reaction rate is observed due to the well-known auto-acceleration or gel-effect. Accordingly, as the reaction proceeds, the movement of macroradicals to find one another to terminate is hindered by the presence of the macromolecular chains. These diffusion-controlled phenomena lead to reduced termination rates of macroradicals, locally increasing their concentration. As a result, this enhanced the reaction rates. During solution polymerization, as the amount of solvent is increased, the auto-acceleration is decreased since macroradicals have more space to move freely and find one another to terminate. Hence, the polymerization rate lowers and more time is needed to complete the reaction. This was observed in both neat PMMA and all nanocomposites (Figure 3). Finally, at conversions higher than 90%, the reaction rate slows down significantly and polymerization almost stops before the full consumption of the monomer. This phenomenon corresponds to the well known glass effect, where the monomer-polymer mixture becomes a "glass". At this point, diffusion-controlled phenomena also affect the propagation and the initiation reaction. The propagation rate constant and the initiator efficiency decrease significantly and even small molecules (i.e., monomer, primary initiator radicals) cannot easily move in space. Thus, unreacted monomers are trapped without being able to react with the macro-radicals, leaving some residual monomer [17–20].

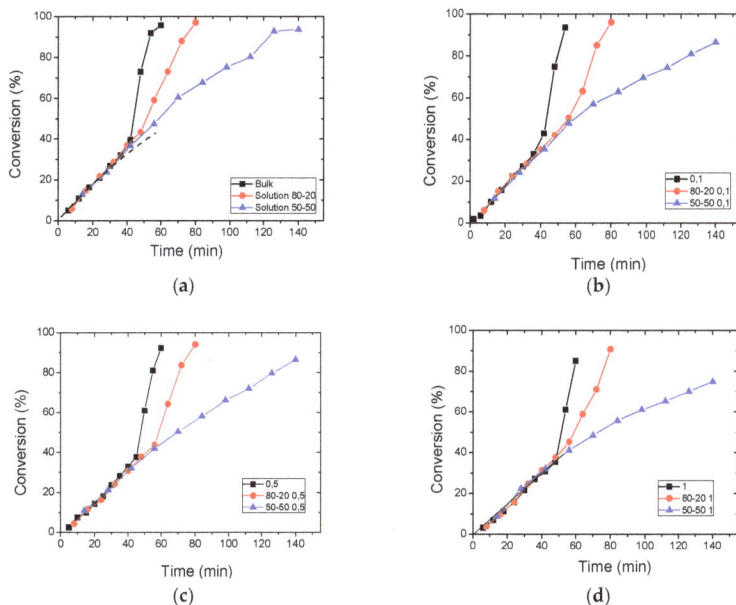

Figure 3. Conversion versus time curves of neat PMMA obtained after bulk and solution polymerization (**a**); PMMA/GO nanocomposites with different relative amounts of GO obtained from in situ bulk polymerization at 80 °C with a 0.03 mol/L initial initiator concentration, 0.1 % (**b**); 0.5% (**c**); 1.0% (**d**).

Furthermore, the effect of adding GO on the PMMA polymerization kinetics is investigated in Figure 4. It is seen that in both bulk and solution polymerization, the behavior of the PMMA with 0.1 wt % GO is very similar to that of neat PMMA, meaning that such a small amount of additive does not influence the reaction kinetics much. However, as the amount of GO added increases, it is clearly observed in all different polymerization conditions that the initial polymerization rate decreases. This has also been observed in the literature in other nano-additives added to PMMA polymerization [14,16].

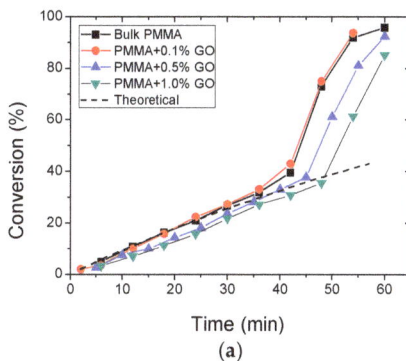

Figure 4. *Cont.*

(b)

(c)

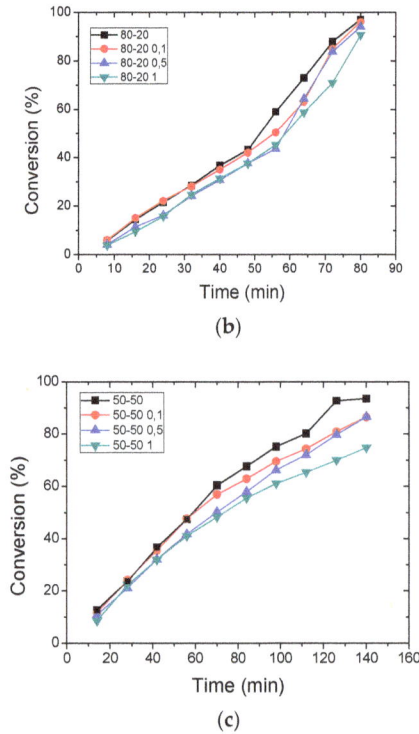

Figure 4. Conversion versus time curves of neat PMMA and PMMA/GO nanocomposites with different relative amounts of GO obtained from in situ polymerization at 80 °C with a 0.03 mol/L initial initiator concentration in bulk (**a**); in solution 80–20 (**b**); 50–50 (**c**).

In order to provide an explanation for the effect of GO on the polymerization kinetics, we used the following Equation (1) for the variation of monomer conversion, X with time, t. Thus, the polymerization rate, dX/dt, assuming the steady-state approximation for the total radical concentration (which has been proven to hold at low monomer conversion), is expressed as [18]:

$$\frac{dX}{dt} = \left(k_p + k_{trM}\right)\left(\frac{fk_d[I]}{k_t}\right)^{1/2}(1 - X) \tag{1}$$

where, k_p, k_{trM}, k_t, and k_d denote the kinetic rate constants of the propagation, chain transfer to monomer, termination, and initiator decomposition reactions, respectively; f is the initiator efficiency; and $[I]$ is the initiator concentration.

Assuming that the initiator concentration remains almost constant at short reaction times and all kinetic rate constants are independent of conversion, Equation (1) can be integrated to give:

$$X = 1 - \exp(-k_{eff}t) \text{ or } -\ln(1 - X) = k_{eff}t \tag{2}$$

with

$$k_{eff} = \left(k_p + k_{trM}\right)\left(\frac{fk_d[I]_0}{k_t}\right)^{1/2} \tag{3}$$

The effective rate constant of PMMA, k_{eff}, can be evaluated from available literature data on the kinetic rate constant at low conversions. A number of different values have been proposed. In the following, we made use of those reported in a very recent paper by Zoller et al. [21], i.e., $k_p = 2.67 \times 10^6 * \exp(-22,360/RT)$, $k_t = 1.984 \times 10^8 * \exp(-5,890/RT)$ L/mol/s, $k_d = 5 \times 10^{16} * \exp(-143,000/RT) \cdot s^{-1}$, $k_{trM} = 5 \times 10^{-5} * k_p$, and $f = 0.5$. The values of the above kinetic rate constants at 80 °C, where the experiments of this study were carried out, are: $k_p = 1314$ L/mol/s, $k_t = 2.668 \times 10^7$ L/mol/s, $k_d = 3.5 \times 10^{-5}$ s^{-1}, $k_{trM} = 0.066$ L/mol/s. Using these values and $[I]_0 = 0.03$ mol/L, the theoretical value of the effective rate constant becomes, $k_{eff} = 1.84 \times 10^{-4}$ s^{-1} or 1.106×10^{-2} min^{-1}. Then, by means of Equation (2), the variation of conversion with time at low conversions can be estimated and is included as a continuous line in Figure 3a. It is seen that the theoretical line simulates the experimental data very well at low conversions. The next step was to identify which kinetic parameter is affected by the addition of GO. From Equation (1), it is unlikely that k_p, k_{trM}, k_t, or k_d would change with the existence of the GO. Then, it seems that the initiator efficiency, f, is affected and particularly decreases with an increasing nano-additive content. Furthermore, the effective rate constant, k_{eff}, can be estimated using Equation (2) from the slope of the curve obtained after plotting $-\ln(1-X)$ vs. t. Such curves for the bulk polymerization of PMMA and its nanocomposites with GO appear in Figure 5. The corresponding curves for the solution polymerizations were similar to those shown in this figure, since as it can be seen in Figure 3, the data at the same amount of GO and at conversions less than 30% (as we used here) are very similar. The estimated values of k_{eff}, together with their standard error and correlation coefficient, appear in Table 1. As it can be seen, very clear straight lines were obtained for all curves. From the values of k_{eff} and using Equation (3) with the aforementioned parameter values, the initiator efficiencies, f, were estimated and are included in Table 1. It can be seen that the value of f for neat PMMA, i.e., 0.47, is very close to the theoretical (literature) value of 0.5. Moreover, a clear decrease of the initiator efficiency with the increasing amount of GO was observed. This is a clear indication that graphene oxide acts as a scavenger of primary initiator radicals at the early stages of polymerization. This is in accordance with the results presented in our previous work [15,16] for in situ MMA homopolymerization in the presence of nano-additives such as organomodified montmorillonites.

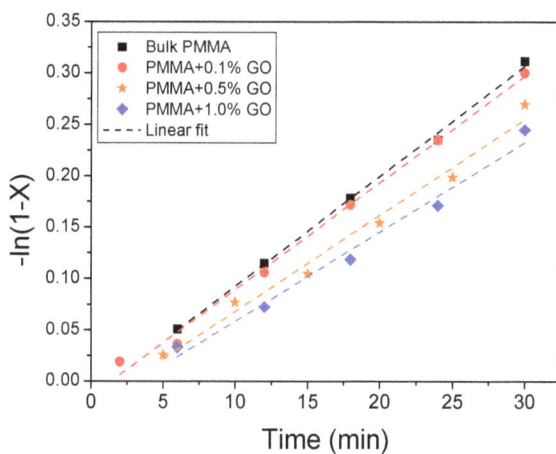

Figure 5. Estimation of the effective rate constant from plots of $-\ln(1-X)$ versus time according to Equation (2) for the neat PMMA and PMMA/GO nanocomposites with different relative amounts of GO obtained from in situ polymerization at 80 °C with a 0.03 mol/L initial initiator concentration in bulk.

Table 1. Effective rate constant, k_{eff}, estimated from the slope of the curves presented in Figure 5 of neat PMMA and its nanocomposites with various amounts of GO obtained in bulk polymerization and corresponding initiator efficiencies estimated from Equation (3).

Sample	k_{eff} (min^{-1})	Standard Error	R^2	f
PMMA	0.0107	2.68×10^{-4}	0.9975	0.47
PMMA/0.1GO	0.01038	3.75×10^{-4}	0.9935	0.44
PMMA/0.5GO	0.00935	5.65×10^{-4}	0.9820	0.36
PMMA/1.0GO	0.0087	6.24×10^{-4}	0.9797	0.31

In order to provide an explanation for the reduced initiator efficiency with the nano-additive content, the decomposition of the initiator used, i.e., BPO, is considered in Scheme 1. Accordingly, two benzoyloxy radicals are initially produced, which can further decompose to phenyl radicals and carbon dioxide. Both primary radicals formed from the decomposition of the initiator may react with the phenolic hydroxyls on the GO surface by abstracting a hydrogen atom. The phenoxy radicals may then scavenge a further radical (Scheme 1). Thus, one or two primary radicals may be terminated for every mole of phenolic OH. As a result, the effective number of primary radicals formed from the fragmentation of the initiator, which can find a monomer molecule and start polymerization, is decreased.

Furthermore, the full molecular weight distribution of neat PMMA and PMMA/GO nanocomposites obtained in bulk and solution with a 80–20 and 50–50 ratio of monomer to solvent, measured via GPC, appears in Figure 6. The average molecular weights of PMMA and all nano-hybrids prepared are illustrated in Table 2. It was observed that the average molecular weight, M_n, of the material formed increases when increased amounts of GO are added. Moreover, from Figure 6, the formation of polymers with a narrower MWD distribution was revealed when GO was added. To explain these measurements we returned again to classical free radical polymerization kinetics. Accordingly, the average molecular weight of a polymer is given by its average degree of polymerization, which in turn is calculated from the average kinetic chain length, ν, knowing the mode of termination by combination or disproportionation. ν can be calculated from the following equation:

$$\frac{1}{\nu} = \frac{k_t[P^\bullet]}{k_p[M]} + \frac{k_{trM}[M]}{k_p[M]} = \frac{(fk_d[I]k_t)^{1/2}}{k_p[M]} + C_M \tag{4}$$

with $C_M = k_{trM}/k_p$.

According to Equation (4), the average kinetic chain length, ν, is inversely proportional to the initiator efficiency, f. Therefore, when f decreases, ν, and as a result, the average molecular weight of the polymer, increases. The physical meaning of this is that macro-radicals with a higher chain length are produced when the number of primary initiator radicals is decreased. Moreover, it was found that higher amounts of GO result in a lower final conversion, which in turn stops the polymer from increasing its high molecular weight tail to higher values, resulting in the reduced polydispersity of the MWD.

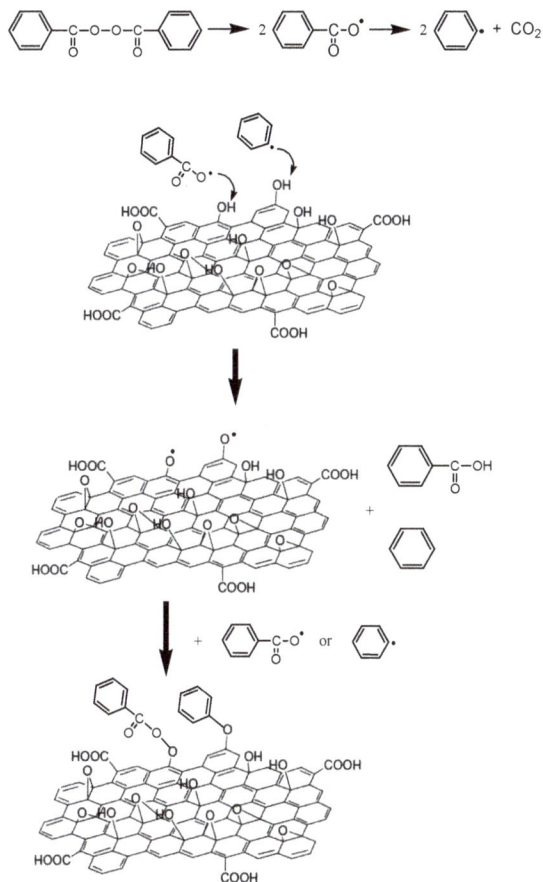

Scheme 1. Schematic illustration of the reaction of primary initiator radicals with GO resulting in their deactivation and reduction of the initiator efficiency.

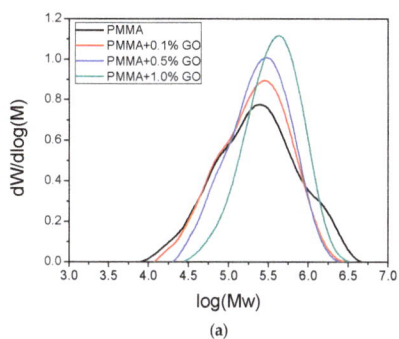

(a)

Figure 6. *Cont.*

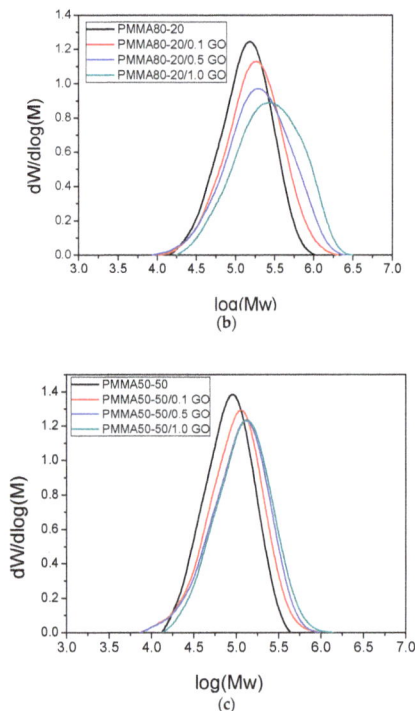

Figure 6. Full Molecular weight distribution of neat PMMA and PMMA/GO nanocomposites obtained in bulk (**a**) and solution with a 80–20 (**b**) and 50–50 (**c**) ratio of monomer to solvent.

Table 2. Number-average, weight-average, and z-average molecular weights ($\overline{M}_n$, $\overline{M}_w$, $\overline{M}_z$), polydispersity of the MWD (PD) and glass transition temperature (T_g) of neat PMMA and its nanocomposites with various amounts of GO obtained in bulk or solution polymerization with a 80–20 and 50–50 ratio of monomer to solvent.

Sample	$\overline{M}_n$	$\overline{M}_w$	$\overline{M}_z$	PD	T_g
PMMA	114,570	427,250	1,121,690	3.73	103.5
PMMA/0.1GO	124,024	321,516	626,860	2.59	105.7
PMMA/0.5GO	160,872	339,196	592,744	2.11	109.9
PMMA/1.0GO	255,553	480,359	781,407	1.88	113.8
PMMA80–20	102,673	173,944	267,436	1.69	86.6
PMMA80–20/0.1GO	119,534	233,037	409,315	1.95	88.0
PMMA80–20/0.5GO	144,446	333,565	634,007	2.31	91.1
PMMA80–20/1.0GO	168,209	390,492	723,507	2.32	92.9
PMMA50–50	65,699	98,390	138,636	1.50	82.2
PMMA50–50/0.1GO	71,766	123,872	193,358	1.73	85.8
PMMA50–50/0.5GO	77,709	140,321	218,256	1.81	88.0
PMMA50–50/1.0GO	91,579	156,504	250,996	1.71	90.9

Moreover, carrying out the polymerization in the presence of solvent results in polymers with significantly reduced average molecular weights (Table 2). As it is seen in Figure 7, the whole MWD shifts to lower values. This is a direct result of Equation (4), where it is clear that as the monomer concentration is decreased (a higher monomer concentration is achieved in bulk polymerization), the kinetic chain length is also decreased. Thus, higher amounts of solvent results in a reduced average molecular weight of the polymer. Another reason is that, as it was observed in Figure 3, the presence of the solvent suppresses the gel-effect. In the presence of strong autoacceleration (effect of diffusion-controlled phenomena), the average molecular weights of the polymer increase with monomer conversion. Since adding a solvent reduces the effect of diffusion controlled phenomena, it also results in a reduced increase of M_n and M_w with monomer conversion, and as a result, in lower final values.

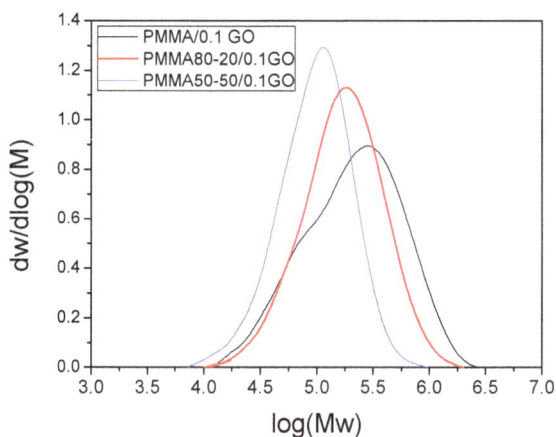

Figure 7. Full Molecular weight distribution of neat PMMA obtained after bulk and solution polymerization.

3.3. Thermal Properties of the Nanocomposites

DSC was used to measure the glass transition temperature, T_g, of neat PMMA and the nano-hybrids, according to the procedure described in the experimental section and the half C_p extrapolation method [9]. Indicative results of DSC traces obtained from the bulk polymerization experiments are shown in Figure 8. The T_g values estimated are given in Table 2. For pristine PMMA, a value near 103.5 °C was recorded, close to that reported in the literature (near 100 °C [9]). When the amount of GO added was increased to 0.1, 0.5, and 1.0 wt %, the T_g of the polymer was increased to 105.7, 109.9, and 113.8 °C, respectively. The same tendency was observed in the solution experiments with the T_g increasing from 86.6 to 92.9 °C and from 82.2 to 90.9 °C for the 80–20 and 50–50 ratios, respectively. The interaction of polymer chains and nano-particles at the surface can alter the chain kinetics by either decreasing or increasing the glass transition temperature of the polymer [12]. The enhancement in the T_g of the nanocomposites could be attributed to the restriction in chain mobility due to the confinement effect of 2D-layered graphene incorporated into the matrix and the strong nanofiller—polymer interactions.

In addition, the thermal stability of pure PMMA and PMMA/GO nanocomposites was examined by thermogravimetric analysis in an inert (nitrogen) atmosphere. Results on the variation of mass loss with an increasing temperature of neat polymer and the nano-hybrids with various amounts of added GO obtained after bulk or solution polymerization are illustrated in Figure 9a–c. The thermal degradation of radically prepared PMMA has been a subject of numerous studies and usually involves multiple steps assigned to: the scission of unsaturated terminal groups, presence of weak head-to-head linkages, and random scission of the carbon-carbon main chain. It is generally considered that most PMMA thermally degrades through depolymerisation. In addition, it has been shown that GO exhibits a significant mass loss (almost 22%) at 210–250 °C, which is due to the pyrolysis of labile oxygen-containing functional groups, such as –COOH, –OH, etc. The incorporation of GO in the polymer matrix results in an enhanced thermal stability of the resulting nanocomposite compared to neat polymer, especially when higher amounts of GO are incorporated [22]. Therefore, the mass loss observed in Figure 9 in the region of 210–250 °C is attributed to the elimination of labile functional groups from the surface of GO. Moreover, the incorporation of 0.1 wt % GO does not seem to show any significant effect. However, the addition of higher GO amounts seems to increase the thermal stability of the polymer formed, which was more pronounced in the PMMA nanocomposites obtained after solution 50–50 polymerization and is shown in Figure 9c. Finally, thermal degradation seems to start earlier in the polymers and nanocomposites formed after solution polymerization compared to corresponding from bulk. This is mainly attributed to the lower average molecular weight of the polymers produced via solution polymerization compared to those from bulk.

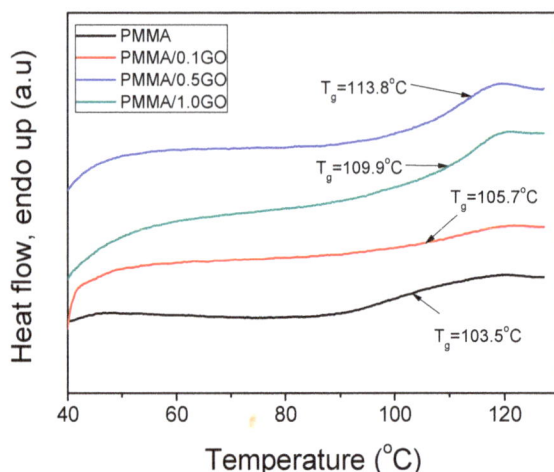

Figure 8. DSC traces of PMMA and all PMMA/GO nanocomposites, to estimate their glass transition temperatures.

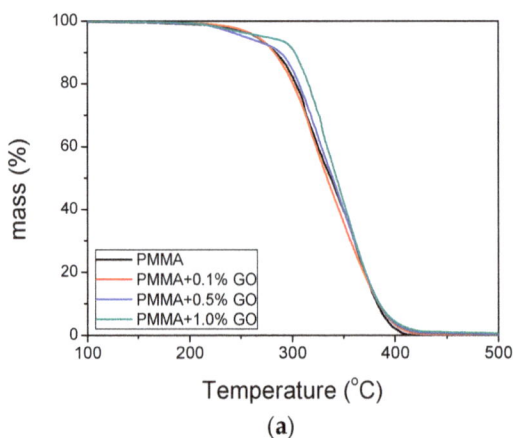

(a)

Figure 9. *Cont.*

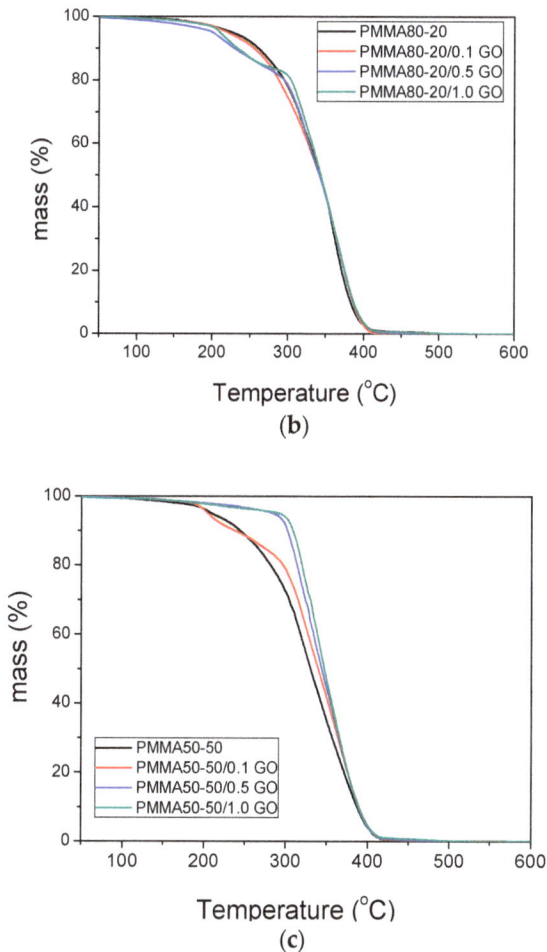

Figure 9. Mass loss of neat PMMA and PMMA/GO nanocomposites with various amounts of added GO obtained in bulk (**a**); solution with 80–20 (**b**); 50–50 (**c**) ratio of monomer to solvent.

4. Conclusions

Nanocomposite materials of poly(methyl methacrylate) with GO were produced using the in situ radical polymerization technique carried out in bulk and solution at different solvent/monomer ratios, from graphite oxide, the MMA monomer, and benzoyl peroxide, the initiator. FTIR data showed that the inclusion of GO in the polymer matrix was rather physical, without a strong chemical bond. XRD data showed that graphite oxide had been transformed to graphene oxide during polymerization. From the study of the reaction kinetics, it was found that the initiator efficiency is reduced by the presence of the GO, resulting in a reduction of the reaction rate and a slight increase in the average molecular weight of the polymer formed. A polymer with a lower average molecular weight was produced when polymerization was carried out in solution, again increasing with the amount of GO added. The polydispersity of the MWD was found to decrease with the amount of solvent added. Moreover, the glass transition temperature of the polymer was increased with the amount of GO added, whereas it was decreased during solution polymerization. Finally, from thermogravimetric analysis, it was verified that the presence of GO results in materials with a slightly higher thermal stability.

Author Contributions: Ioannis Tsagkalias and Dimitris S. Achilias conceived and designed the experiments; Triantafyllos Manios performed the experiments; Ioannis Tsagkalias and Dimitris S. Achilias analyzed the data; Dimitris S. Achilias and Ioannis Tsagkalias wrote the paper.

Conflicts of Interest: The authors declare no conflict of interest.

References

1. Kim, H.; Abdala, A.A.; Macosko, C.W. Graphene/Polymer Nanocomposites. *Macromolecules* **2010**, *43*, 6515–6530. [CrossRef]
2. Park, S.; Ruoff, R.S. Chemical methods for the production of graphenes. *Nat. Nano* **2009**, *4*, 217–224. [CrossRef] [PubMed]
3. Dreyer, D.R.; Park, S.; Bielawski, C.W.; Ruoff, R.S. The chemistry of graphene oxide. *Chem. Soc. Rev.* **2010**, *39*, 228–240. [CrossRef] [PubMed]
4. Park, S.; An, J.; Jung, I.; Piner, R.D.; An, S.J.; Li, X.; Velamakanni, A.; Ruoff, R.S. Colloidal suspensions of highly reduced graphene oxide in a wide variety of organic solvents. *Nano Lett.* **2009**, *9*, 1593–1597. [CrossRef] [PubMed]
5. Aldosari, M.A.; Othman, A.A.; Alsharaeh, E.H. Synthesis and characterization of the in situ bulk polymerization of PMMA containing graphene sheets using microwave irradiation. *Molecules* **2013**, *18*, 3152–3167. [CrossRef] [PubMed]
6. Yuan, X.Y.; Zou, L.L.; Liao, C.C.; Dai, J.W. Improved properties of chemically modified graphene/poly(methyl methacrylate) nanocomposites via a facile in-situ bulk polymerization. *Express Polym. Lett.* **2012**, *6*, 847–858. [CrossRef]
7. Kuila, T.; Bose, S.; Hong, C.E.; Uddin, M.E.; Khanra, P.; Kim, N.H.; Lee, J.H. Preparation of functionalized graphene/linear low density polyethylene composites by a solution mixing method. *Carbon* **2011**, *49*, 1033–1051. [CrossRef]
8. Achilias, D.S.; Nikolaidis, A.K.; Karayannidis, G.P. PMMA/organomodified montmorillonite nanocomposites prepared by in situ bulk polymerization: Study of the reaction kinetics. *J. Therm. Anal. Calorim.* **2010**, *102*, 451–460. [CrossRef]
9. Achilias, D.S.; Siafaka, P.; Nikolaidis, A.K. Polymerization kinetics and thermal properties of poly(alkyl methacrylate)/organomodified montmorillonite nanocomposites. *Polym. Int.* **2012**, *61*, 1510–1518. [CrossRef]
10. Siddiqui, M.N.; Redhwi, H.H.; Gkinis, K.; Achilias, D.S. Synthesis and characterization of novel nanocomposite materials based on poly(styrene-co-butyl methacrylate) copolymers and organomodified clay. *Eur. Polym. J.* **2013**, *49*, 353–365. [CrossRef]
11. Siddiqui, M.N.; Redhwi, H.H.; Charitopoulou, D.; Achilias, D.S. Effect of organomodified clay on the reaction kinetics, properties and thermal degradation of nanocomposite based on poly(styrene-*co*-ethyl methacrylate). *Polym. Int.* **2014**, *63*, 766–777. [CrossRef]
12. Nikolaidis, A.K.; Achilias, D.S.; Karayannidis, G.P. Effect of the type of organic modifier on the polymerization kinetics and the properties of poly(methyl methacrylate)/organomodified montmorillonite nanocomposites. *Eur. Polym. J.* **2012**, *48*, 240–251. [CrossRef]
13. Paredes, J.I.; Villar-Rodil, S.; Martínez-Alonso, A.; Tascón, J.M.D. Graphene Oxide Dispersions in Organic Solvents. *Langmuir* **2008**, *24*, 10560–10564. [CrossRef] [PubMed]
14. Siddiqui, M.N.; Redhwi, H.H.; Verros, G.D.; Achilias, D.S. Evaluating the role of nanomontmorillonite in bulk in situ radical polymerization kinetics of butyl methacrylate through a simulation model. *Ind. Eng. Chem. Res.* **2014**, *53*, 11303–11311. [CrossRef]
15. Verros, G.D.; Achilias, D.S. Towards the development of a mathematical model for the bulk in situ radical polymerization of methyl methacrylate in the presence of nano-additives. *Can. J. Chem. Eng.* **2016**, *94*, 1783–1791. [CrossRef]
16. Siddiqui, M.N.; Redhwi, H.H.; Vakalopoulou, E.; Tsagkalias, I.; Ioannidou, M.D.; Achilias, D.S. Synthesis, characterization and reaction kinetics of PMMA/silver nanocomposites prepared via in situ radical polymerization. *Eur. Polym. J.* **2015**, *72*, 256–269. [CrossRef]
17. Verros, G.D.; Latsos, T.; Achilias, D.S. Development of a unified framework fr calculating molecular weight distribution in diffusion controlled free radical bulk homo-polymerization. *Polymer* **2005**, *46*, 539–552. [CrossRef]

18. Achilias, D.S. A review of modelling of diffusion controlled polymerization reactions. *Macromol. Theory Simul.* **2007**, *16*, 319–347. [CrossRef]

19. Verros, G.D.; Achilias, D.S. Modeling gel effect in branched polymer systems: Free radical solution homopolymerization of vinyl acetate. *J. Appl. Polym. Sci.* **2009**, *111*, 2171–2185. [CrossRef]

20. Achilias, D.S.; Verros, G.D. Modelling of diffusion controlled reactions in free radical solution and bulk polymerization: Model validation by DSC experiments. *J. Appl. Polym. Sci.* **2010**, *116*, 1842–1856.

21. Zoller, A.; Gigmes, D.; Guillaneuf, Y. Simulation of "cold" free radical polymerization of methyl methacrylate using a tertiary amine/BPO initiating system. *Polym. Chem.* **2015**, *6*, 5719–5727. [CrossRef]

22. Michailidis, M.; Verros, G.D.; Deliyanni, E.A.; Andriotis, E.G.; Achilias, D.S. An Experimental and Theoretical Study of Butyl Methacrylate In Situ Radical Polymerization Kinetics in the Presence of Graphene Oxide Nanoadditive. *J. Polym. Sci. A* **2017**, *55*, 1433–1441. [CrossRef]

Article

Imidazolium Ionic Liquid Modified Graphene Oxide: As a Reinforcing Filler and Catalyst in Epoxy Resin

Qing Lyu [1], Hongxia Yan [1,*], Lin Li [1], Zhengyan Chen [1], Huanhuan Yao [1] and Yufeng Nie [2]

[1] Department of Applied Chemistry, School of Natural and Applied Sciences, Northwestern Polytechnical University, Xi'an 710129, China; lq429216099@live.com (Q.L.); lilin20172017@outlook.com (L.L.); chenzhengyan@mail.nwpu.edu.cn (Z.C.); huanhuanyao@outlook.com (H.Y.)

[2] Department of Applied Mathematics, School of Natural and Applied Sciences, Northwestern Polytechnical University, Xi'an 710129, China; yfnie@nwpu.edu.cn

* Correspondence: hongxiayan@nwpu.edu.cn; Tel.: +86-029-8843-1657

Received: 18 July 2017; Accepted: 11 September 2017; Published: 14 September 2017

Abstract: Surface modification of graphene oxide (GO) is one of the most important issues to produce high performance GO/epoxy composites. In this paper, the imidazole ionic liquid (IMD-Si) was introduced onto the surface of GO sheets by a cheap and simple method, to prepare a reinforcing filler, as well as a catalyst in epoxy resin. The interlayer spacing of GO sheets was obviously increased by the intercalation of IMD-Si, which strongly facilitated the dispersibility of graphene oxide in organic solvents and epoxy matrix. The addition of 0.4 wt % imidazolium ionic liquid modified graphene oxide (IMD-Si@GO), yielded a 12% increase in flexural strength (141.3 MPa), a 26% increase in flexural modulus (4.69 GPa), and a 52% increase in impact strength (18.7 kJ/m^2), compared to the neat epoxy. Additionally the IMD-Si@GO sheets could catalyze the curing reaction of epoxy resin-anhydride system significantly. Moreover, the improved thermal conductivities and thermal stabilities of epoxy composites filled with IMD-Si@GO were also demonstrated.

Keywords: graphene oxide; epoxy; ionic liquid; imidazole

---

1. Introduction

As a kind of classic thermosetting polymers, epoxy resins are of particular interest to structural engineers due to their good stiffness, high strength, excellent dimensional stability, low curing shrinkage, and unique chemical resistance. Nevertheless, inherent brittleness strongly hinders the applications of epoxy resins in many fields. A large number of researchers have focused on the reinforcement of epoxy matrices with nano-fillers, such as SiO_2, Al_2O_3, Fe_3O_4, carbon nanotubes, etc. [1–4]. As is well known, graphene, one of the most effective additives due to its exceptional physical properties, is widely used to improve the comprehensive performances of epoxy composites, such as mechanical properties, electrical properties and thermal properties [5]. However, the lack of reactive functional groups on the graphene surface inhibits the use of graphene, due to the poor dispersibility and extremely weak interfacial bonding between graphene and epoxy matrix [6].

Graphene oxide (GO), as an oxide form of graphene, contains a large number of oxygenated groups on the basal planes and along the edges. As a result, GO has high organic compatibility with matrices and can be a suitable nano-filler to reinforce epoxy resins [7]. However, strong van der Waals forces exist between GO sheets, inducing the aggregation of GO sheets and causing an uneven stress concentration in epoxy matrix [8]. This can seriously limit the reinforcing effect of GO in epoxy resins. Additionally only a few active groups on GO surface can react with epoxy matrix, resulting in weak interfacial interactions between GO sheets and matrix. These problems can be addressed by functionalizing GO with some small molecules or polymers that can introduce some organic groups such as epoxide groups, amino groups, and isocyanate groups [9–11]. For example, Wan et al. [12]

reported a modification method of GO by covalent grafting of diglycidyl ether of bisphenol A chains. The surface functionalization was found to obviously improve the compatibility and dispersibility of GO sheets in epoxy matrix, with the tensile strength and fracture toughness increased by 75% and 41% at 0.5 wt % filler loading, compared to the neat epoxy. Ryu et al. [13] prepared the hexamethylene diamine functionalized graphene oxide successfully. The modification could significantly improve the dispersibility of GO sheets and interfacial interactions with the matrix, leading to promote the curing reaction well.

Ionic liquids have attracted a great deal of attention for their unique properties, such as excellent chemical and thermal stabilities, good electrical conductivity, and high ionic mobility [14,15]. Yang et al. [16] synthesized polydisperse graphene nanosheets stabilized by amine-terminated ionic liquids, which could be dispersed well in water, *N,N*-dimethylformamide (DMF), and dimethyl sulphoxide (DMSO). They thought the good dispersibility was attributed to the improved solubility and electrostatic repulsion of modified graphene nanosheets because of the ionic liquid incorporation. Due to the good dispersibility provided by ionic liquids, we think that the ionic liquid modified graphene or graphene oxide is of great value for reinforcing polymers. However, most reports about ionic liquid modified graphene or graphene oxide just focus on the electrochemical applications, such as electrodes, sensors, and supercapacitors [17,18]. We think it is necessary to investigate the reinforcing effects of ionic liquid modified graphene or graphene oxide in epoxy resin.

Meanwhile, it is worth mentioning that a majority of ionic liquids contain imidazole groups. As is well known, the imidazole rings can obviously catalyze the curing reactions of epoxy resins [19]. Pour et al. [20] prepared the poly (vinyl imidazole) grafted GO nanosheets and found that imidazole modified GO could enhance the curing rate by decreasing the activation energy of epoxy-amine curing system. Consequently, it might be a good choice to modify GO sheets by imidazolium ionic liquid, in order to enhance the dispersibility of GO sheets, as well as introduce imidazole groups to the epoxy curing system. In this way, it is possible to prepare a multifunctional nanoparticle with catalytic and reinforcing effects in epoxy resins.

Herein, we fabricated the imidazolium ionic liquid (IMD-Si) modified graphene oxide (IMD-Si@GO) by a cheap and simple method, based on the silanization reaction. The IMD-Si@GO nanosheets were incorporated into an epoxy resin-anhydride system to prepare the novel epoxy composites with high performances. The dispersibility of IMD-Si@GO, and the reinforcing and catalytic effects of IMD-Si@GO in the epoxy resin-anhydride system, have been investigated. Additionally, the thermal conductivities and thermal stabilities of IMD-Si@GO/epoxy (IMD-Si@GO/EP) composites were studied.

2. Materials and Methods

2.1. Materials

Diglycidyl ether of bisphenol A (DGEBA) epoxy with an epoxy value of 0.48–0.54 mol/100 g was purchased from Wuxi Resin Factory of Bluestar New Chemical Materials Co., Ltd., Wuxi, China. Methyltetrahydrophthalic anhydride (MTHPA, Puyang Huicheng Chemicals Co., Ltd., Puyang, China) was used as the curing agent. GO sheets were prepared from natural graphite flakes (500 mesh, Qingdao Hensen Graphite Co., Ltd., Qingdao, China) by a modified Hummers method [21]. Imidazole and 3-chloromethoxypropylsilane were purchased from (Alading Reagent Co., Ltd., Shanghai, China). Ethyl acetate, anhydrous alcohol and acetone were purchased from (Tianjin Tianda Chemical Co., Ltd., Tianjin, China).

2.2. Synthesis of IMD-Si@GO

A schematic representation of the synthesis of IMD-Si@GO is shown in Scheme 1. In the first step, 5.5 mL (30 mmol) 3-chloromethoxypropylsilane and 2.04 g (30 mmol) imidazole were mixed at 110 °C with magnetic stirring for 24 h under N_2 atmosphere [22]. The ionic liquid was washed by ethyl acetate for several times for purification. Then 500 mg GO was dispersed into 250 mL anhydrous alcohol by sonication for 30 min. IMD-Si (0.5 g) was added into the dispersion purged with N_2 at 50 °C. After refluxing for 4 h, the mixture was filtered, washed with anhydrous alcohol, and dried in a vacuum oven at 60 °C.

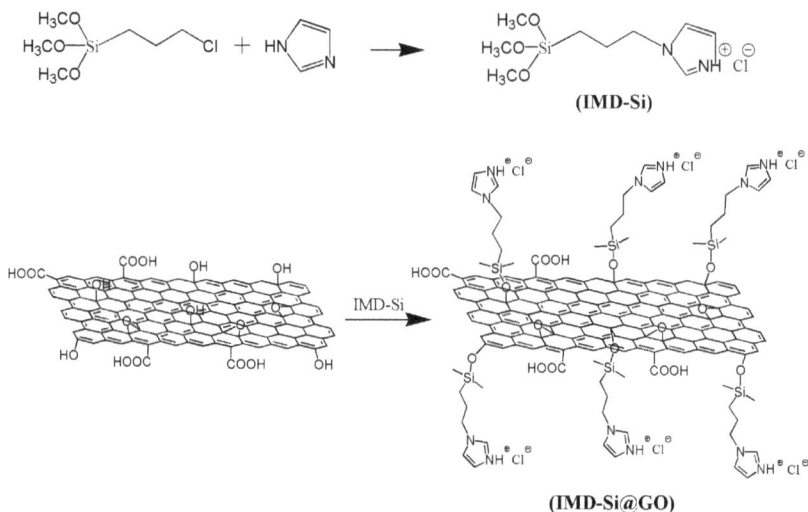

Scheme 1. Schematic of synthesis of the IMD-Si@GO.

2.3. Fabrication of Epoxy Composites

The IMD-Si@GO/EP were prepared by casting method with different contents of fillers (0.1, 0.2, 0.4, 0.6, and 0.8 wt %). The IMD-Si@GO sheets were first dispersed in acetone by sonication and then mixed with DGEBA epoxy. Afterwards, the mixture was put into a vacuum oven at 70 °C to remove solvent. Then, a stoichiometric amount of the curing agent MTHPA was added to the above mixture, followed by stirring for 10 min. Finally, the mixture was poured into a mold, cured following the schedule of 140 °C/2 h + 160 °C/4 h + 180 °C/3 h. As a reference, the pure epoxy resin was also prepared in similar procedures to perform contrast experiments.

2.4. Characterization

^{1}H Nuclear magnetic resonance (^{1}H NMR, 400 MHz) spectra were recorded in deuterated DMSO solvent using a Bruker Avance spectrometer (Bruker Instrument, Billerica, MA, USA). Atomic force microscope (AFM) images were taken in a tapping mode using Hitachi Nanonavi E-sweep (Hitachi High-Tech Science Co., Tokyo, Japan) with the size of 6 μm × 6 μm. X-ray photoelectron spectra (XPS) were performed on a PHI Quantum 2000 Scanning ESCA Microprobe system (ULVAC-PHI Inc., Kanagawa, Japan). All XPS spectra were corrected using the C 1s line at 284.8 eV. Fourier transform infrared (FTIR) spectra were characterized by a PerkinElmer-283B FTIR (Perkin Elmer Inc., Waltham, MA, USA) spectrometer ranging from 4000 to 400 cm^{-1}. Elemental analysis was carried out with a VarioEL III Analyser (Elementar Analysensyteme GmbH, Hanau, Germany). X-ray diffraction (XRD) patterns were recorded with a Bruker D8 ADVANCE X-ray diffractometer (Bruker AXS GmbH, Karlsruhe, Germany) (Cu Kα radiation, 0.1542 nm). The XRD data were collected from 5° to 85° with the scanning rate of 0.02°/s. Thermogravimetric analysis (TGA) was performed on a TGA Q50 (TA Instrument, New Castle, DE, USA) at a heating rate of 10 °C/min, in an argon atmosphere.

The dispersion of IMD-Si@GO in epoxy matrix was verified by using transmission optical microscopy (TOM, Pudan MM-8, Pu Dan Optical Instrument Co., Ltd., Shanghai, China). Bending

Polymers **2017**, *9*, 447

tests and impact tests were carried out in a universal testing machine (CMT-6303, Shenzhen SANS Testing Machine Co., Ltd., Shenzhen, China) following the Chinese standard GB/T2567-2008, and at least five tests were carried out for each sample. The samples' dimensions for bending tests and impact tests were $(80 \pm 0.2) \times (15 \pm 0.2) \times (4 \pm 0.2)$ mm^3 and $(80 \pm 0.2) \times (10 \pm 0.2) \times (4 \pm 0.2)$ mm^3, respectively. The fractured surface of epoxy composites obtained from the impact test was observed using a scanning electronic microscope (SEM, Hitachi S-570, Hitachi High-Tech Science Co., Tokyo, Japan) and the samples were sputter-coated with gold. Differential scanning calorimetry (DSC) was conducted on a TA Instruments DSC 2920 (TA Instrument, New Castle, DE, USA) using N$_2$ as a purge gas from 30 to 300 °C with a heating rate of 10 °C/min. Thermal conductivities of the samples were measured at room temperature by a Hot Disk instrument (AB Co., Uppsala, Sweden).

3. Results and Discussion

3.1. Characterization of the IMD-Si@GO

The chemical structure of IMD-Si was characterized by ^{1}H NMR (Figure 1). The H1, H2, and H3 related to methoxyl group, are observed at 3.53–3.38 ppm. The signals of the protons linked to –CH$_2$–CH$_2$–CH$_2$– at 0.61–0.47, 1.91–1.77, and 4.25–4.08 ppm, are respectively corresponding to H4, H5, and H6. The protons marked by 7 and 8, 10 and 9, corresponding to protons in the imidazole ring observed at 7.72–7.50, 7.95–7.83, and 9.03–8.89 ppm, respectively. Additionally, a sharp peak can be seen at 3.29 ppm, which is most likely to correlate to moisture in IMD-Si. It is notable that the peak at 9.03–8.89 ppm shifts to lower field significantly, for H9 is an active hydrogen.

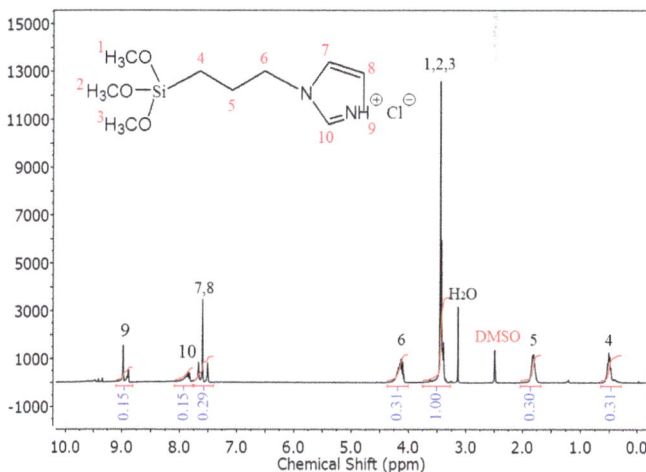

Figure 1. ^{1}H NMR spectra of IMD-Si.

AFM measurement was used to observe the morphology of GO and IMD-Si@GO, after deposition on mica surface. It can be seen from Figure 2 that single layer GO and IMD-Si@GO sheets of varying size are deposed on the substrates with overlaps. The single layer GO sheets are varied in the thickness range of 1.0–1.3 nm, which is in agreement with the values reported previously [23]. After functionalized by ionic liquid, the thicknesses of single layer sheets are obviously increased to nearly 2.8–3.8 nm, indicating the successful presence of the ionic liquid grafted onto GO sheets.

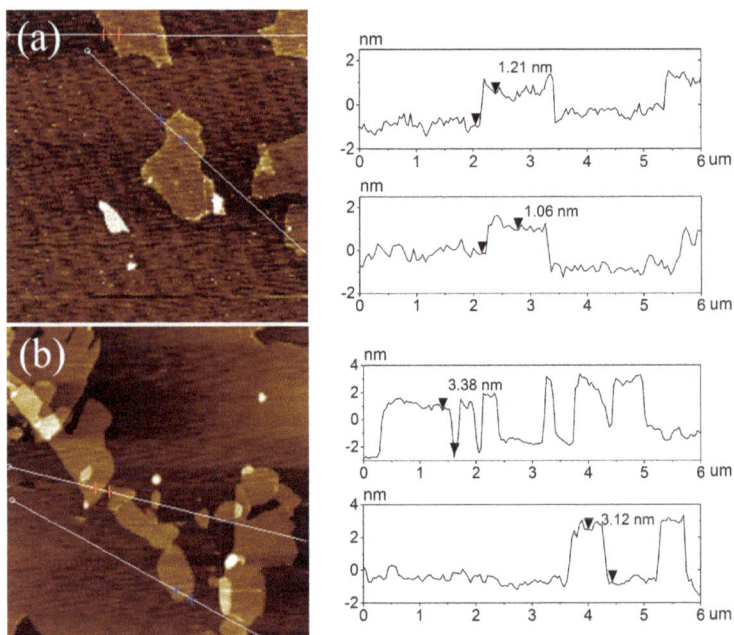

Figure 2. AFM images of GO (**a**) and IMD-Si@GO (**b**).

XPS spectra are very efficient to characterize the chemical composition and electronic structure of nanomaterials. Figure 3a shows the XPS survey spectra of GO and IMD-Si@GO. As for GO, there are two obvious peaks at 284.8 and 531.8 eV, corresponding to C 1s and O 1s, respectively. After functionalization, new peaks appear at 101.8, 152.8, 198.8, and 400.8 eV attributed to Si 2p, Si 2s, Cl 1s, and N 1s, respectively. The high resolution C 1s XPS spectrum of GO (Figure 3b) shows three types of carbon, located at 284.8 (C=C), 287.0 (C–O), and 288.4 eV (C=O) [24]. Compared with GO, the C 1s band of IMD-Si@GO (Figure 3c) can be fitted to five components: 284.3 (C–Si), 284.8 (C=C), 285.6 (C=N), 286.6 (C–O, C–N), and 288.6 eV (C=O). All these proofs can demonstrate the imidazolium ionic liquid grafted onto GO sheets successfully.

The FTIR spectra of GO and IMD-Si@GO are shown in Figure 4, which are consistent with XPS data. In the spectrum of GO, the absorption peaks at 3400, 1724, 1626, 1228, and 1055 cm^{-1} are attributed to –OH, –COOH, C=C, epoxy C–O, and alkoxy C–O, respectively [9]. As for IMD-Si@GO, a new peak appears at 3147 cm^{-1} due to the stretching vibrations of C–H in imidazole rings. The bands at 1564 and 1448 cm^{-1} are assigned to C–N and C=N, indicating the imidazole rings introduced onto the GO surface. In addition, a broad absorption band is observed in the range of 1100–1000 cm^{-1} related to Si–O–C and Si–O–Si, which also confirms the presence of imidazolium ionic liquid.

From the XPS and FTIR spectra, it can be concluded that the imidazolium ionic liquid has been successfully grafted onto GO surface. As we know, elemental analysis is an efficient method to calculate the grafting ratio of carbon materials [25]. The carbon, hydrogen, and nitrogen content of IMD-Si@GO are listed in Table 1. The grafting ratio of ionic liquid is defined as the weight percent of IMD-Si to IMD-Si@GO, which is calculated as the nitrogen weight content of IMD-Si@GO divided by that of IMD-Si (calc. 12.6%, not including three methyl groups). The grafting ratio of IMD-Si is calculated to be about 22.8%, that is, the content of IMD-Si groups is 1.03 mmol/g in IMD-Si@GO.

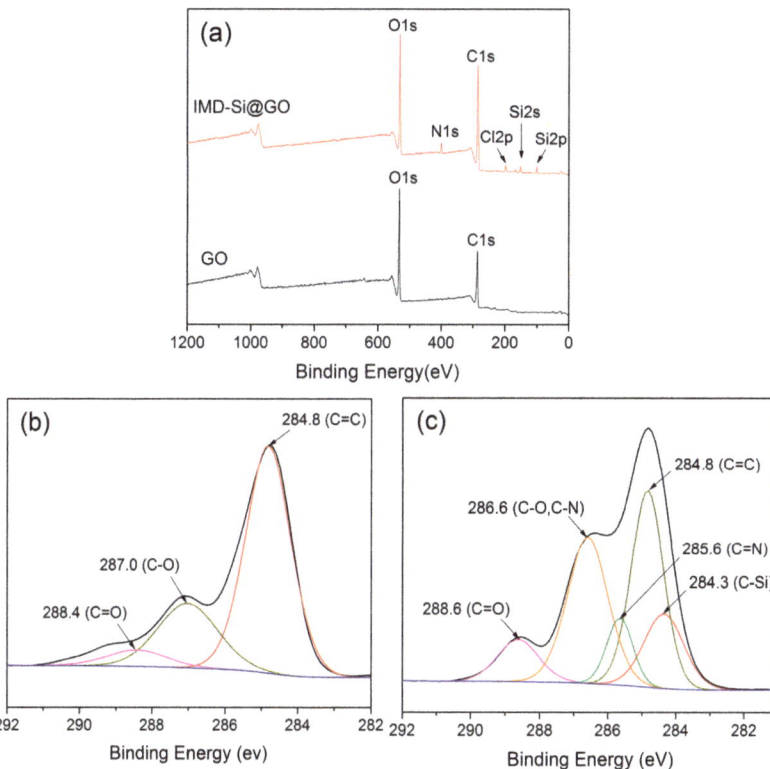

Figure 3. XPS spectra of GO and IMD-Si@GO: (**a**) survey spectra; (**b**) C 1s of GO; (**c**) C 1s of IMD-Si@GO.

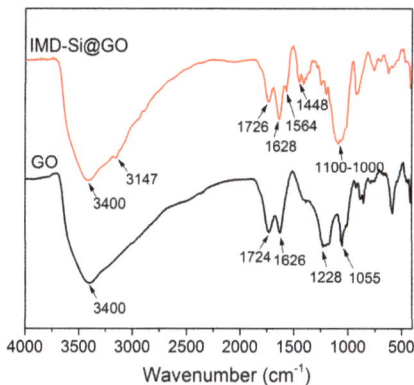

Figure 4. FTIR spectra of GO and IMD-Si@GO.

XRD patterns of GO and IMD-Si@GO are showed in Figure 5. The sharp peak at 9.68° for GO corresponds to an interlayer space of 0.91 nm. After functionalization, the diffraction peak shifts to 8.10°, indicating the interlayer distance of 1.09 nm, which can confirm the intercalation of GO sheets by imidazolium ionic liquid. The intercalation of GO makes it easy to dispersed in polymer matrix and prevents the aggregation efficiently, providing a good potential to GO for the preparation of polymer composites [26,27].

Table 1. Elemental analysis for IMD-Si@GO.

Element name	C	H	N
IMD-Si@GO (wt %)	43.64	3.07	2.87

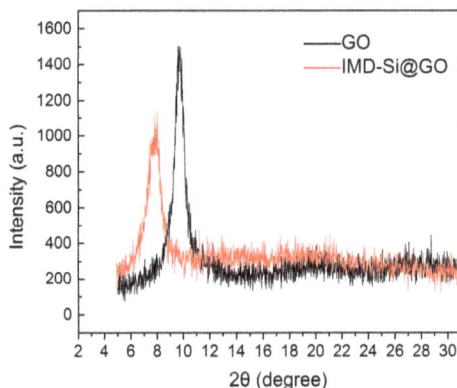

Figure 5. XRD patterns of GO and IMD-Si@GO.

We can see the TGA curves of IMD-Si, GO and IMD-Si@GO from Figure 6. GO has a slight mass loss below 100 °C because of the water evaporation. The main weight reduction for GO at the range of 100–310 °C is 47.7%, attributed to the pyrolysis of the oxygenated groups [28]. As for IMD-Si@GO, the mass loss at 100–310 °C (21.9%) is much lower than that for GO, indicating an enhancement in thermal stability. The weight reduction of IMD-Si@GO is mainly attributed to the pyrolysis of residual oxygenated groups and the decomposition of surface-attached ionic liquid [29]. Additionally the residual weight of IMD-Si, GO and IMD-Si@GO at 800 °C is 28.2%, 38.5%, and 47.1%, respectively. It is worth mentioning that the residual weight of IMD-Si@GO is bigger than that of ionic liquid or GO sheets, also demonstrating the improvement of thermal stability by the grafted ionic liquid. The reason may be that the imidazolium ionic liquid could generate a large amount of Si–O–Si bonds at high temperature, which could greatly hinder the thermal decomposition. In addition, the introduction of imidazole rings onto the GO surface can also slow down the rate of thermal decomposition.

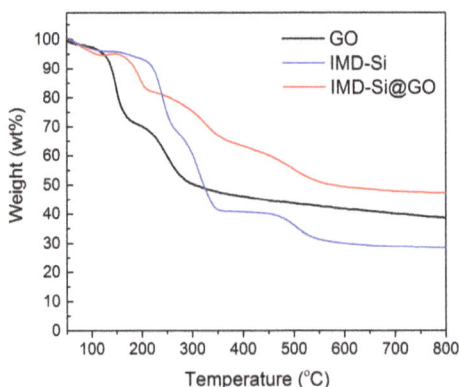

Figure 6. TGA curves of IMD-Si, GO and IMD-Si@GO.

3.2. Dispersion and Exfoliation of IMD-Si@GO

To test the dispersibility of GO and IMD-Si@GO in organic solvents, both of them were dispersed in acetone and ethanol with a concentration of 0.5 mg/mL by sonication for 10 min. Figure 7 shows the photos of GO and IMD-Si@GO dispersions in acetone and ethanol after storage for 2 h. By way of comparison, the unmodified graphene oxide sheets have a complete settlement in acetone and a significant settlement in ethanol, while the IMD-Si@GO dispersions are completely stable in both solvents. This verifies that covalent grafting of imidazolium ionic liquid can significantly facilitate the dispersibility of graphene oxide, due to the intercalation of GO sheets by ionic liquid and the electrostatic repulsion provided by IMD-Si. The good dispersibility of modified GO sheets in organic solvents makes it easy to fabricate nanomaterials/epoxy composites by solution blending technique, and also provides the potential for various applications.

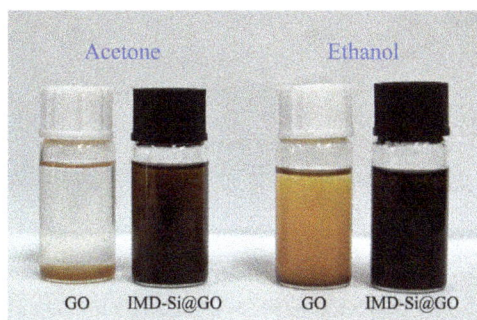

Figure 7. Photos of GO and IMD-Si@GO dispersions in acetone and ethanol.

The exfoliation degree of the IMD-Si@GO sheets in epoxy matrix was studied by XRD. Figure 8 shows XRD patterns of neat epoxy and the epoxy composites containing 0.4 and 0.8 wt % of fillers. Obviously, there is no apparent difference in terms of the broad diffraction peaks for all these nanocomposites from 11° to 28°. The wide diffraction originates from the scattering of cured epoxy resin. Furthermore, the characteristic diffraction peak of IMD-Si@GO at 8.1° does not appear in both XRD curves of the IMD-Si@GO/EP composites, indicating that IMD-Si@GO nanosheets are well exfoliated in epoxy matrix [12]. It is worth mentioning that good exfoliation degree of nano-fillers does not mean uniform dispersion in matrix, which needs to be further confirmed by TOM photos of cured epoxy resin and SEM photos of the fracture surfaces.

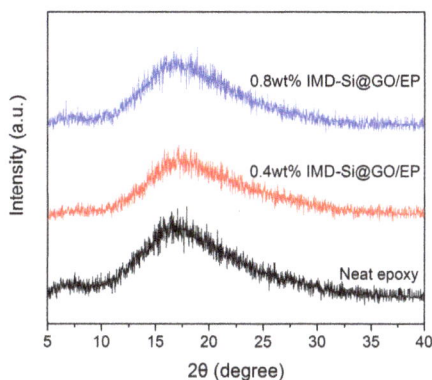

Figure 8. XRD patterns of epoxy composites with different contents of IMD-Si@GO fillers.

In order to evaluate the dispersibility of IMD-Si@GO sheets in epoxy matrix, TOM images of cured epoxy composites with 0.4 and 0.8 wt % fillers were carried out first. As shown in Figure 9a, the IMD-Si@GO sheets present a good dispersion in the matrix at 0.4 wt % filler content, although some small aggregations can been observed. When the filler content is increased to 0.8 wt %, some big clusters of IMD-Si@GO sheets can be easily found. These obvious aggregations may produce stress concentration regions in epoxy matrix and lead to decreased mechanical properties.

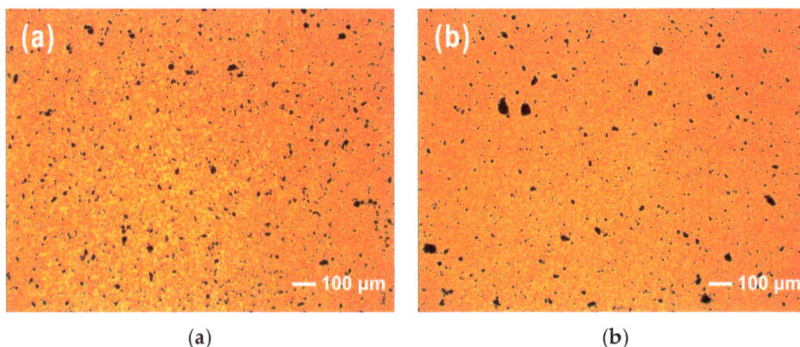

(a)   (b)

Figure 9. TOM images of epoxy composites with different contents of fillers: (**a**) 0.4 wt %; and (**b**) 0.8 wt %.

3.3. Mechanical Properties of IMD-Si@GO/EP

The mechanical properties of epoxy resin with different contents of IMD-Si@GO fillers were experimentally studied. The flexural strength, flexural modulus, and impact strength of IMD-Si@GO/EP composites are shown in Figure 10. It is interesting that the flexural and impact strength of epoxy composites reach the maximum values of 141.3 MPa and 18.7 kJ/m^2 at 0.4 wt % IMD-Si@GO, increased by 12% and 52% compared with the neat epoxy (126.5 MPa and 12.3 kJ/m^2), respectively. Additionally it shows a maximum increase of 26% in flexural modulus (4.69 GPa) for epoxy composites containing 0.4 wt % IMD-Si@GO. The enhanced strength and modulus of epoxy by adding a small amount of IMD-Si@GO is considered to be due to the strong interfacial bonding between fillers and epoxy matrix. The imidazole rings and residual oxygenated groups on the IMD-Si@GO surface can participate in the curing reaction. Consequently, the IMD-Si@GO is not only a nano-filler, but also a co-curing agent in epoxy matrix, which can strongly improve the interfacial interactions [30]. Meanwhile, in the presence of IMD-Si, GO sheets can be well exfoliated and uniformly dispersed in the matrix, promoting the reinforcing effect, obviously [31]. By increasing the filler content continuously, the flexural strength, flexural modulus and impact strength starts to decrease. This may impute to the following two points: (i) too high a filler loading induces the aggregation of IMD-Si@GO sheets in the matrix, which can be seen in Figure 9b, leading to weakening of the mechanical properties of epoxy composites; and (ii) too much IMD-Si@GO hinders the curing reaction and may decrease the crosslinking density [32,33].

To further study the reinforcing mechanism of IMD-Si@GO in epoxy matrix, the fracture surfaces of epoxy and its composites were investigated by SEM. In Figure 11a, the neat epoxy specimen shows a river-like fracture surface, indicating the characteristics of brittle fracture without stress dispersion [34]. Figure 11b shows a rough and uneven fracture surface of 0.4 wt % IMD-Si@GO/EP, exhibiting a typical tough feature. This proves a significant enhancement of the interfacial adhesion between the epoxy matrix and fillers [35]. In addition, almost no aggregation of IMD-Si@GO sheets can be seen on the surface, demonstrating a good dispersibility of IMD-Si@GO in the epoxy matrix. When the filler content is increased to 0.8 wt %, some aggregations can be found on the fracture surface from

Figure 11c, which is in good agreement with the observation of TOM images and analysis of the mechanical properties.

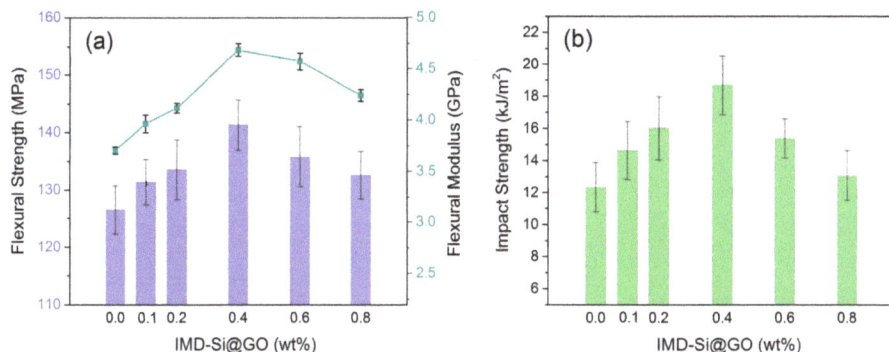

Figure 10. Flexural strength, flexural modulus (**a**) and impact strength (**b**) of epoxy composites with different contents of IMD-Si@GO fillers.

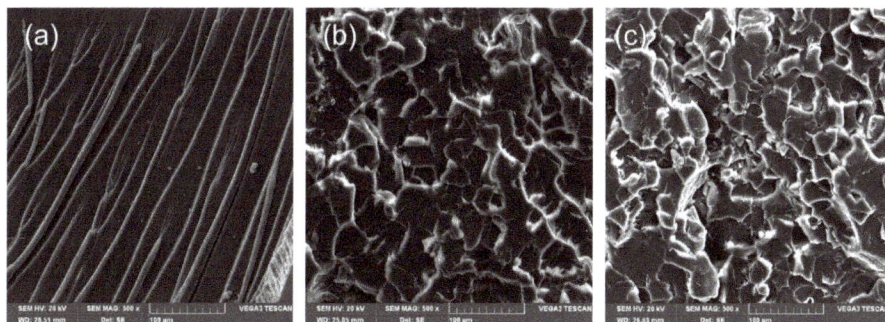

Figure 11. SEM images of the fracture surfaces of epoxy composites with different contents of IMD-Si@GO fillers: (**a**) neat epoxy; (**b**) 0.4 wt %; (**c**) 0.8 wt %.

3.4. Catalytic Effect of IMD-Si@GO in Epoxy Resin-Anhydride System

Imidazole and its derivatives are widely used as a type of accelerators in epoxy resin-anhydride systems, which can initiate alternating copolymerization between anhydride and epoxy groups, as a result, initiate the esterification reaction and improve the curing speed dramatically. Thence, the catalytic effect of IMD-Si@GO in epoxy resin-anhydride system was investigated by DSC (Figure 12). For the neat DGEBA-MTHPA curing system, a broad exothermic peak with a peak temperature at 289 °C has been seen for a heating rate of 10 °C/min. As for DGEBA-MTHPA with 0.4 wt % IMD-Si@GO, the exothermic peak moves to around 268 °C and becomes sharper. By increasing the filler content to 0.8 wt %, the broad peak continues to shift to the left with a peak temperature at 259 °C. This indicates that imidazole modified GO can indeed play the role in catalyzing epoxy resin-anhydride curing system, and the increase in the content of IMD-Si@GO results in a higher reaction rate [36,37]. Additionally it is obvious that a small peak appears at around 191 °C by adding IMD-Si@GO nanosheets, without movement when the filler content rises. This peak may originate from the reaction between the pyridine nitrogen and epoxy groups.

The main catalytic mechanism of IMD-Si@GO in epoxy resin-anhydride system is anionic ring-opening alternating copolymerization, which is shown in Scheme 2. The imidazole rings on the IMD-Si@GO surface attack anhydride groups to generate carboxylate anions. Then the carboxylate anions initiate the ring-opening reaction to yield oxygen anions. Oxygen anions on the IMD-Si@GO surface subsequently transform to carboxylate anions by reacting with other anhydride groups. Thence, alternating copolymerization continues to take place between the anhydride and epoxy groups to form cross-linked polymer networks [38]. It can be concluded that the IMD-Si@GO can initiate esterification reaction, that is to say, the functionalized graphene oxide can be a catalyst in epoxy resin-anhydride systems.

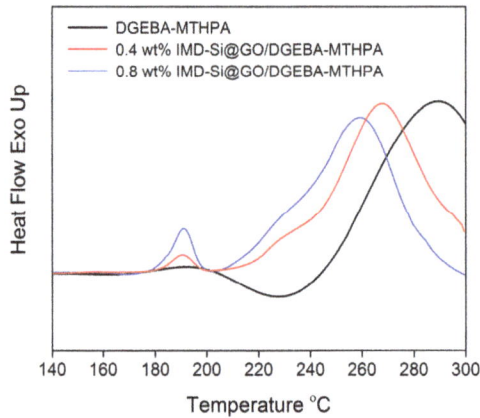

Figure 12. DSC curing curves of epoxy resin-anhydride with different contents of IMD-Si@GO with a heating rate of 10 °C/min.

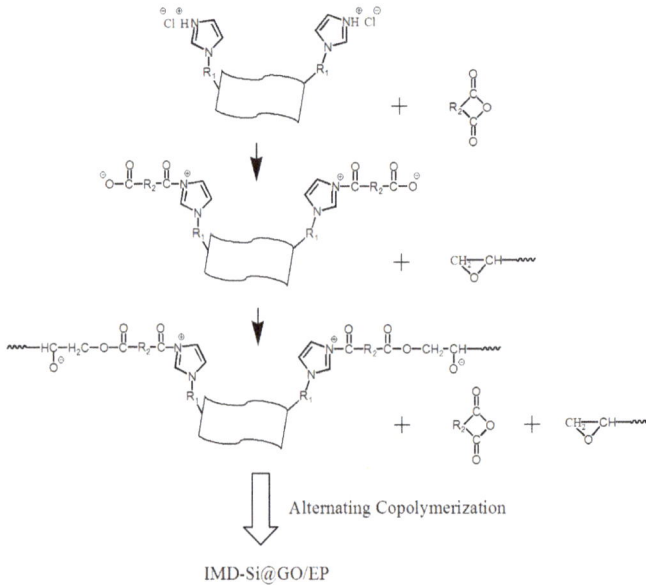

Scheme 2. Catalytic mechanism of IMD-Si@GO/epoxy resin-anhydride system.

3.5. Thermal Conductivities of IMD-Si@GO/EP

Thermal conductivity is also effectively affected by carbon fillers within the epoxy matrices [39]. Figure 13 shows the thermal conductivities of the IMD-Si@GO/EP composites prepared with 0.1, 0.2,

0.4, 0.6 and 0.8 wt % of fillers. We can see that the thermal conductivities of IMD-Si@GO/EP composites are increased with increased contents of fillers. The thermal conductivity is greatly improved to 0.294 W/mK with only a little amount of IMD-Si@GO (0.8 wt %), getting a 15.7% increase compared to that of unfilled epoxy (0.244 W/mK) [40]. From the enhancement of thermal conductivities, we can deduce at least following two points: (i) there are strong interfacial interactions between epoxy and IMD-Si@GO sheets, which could reduce the thermal interfacial resistance effectively; and (ii) the uniform dispersion of IMD-Si@GO in epoxy matrix results in an increased contact area between IMD-Si@GO and matrix and promotes the phonon diffusion efficiency [41].

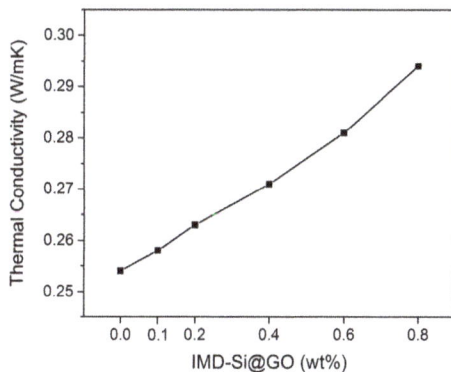

Figure 13. Thermal conductivities of epoxy composites with different contents of IMD-Si@GO fillers.

3.6. Thermal Stabilities of IMD-Si@GO/EP

Thermal stabilities of the neat epoxy and IMD-Si@GO/EP composites loading 0.8 wt % of nano-fillers were investigated by TGA analysis. TGA and derivative thermogravimetry (DTG) curves are shown in Figure 14. The neat epoxy has T_d (the temperatures at 5 wt % weight loss) and T_{max} (the temperatures at maximum weight loss) of 225.2 and 399.0 °C, respectively. By adding 0.8 wt % IMD-Si@GO, the T_d and T_{max} (267.3 and 408.1 °C) are both higher than these of neat epoxy, suggesting that IMD-Si@GO can efficiently inhibit the mass loss during thermal degradation process. As for the char yield at 800 °C, the char residue of 0.8 wt % IMD-Si@GO/EP dramatically increases to 6.41 wt %, 2.2 times compared to that of unfilled epoxy (2.91 wt %). The enhancement of thermal stabilities can be explained by the barrier effect of modified GO sheets, uniform dispersion of IMD-Si@GO in matrix, and also strong interfacial bonding between IMD-Si@GO sheets and epoxy matrix [42].

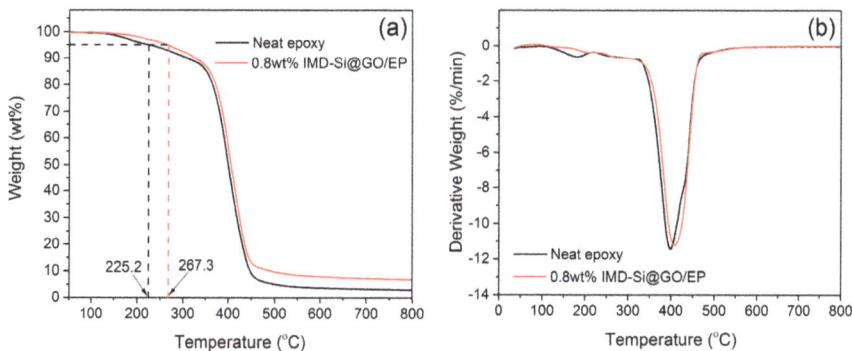

Figure 14. TGA (**a**) and DTG (**b**) curves of the neat epoxy and 0.8 wt % IMD-Si@GO/EP composites.

4. Conclusions

In this study, imidazolium ionic liquid modified graphene oxide was prepared by a cheap and simple method. The results from AFM, XPS, FTIR, XRD, and TGA indicate the successful preparation of IMD-Si@GO, and the grafting ratio is calculated to be 22.7% from the elemental analysis. The interlayer spacing of GO sheets is increased by intercalation of IMD-Si, which strongly facilitates the dispersibility of graphene oxide in organic solvents and epoxy matrix. The flexural strength, flexural modulus and impact strength of the IMD-Si@GO/EP composites are optimal with 0.4 wt % fillers, increased by 12%, 26%, and 52%, respectively, compared to the neat epoxy. Additionally, DSC reveals that the IMD-Si@GO sheets can catalyze the curing reaction of epoxy resin-anhydride system at very low IMD-Si@GO loadings. Moreover, a little inclusion of IMD-Si@GO into epoxy matrix can remarkably improve the thermal conductivities and thermal stabilities. These enhancements can be attributed to the uniform dispersion of IMD-Si@GO and strong interfacial interactions with the epoxy matrix. Consequently, the modification of GO sheets by covalent grafting of ionic liquid could be a promising method in the application of reinforced polymers by nanomaterials.

Acknowledgments: This work is financially sponsored by the Doctoral Program of Higher Education of China (20136102110049).

Author Contributions: Qing Lyu conceived and designed the experiments; Qing Lyu, Lin Li, Zhengyan Chen, and Huanhuan Yao performed the experiments; Hongxia Yan and Yufeng Nie supervised the project; Qing Lyu and Lin Li analyzed the data; and Qing Lyu wrote the paper. All the authors contributed to the realization of the manuscript.

Conflicts of Interest: The authors declare no conflict of interest.

References

1. Gao, J.; Li, J.; Benicewicz, B.C.; Zhao, S.; Hillborg, H.; Schadler, L.S. The mechanical properties of epoxy composites filled with rubbery copolymer grafted SiO_2. *Polymers* **2012**, *4*, 187–210. [CrossRef]
2. Hu, Y.; Du, G.; Chen, N. A novel approach for Al_2O_3/epoxy composites with high strength and thermal conductivity. *Compos. Sci. Technol.* **2016**, *124*, 36–43. [CrossRef]
3. Pour, Z.S.; Ghaemy, M. Fabrication and characterization of superparamagnetic nanocomposites based on epoxy resin and surface-modified γ-Fe_2O_3, by epoxide functionalization. *J. Mater. Sci.* **2014**, *49*, 4191–4201.
4. Cha, J.; Jin, S.; Shim, J.H.; Chong, S.P.; Ryu, H.J.; Hong, S.H. Functionalization of carbon nanotubes for fabrication of CNT/epoxy nanocomposites. *Mater. Des.* **2016**, *95*, 1–8.
5. Hu, K.; Kulkarni, D.D.; Choi, I.; Tsukruk, V.V. Graphene-polymer nanocomposites for structural and functional applications. *Prog. Polym. Sci.* **2014**, *39*, 1934–1972. [CrossRef]
6. Atif, R.; Shyha, I.; Inam, F. Mechanical, thermal, and electrical properties of graphene-epoxy nanocomposites-a review. *Polymers* **2016**, *8*, 281. [CrossRef]
7. Zhu, Y.; James, D.K.; Tour, J.M. ChemInform abstract: new routes to graphene, graphene oxide and their related applications. *Adv. Mater.* **2012**, *24*, 4924–4955. [CrossRef] [PubMed]
8. Pourhashem, S.; Vaezi, M.R.; Rashidi, A.; Bagherzadeh, M.R. Distinctive roles of silane coupling agents on the corrosion inhibition performance of graphene oxide in epoxy coatings. *Prog. Org. Coat.* **2017**, *111*, 47–56. [CrossRef]
9. Wan, Y.; Gong, L.; Tang, L.; Wu, L.; Jiang, J. Mechanical properties of epoxy composites filled with silane-functionalized graphene oxide. *Compos. Part A* **2014**, *64*, 79–89. [CrossRef]
10. Sharif, F.; Gudarzi, M.M. Enhancement of dispersion and bonding of graphene-polymer through wet transfer of functionalized graphene oxide. *Express. Polym. Lett.* **2012**, *6*, 1017–1031.
11. Jiang, T.; Kuila, T.; Kim, N.H.; Lee, J.H. Effects of surface-modified silica nanoparticles attached graphene oxide using isocyanate-terminated flexible polymer chains on the mechanical properties of epoxy composites. *J. Mater. Chem. A* **2014**, *2*, 10557–10567. [CrossRef]
12. Wan, Y.; Tang, L.; Gong, L.; Yan, D.; Li, Y.; Wu, L.; Jiang, J.; Lai, G. Grafting of epoxy chains onto graphene oxide for epoxy composites with improved mechanical and thermal properties. *Carbon* **2014**, *69*, 467–480. [CrossRef]

13. Ryu, S.H.; Sin, J.H.; Shanmugharaj, A.M. Study on the effect of hexamethylene diamine functionalized graphene oxide on the curing kinetics of epoxy nanocomposites. *Eur. Polym. J.* **2014**, *52*, 88–97. [CrossRef]

14. Dupont, J.; Souza, R.F.; Suarez, P.A. Ionic liquid (molten salt) phase organometallic catalysis. *Chem. Rev.* **2002**, *102*, 3667–3692. [CrossRef] [PubMed]

15. Liu, N.; Luo, F.; Wu, H.; Liu, Y.; Zhang, C.; Chen, J. One-step ionic-liquid-assisted electrochemical synthesis of ionic-liquid-functionalized graphene sheets directly from graphite. *Adv. Funct. Mater.* **2008**, *18*, 1518–1525. [CrossRef]

16. Yang, H.; Shan, C.; Li, F.; Han, D.; Zhang, Q.; Niu, L. Covalent functionalization of polydisperse chemically-converted graphene sheets with amine-terminated ionic liquid. *Chem. Commun.* **2009**, *45*, 3880–3882. [CrossRef] [PubMed]

17. Ye, Y.; Wang, H.; Bi, S.; Xue, Y.; Xue, Z.; Zhou, X.; Xie, X.; Mai, Y. High performance composite polymer electrolyte using polymeric ionic liquid-functionalized graphene molecular brushes. *J. Mater. Chem. A* **2015**, *3*, 18064–18073. [CrossRef]

18. Wu, W.; Wang, J.; Liu, J.; Chen, P.; Zhang, H.; Huang, J. Intercalating ionic liquid in graphene oxide to create efficient and stable anhydrous proton transfer highways for polymer electrolyte membrane. *Int. J. Hydrog. Energy* **2017**, *42*, 11400–11410. [CrossRef]

19. Bouillon, N.; Pascault, J.P.; Lan, T. Influence of different imidazole catalysts on epoxy-anhydride copolymerization and on their network properties. *J. Appl. Polym. Sci.* **2010**, *38*, 2103–2113. [CrossRef]

20. Pour, Z.S.; Ghaemy, M. Polymer grafted graphene oxide: For improved dispersion in epoxy resin and enhancement of mechanical properties of nanocomposite. *Compos. Sci. Technol.* **2016**, *136*, 145–157. [CrossRef]

21. Zhu, J.; Chen, M.; Qu, H.; Zhang, X.; Wei, H.; Luo, Z.; Colorado, H.A.; Wei, S.; Guo, Z. Interfacial polymerized polyaniline/graphite oxide nanocomposites toward electrochemical energy storage. *Polymers* **2012**, *53*, 5953–5964. [CrossRef]

22. Kooti, M.; Afshari, M. Phosphotungstic acid supported on magnetic nanoparticles as an efficient reusable catalyst for epoxidation of alkenes. *Mater. Res. Bull.* **2012**, *47*, 3473–3478. [CrossRef]

23. Xu, J.; Xu, M.; Wu, J.; Wu, H.; Zhang, W.; Li, Y. Graphene oxide immobilized with ionic liquids: Facile preparation and efficient catalysis for solvent-free cycloaddition of CO_2 to propylene carbonate. *RSC Adv.* **2015**, *5*, 72361–72368. [CrossRef]

24. Paredes, J.I.; Villar-Rodil, S.; Martínez-Alonso, A.; Tascón, J.M.D. Graphene oxide dispersions in organic solvents. *Langmuir* **2008**, *24*, 10560–10564. [CrossRef] [PubMed]

25. Deng, Y.; Li, Y.; Dai, J.; Lang, M.; Huang, X. An efficient way to functionalize graphene sheets with presynthesized polymer via ATNRC chemistry. *J. Polym. Sci. Polym. Chem.* **2015**, *49*, 1582–1590. [CrossRef]

26. Acik, M.; Dreyer, D.R.; Bielawski, C.W.; Chabal, Y.J. Impact of ionic liquids on the exfoliation of graphite Oxide. *J. Phys. Chem. C* **2012**, *116*, 7867–7873. [CrossRef]

27. Tasviri, M.; Ghasemi, S.; Ghourchian, H.; Gholami, M.R. Ionic liquid/graphene oxide as a nanocomposite for improving the direct electrochemistry and electrocatalytic activity of glucose oxidase. *J. Solid State Electrochem.* **2013**, *17*, 183–189. [CrossRef]

28. Layek, R.K.; Das, A.K.; Min, J.P.; Kim, N.H.; Lee, J.H. Enhancement of physical, mechanical, and gas barrier properties in noncovalently functionalized graphene oxide/poly(vinylidene fluoride) composites. *Carbon* **2015**, *81*, 329–338. [CrossRef]

29. Afshari, M.; Gorjizadeh, M.; Nazari, S.; Naseh, M. Cobalt salophen complex supported on imidazole functionalized magnetic nanoparticles as a recoverable catalyst for oxidation of alkenes. *J. Magn. Magn. Mater.* **2014**, *363*, 13–17. [CrossRef]

30. Ahmadi-Moghadam, B.; Sharafimasooleh, M.; Shadlou, S.; Taheri, F. Effect of functionalization of graphene nanoplatelets on the mechanical response of graphene/epoxy composites. *Mater. Des.* **2015**, *66*, 142–149. [CrossRef]

31. Kleinschmidt, A.C.; Donato, R.K.; Perchacz, M.; Benes, H.; Stengl, V.; Amico, S.C.; Schrekker, H.S. "Unrolling" multi-walled carbon nanotubes with ionic liquids: Application as fillers in epoxy-based nanocomposites. *RSC Adv.* **2014**, *4*, 43436–43443. [CrossRef]

32. Marra, F.; D'Aloia, A.G.; Tamburrano, A.; Ochando, I.M.; Bellis, G.D.; Ellis, G.; Sarto, M.S. Electromagnetic and dynamic mechanical properties of epoxy and vinylester-based composites filled with graphene nanoplatelets. *Polymers* **2016**, *8*, 272. [CrossRef]

33. Chen, Z.; Yan, H.; Liu, T.; Niu, S. Nanosheets of MoS_2 and reduced graphene oxide as hybrid fillers improved the mechanical and tribological properties of bismaleimide composites. *Compos. Sci. Technol.* **2016**, *125*, 47–54. [CrossRef]

34. Umboh, M.K.; Adachi, T.; Nemoto, T.; Higuchi, M.; Major, Z. Non-stoichiometric curing effect on fracture toughness of nanosilica particulate-reinforced epoxy composites. *J. Mater. Sci.* **2014**, *49*, 7454–7461. [CrossRef]

35. Liu, C.; Yan, H.; Lv, Q.; Li, S.; Niu, S. Enhanced tribological properties of aligned reduced graphene oxide-Fe_3O_4@polyphosphazene/bismaleimides composites. *Carbon* **2016**, *102*, 145–153. [CrossRef]

36. Monteserín, C.; Blanco, M.; Aranzabe, E.; Aranzabe, A.; Vilas, J.L. Effects of graphene oxide and chemically reduced graphene oxide on the curing kinetics of epoxy amine composites. *J. Appl. Polym. Sci.* **2017**, *134*, 44803–44815. [CrossRef]

37. Park, S.; Kim, D.S. Curing behavior and physical properties of an epoxy nanocomposite with amine-functionalized graphene nanoplatelets. *Compos. Interfaces* **2016**, *23*, 1–13. [CrossRef]

38. Park, W.H.; Lee, J.K.; Kwon, K.J. Cure behavior of an epoxy-anhydride-imidazole system. *Polym. J.* **1996**, *28*, 407–411. [CrossRef]

39. Gu, J.; Yang, X.; Lv, Z.; Li, N.; Liang, C.; Zhang, Q. Functionalized graphite nanoplatelets/epoxy resin nanocomposites with high thermal conductivity. *Int. J. Heat. Mass Transf.* **2016**, *92*, 15–22. [CrossRef]

40. Wang, R.; Zhuo, D.; Weng, Z.; Wu, L.; Cheng, X.; Zhou, Y.; Wang, J.; Xuan, B. A novel nanosilica/graphene oxide hybrid and its flame retarding epoxy resin with simultaneously improved mechanical, thermal conductivity, and dielectric properties. *J. Mater. Chem. A* **2015**, *3*, 9826–9836. [CrossRef]

41. Teng, C.; Ma, C.M.; Lu, C.; Yang, S.; Lee, S.; Hsiao, M.; Yen, M.; Chiou, K.; Lee, T. Thermal conductivity and structure of non-covalent functionalized graphene/epoxy composites. *Carbon* **2011**, *49*, 5107–5116. [CrossRef]

42. Wan, Y.; Yang, W.; Yu, S.; Sun, R.; Wong, C.; Liao, W. Covalent polymer functionalization of graphene for improved dielectric properties and thermal stability of epoxy composites. *Compos. Sci. Technol.* **2016**, *122*, 27–35. [CrossRef]

polymers

MDPI

Article

Preparation of Electrospun Nanocomposite Nanofibers of Polyaniline/Poly(methyl methacrylate) with Amino-Functionalized Graphene

Hanan Abdali [1,2] and Abdellah Ajji [1,*]

[1] CREPEC, Department of Chemical Engineering, Polytechnique Montréal,
P.O. Box 6079, Station Centre-Ville, Montreal, QC H3C 3A7, Canada; hanan.abdali@polymtl.ca
[2] Ministry of Higher Education, P.O. Box 225085, Riyadh 11153, Saudi Arabia
* Correspondence: abdellah.ajji@polymtl.ca; Tel.: +1-514-340-4711 (ext. 3703)

Received: 31 August 2017; Accepted: 12 September 2017; Published: 16 September 2017

Abstract: In this paper we report upon the preparation and characterization of electrospun nanofibers of doped polyaniline (PANI)/poly(methyl methacrylate) (PMMA)/amino-functionalized graphene (Am-rGO) by electrospinning technique. The successful functionalization of rGO with amino groups is examined by Fourier transforms infrared (FTIR), X-ray photoelectron spectroscopy (XPS) and Raman microspectrometer. The strong electric field enables the liquid jet to be ejected faster and also contributes to the improved thermal and morphological homogeneity of PANI/PMMA/Am-rGO. This results in a decrease in the average diameter of the produced fibers and shows that these fibers can find promising uses in many applications such as sensors, flexible electronics, etc.

Keywords: electrospun nanofibers; electrospinning; polyaniline; nanocomposites; amino-functionalized graphene

1. Introduction

Electrospinning is an efficient, relatively simple and low-cost process used to produce continuous nanofibers on a large scale, where the fiber diameter can be adjusted from nanometers to microns by applying a high voltage to a polymer solution from a micro-syringe pump [1–5]. Polymer nanofibers produced via electrospinning have specific surface areas approximately one to two orders of magnitude larger than flat films, making them the most promising candidates for applications in filtrations, engineering tissue scaffolds, wound healing, release control, energy storage and sensors [5–8].

Polyaniline (PANI) is one of the most conductive polymers, that has been used in many electronic, optical and electrochemical applications, due to its low cost, good environmental stability, redox reversibility, and electrical conductivity [9,10]. However, processing PANI into nanofibers by using the electrospinning is a challenge, mainly due to its rigid backbone that is related to its high degree of aromaticity making the elastic properties of the solution insufficient for electrospinning [11,12]. In this regard, non-conductive hosting polymers such as poly(methyl methacrylate) (PMMA), is blended to assist polyaniline to form composite fibers [13,14]. Consequently, the nanofibers of PANI have garnered much interest because of their properties as candidates for chemical sensors [15], light-emitting and electronic devices [16]. Yet, some disadvantages such as poor mechanical properties do exist, although combining PANI with carbon materials reinforces its stability and enhances some of its properties, such as capacitance [17,18].

Graphene is a potential nanofiller that can efficiently enhance the mechanical, thermal and electrical properties of polymer-based nanocomposites at a very low loading, useful for various novel applications due to its high thermal conductivity, superior mechanical strength, high specific surface area, excellent mobility of charge carriers and high chemical stability [19–21].

However, the homogenous dispersion of graphene in a polymer matrix is a necessary feature when it is used as a nanofiller. Graphene is predisposed to agglomerate because of its hydrophobic nature, high surface energy, and intrinsic van der Waals forces preventing its uniform distribution in the polymer matrix [21–23]. This reduces its beneficial effects and therefore dispersing the graphene in an electrospinning solution is an important step in forming the nanofibers [21]. The problems can be overcome by functionalizing the graphene. This procedure provides multiple bonding sites to the resin matrix where the remarkable properties of graphene can be successfully transferred to a polymer composite [23–25]. Amine groups are attributed with high reactivity enabling them to react with other chemical groups easily and providing a favorable approach in the preparation and applications of graphene/polymer nanocomposites [23]. As the Nitrogen atom in amine is more nucleophilic than the oxygen atom, it can be expected that substituting graphene or graphene oxide with amine will increase the nucleophilic properties of graphene. Consequently, interfacial binding can result between graphene and the materials of interest. These interactions will improve the performance and functionality of the intended applications of graphene [23,24]. Former studies about the functionalization of graphene with amine groups have indicated that it could be a promising strategy to improve the electrical conductivity of graphene, due to the electron donating effect of amine groups [23].

Herein, the ethylenediamine (NH_2–$(CH_2)_2$–NH_2) was utilized to functionalize graphene surfaces, which produced stitched graphene owing to the presence of two amine (–NH_2) functionalities on both sides of the ethylene moiety [24]. Therefore, due to the intriguing properties of graphene and the advantages of PANI, composites of graphene and PANI fibers are eminently suitable for many applications such as organic photovoltaics, supercapacitors and resistance-based sensors. In this article, the preparation of electrospun fiber mats of doped polyaniline/poly(methyl methacrylate)/amine-functionalized graphene using the electrospinning process is studied. Literature has reported on graphene/polyaniline composites [26–28] but, in this study, for the first time, amino-functionalized graphene/polyaniline nanofibers are investigated using the electrospinning process.

More specifically, the objectives of this study were to identify and detail the primary materials and process factors necessary to produce amino-graphene/polyaniline nanofibers using the electrospinning process. The details of the amine functionalization of graphene are presented and the morphology and the thermal stability of the PANI/PMMA/Am-rGO nanofibers are investigated.

2. Materials and Methods

2.1. Materials

Graphene oxide (GO), poly(methyl methacrylate) (PMMA) Mw ~996,000 g·mol^{-1}, polyaniline (PANI, emeraldine base) Mw ~100,000 g·mol^{-1}, camphor-10-sulfonic acid (HCSA, 98%), N,N dimethylformamide (DMF, 99.8%), chloroform (CHCl$_3$, ≥99%), ethylenediamine (EDA, ≥99%), were all purchased from Sigma-Aldrich, (Oakville, ON, Canada). Deionized (DI) water was used for all the experiments.

2.2. Reduction of Graphene Oxide to Graphene

Commercial graphene oxide (GO) was reduced by thermal annealing treatment [29]. First, the GO powder was heated in a tubular quartz furnace (High Temperature Tube Furnace (HTF), GSL-1300-40X, MTI Corporation, CA, USA) from room temperature to 400 °C at the rate of 5 °C·min^{-1}, and kept at 400 °C for 30 min under an argon (Ar) gas flow of 80 mL·min^{-1}; secondly, the temperature was increased from 400 to 650 °C at a rate of 5 °C·min^{-1}, and kept at 650 °C for 30 min under an Ar gas flow of 80 mL·min^{-1}. Finally, the reduced GO samples were naturally cooled to room temperature without argon.

2.3. Surface Modification of Graphene with Amines

The rGO made in the previous step, where 150 mg of rGO was mixed with 10 mL of EDA, in a vessel under vigorous stirring. The reaction was continued for 24 h under reflux at 80 °C. Afterwards, the aminated-rGO (Am-rGO) was centrifuged at 10,000 rpm for 1 h and was thoroughly washed with deionized water, filtered, and dried in a vacuum oven at 80 °C for 24 h [23].

2.4. Preparation of the PANI/PMMA/Am-rGO Solution

10 mg of the Am-rGO was dispersed in 2.96 g of DMF by sonication for 1 h. 100 mg of PANI was mixed with 130 mg of HCSA to dope it and dissolving it in 14.78 g of chloroform. The solution was stirred constantly for 6 h and subsequently filtered using Whatman Puradisk PTFE syringe filter (pore size—0.45 µm, GE Healthcare, Buckinghamshire, UK) to remove the particulate matter. Then, the Am-rGO solution was mixed with PANI solution and subsequently 85 mg of PMMA (as a carrier polymer) was added to the solution and stirred for 24 h to form the solution for electrospinning (see Table 1). The PANI/PMMA solution was similarly prepared without the addition of functionalized graphene for comparative analysis.

Table 1. Composition of electrospun PANI/PMMA/Am-rGO and PANI/PMMA solutions.

PANI (mg)	HCSA (mg)	Am-rGO (mg)	CHCl₃ (g)	DMF (g)	PANI:PMMA (wt %)	Am-rGO:PANI (wt %)
100	130	10	14.78	2.96	54.05	9.09
100	130	-	14.78	2.96	54.05	-

2.5. Electrospinning Setup

Figure 1 shows the schematic diagram of the fabrication of graphene-polymer nanofiber composite by electrospinning. The homogeneous dispersed solutions were electrospun using a homemade horizontal set up containing a programmable micro-syringe pump (Harvard Apparatus, PHD 2000, Holliston, MA, USA) and a variable high DC voltage power supply (ES60P-5W Gamma High Voltage Research Inc, Omaha Beach, FL, USA). Parameters such as the flow rate, voltage and distance were harnessed at peak efficiency to obtain the desired nanofibers with the least beading to perform the subsequent experiments. The PANI/PMMA/Am-rGO solution was loaded into a 3 mL syringe with Luer-Lock connection to an 18-gauge blunt tip needle (Cadence Science, Cranston, RI, USA). The syringe was mounted on the pump with a grip and grounded by use of an alligator clip. The applied voltage was in the range of 18–20 kV between the needle tip and the collector. A syringe pump was utilized to control the flow rate of the solution which was 0.3 mL/h and the distance between the needle and the collector was 15 cm. The spun nanofiber mats were collected on an aluminum foil attached to a stationary collector plate. All experiments were conducted in a chamber at a relative humidity of 19–25%.

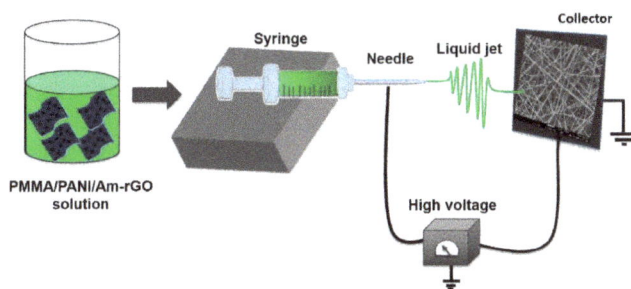

Figure 1. Schematic illustration of the electrospinning setup.

2.6. Characterization of Amino Functionalized Graphene

In order to ensure the presence of functional groups that should be present from the successful reduction of GO to rGO and functionalization of rGO with amino group, the Fourier transform infrared (FTIR) spectroscopy analysis was undertaken using Perkin Elmer 65 FTIR-ATR instrument (PerkinElmer, Woodbridge, ON, Canada). A total of 128 scans were accumulated for signal averaging of each IR spectral measurement with a 4 cm^{-1} resolution. The spectra of the samples were recorded over a wavenumber range of 4000–650 cm^{-1}. Raman microspectrometer and the electron diffraction (SAED) patterns were utilized to investigate the structural changes of GO, rGO, and Am-rGO. Characterization of graphene specimens were performed by Raman microspectrometer with a Renishaw InVia Raman microscope (Renishaw, Mississauga, ON, Canada) at an excitation laser wavelength of 514 nm. Raman spectroscopy is a powerful non-destructive tool for studying disorder and defects in crystal structure, it is often used to characterize microstructure of carbon materials. All specimens were deposited on glass slides in powder form without any solvent. For SAED patterns characterization, the graphene specimens were scooped onto a transmission electron microscopy (TEM) copper grid with supporting carbon film (CF400-Cu, Electron Microscopy Sciences) directly. The dispersion characteristics of rGO and Am-rGO were measured by ultraviolet visible (UV–Vis) spectrophotometer at ambient temperature utilizing Infinite 200 PRO (Tecan, Männedorf, Switzerland) cuvette reader. The chemical composition of the samples were determined by X-ray photoelectron spectroscopic (XPS) analysis using a VG Scientific ESCALAB 3 MK II X-ray photoelectron spectrometer (VG Escalab 3 Mk, East Grinstead, England) using an Mg Kα source (15 kV, 20 mA).

2.7. Characterization of the PMMA/PANI/Am-rGO Nanofibers

The scanning electron microscope (SEM, JSM-7600TFE, FEG-SEM, Calgary, AB, Canada) at an operational voltage of 2 kV was used to study the morphology of electrospun fibers. Fiber diameters were calculated using Image-Pro Plus®software by taking an average of about 300 fibers. To confirm the presence of graphene sheets in the nanofibers of PANI/PMMA/Am-rGO, transmission electron microscopy (TEM, JEOL, JEM 2100F, JEOL, Pleasanton, CA, USA) was used. For TEM observation, fibers were directly deposited onto a TEM copper grid with supporting carbon film (CF400-Cu, Electron Microscopy Sciences). Thermogravimetric analysis (TGA) was conducted under nitrogen atmosphere using Q5000 TGA (TA Instruments, New Castle, DE, USA) in the temperature range of 20–900 °C, with a heating ramp of 10 °C·min^{-1}.

3. Results and Discussion

3.1. Morphology and Structure Analysis of Am-rGO

The UV–Vis spectrum of rGO suspension showed an absorption peak at around 265 nm. This observation confirms the formation of C=C conjugated graphene structure after the thermal reduction process (see Figure 2b) [24]. The UV–Vis spectroscopy was used also for monitoring the stability of the rGO and Am-rGO in mixture of CHCl$_3$/DMF (5:1). As shown in Figure 2b,c, there is a very slight decrease in the absorbance spectra of Am-rGO over five days in comparison with rGO, indicating a good stability of the Am-rGO dispersion. Visually, dispersion of Am-rGO is more stable, whereas in comparison with dispersions of rGO in the same solution, which means that the dispersion of Am-rGO is greatly improved within CHCl$_3$/DMF in the presence of amino groups (in the inset of Figure 2c).

FTIR spectroscopy was performed on GO, rGO and Am-rGO in order to ensure the presence of functional groups that should be present from the successful reduction of GO to rGO and functionalization of rGO with amino groups (see Figure 2a). Different oxygen containing functional groups were observed on the GO spectrum bands as shown in Figure 3a. The C=O stretching vibrations in the carboxyl groups at 1700 cm^{-1}; the C–OH deformation from the hydroxyl groups attached to the aromatic graphene network at 1409 cm^{-1}; the C–O (hydroxyl) stretching at 1601 cm^{-1}; the C–O (epoxy) stretching at 1040 cm^{-1} and the C–O (phenolic) stretching at 1213 cm^{-1} [18,30].

The band at 1620 cm^{-1} is ascribed to the skeletal vibration of unoxidized sp^2 graphitic domains. After the thermal reduction of GO, the skeletal vibration of sp^2 graphitic domain still remains shifts to 1573 cm^{-1}. Besides, some residual presence of bands at 1710, 1150, and 1280 cm^{-1} were detected providing evidence of the different types of oxygen functionalities in the rGO and their decreases in intensity and others vanished due to thermal reduction [18,30]. In the spectrum of Am-rGO, the N–H deformation peaks at 1565 cm^{-1}; the C–N stretching vibrations at 1180 and 1120 cm^{-1}; and the C=O stretching vibrations of carboxyl group at 1725 cm^{-1}. Furthermore, the Am-rGO has peaks of the C–H stretch of alkyl chain at 2918 and 2854 cm^{-1}; and the C–O stretching in hydroxyl groups at 1015 cm^{-1}. The FTIR spectroscopy results indicate that the chemically functionalized graphene (Am-rGO) was successfully synthesized. Similar results for functionalization of graphene with amino groups have been previously reported in literature [23,30].

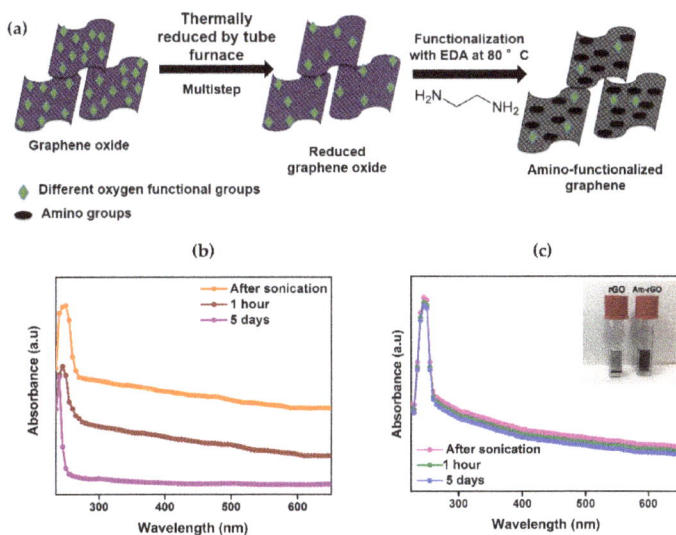

Figure 2. (**a**) Schematic illustration of the non-covalent functionalization of graphene surfaces with amino groups. Time evolution of UV-Vis absorption spectra of (**b**) rGO and (**c**) Am-rGO dispersed in CHCl$_3$/DMF (inset shows photograph of GO, rGO and Am-rGO).

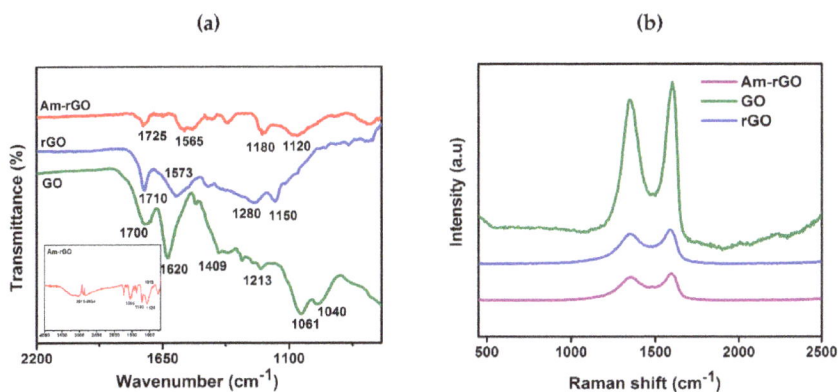

Figure 3. (**a**) FTIR and (**b**) Raman spectra of GO, rGO and Am-rGO.

Raman spectra was employed to analysis the graphitic structures of of GO, rGO, Am-rGO, as shown in Figure 3b. The G band is derived from stretching the C–C bond, and is usual for all sp^2 carbon forms system, and it is obtained from the first order Raman scattering [31,32], and the D band is due to disordered carbon atoms [31,32]. As expected, the GO revealed an intensive G band at 1600 cm^{-1} owing to the oxygenation of graphite, which results in the formation of sp^3 while, the D band is presented at 1350 cm^{-1} because of the reduction in size of the sp^2 domains by the creation of defects, and distortions during oxidation [32,33]. Moreover, in rGO and Am-rGO, the G band were shifted to lower wavenumber exhibited at 1587 and 1595 cm^{-1} and the D band positions remained unchanged at 1350 cm^{-1}, respectively. The G band appears in lower frequency due to the increased number of sp^2 carbon atoms following reduction of GO. The intensity ratio of D and G band I_D/I_G is slightly increased from (0.89) in GO to (0.93) in rGO, demonstrating a decrease in the size of the in-plane sp^2 domains after reduction, and can be explained that the thermal redaction creates many new graphitic domains, that are more numerous in number and smaller in size [32–34]. Whereas, the Am-rGO showed a higher I_D/I_G intensity ratio (0.97) than the rGO, which is attributed to the formation of the chemical bond between amino groups and basal planes of the rGO. This corresponds to other results reported for functional graphene [34–36].

The XPS was also applied as an effective tool to characterize the presence of different elements such as carbon, oxygen and nitrogen in GO, rGO and Am-rGO samples. Table 2 shows that the elemental analysis of GO, rGO and Am-rGO. The results confirm the successful functionalization of rGO with amino groups. The increase in C/O atomic ratio in Am-rGO indicates that EDA can successfully functionalized graphene sheets. The presence of N containing groups in Am-rGO can also be demonstrated from its XPS spectrum, where three peaks corresponding to N, C and O elements can be clearly visualized. As shown in Figure 4a, only the carbon (C1s at 284.8 eV) and oxygen (O1s at 531.2 eV) appeared in the wide-scan spectrum in the GO, rGO and Am-rGO. After the functionalization of rGO with amino groups, as expected, the nitrogen (N1s at 400.1 eV) was clearly evident in the wide-scan spectrum in the Am-rGO [30,37]. The appearance of the N1s peak and the greatly decreased intensity of the O1s peak in the XPS spectrum of Am-rGO indicate the efficient displacement of oxygen moieties by amino groups during the chemical amination of rGO [30,37]. Peak fitting of C1s and N1s high resolution C1s and N1s XPS spectrum reveals the diverse carbon and nitrogen components in the Am-rGO framework. As shown in Figure 4b, carbon atoms exists in different functional groups: at 284.6 (C–C/C=C), 285.5 (C–N/C=N), 286.5 (C–O), 287.9 (O=C–N) and 289.3 eV (O–C=O). The amination process led to the formation of (N=C) at 398.5 eV, (C–NH$_2$) at 399.7 eV, (C–N–C) at 400.4 eV, and (C–N$^+$–C) at 401.6 eV (see Figure 4c) [29,30]. As it can be seen in Figure 4c, the most intense peak is assigned to the C–NH$_2$, indicating that amino functionalized-rGO was successfully prepared. These XPS results were consistent with other studies presented in the literature [30,37].

The morphology and microstructure of the GO, rGO and Am-rGO and the electron diffraction (SAED) patterns in the selected area were analyzed by TEM (see Figure 5a–f). The TEM images of the GO, rGO and Am-rGO in Figure 5a–c, respectively, clearly show the presence of wrinkles, ripples and scrolls in the GO, rGO and Am-rGO indicating the occurrence of few-layered graphene sheets [38,39]. The SAED patterns of GO, rGO and Am-rGO (Figure 5d–f) were compared in order to understand the successful reduction and fictionalization with amino groups. Only diffraction rings are found in the SAED pattern of the GO, demonstrating the disordered structure of GO, while the diffraction spots in the rGO confirm the crystalline structure obtained after the thermal reduction of GO [38–40], as shown in Figure 5d–e. Moreover, owing to the addition of amino functional groups, the bright spots were not fully restored into the hexagonal graphene framework [40] (see Figure 5f). These results show that functionalization caused less damage to the graphene structure.

Table 2. Elemental composition of GO, rGO and Am-rGO samples extracted based on the XPS results.

Elements	Relative atomic percent (%)			C/O Ratio
	C	O	N	
GO	66.4	32.5	0.3	2.0
rGO	86.1	12.7	0.3	6.7
Am-rGO	83.2	10.2	6.6	8.2

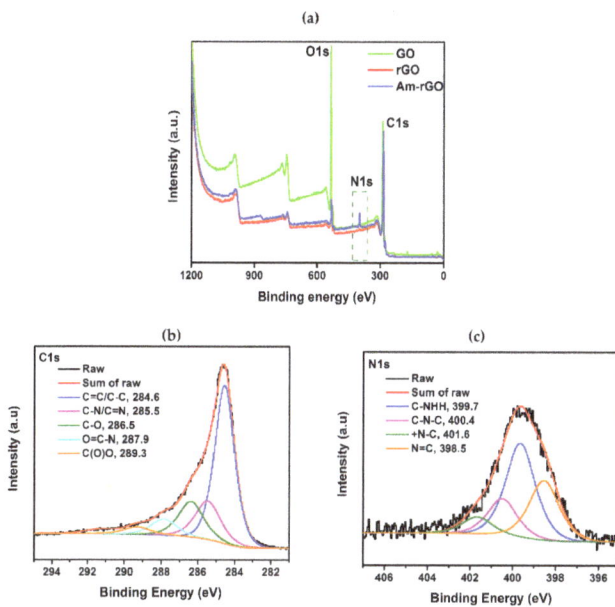

Figure 4. (**a**) XPS survey of GO, rGO and Am-rGO. High resolution XPS spectra of (**b**) C1s and (**c**) N1s of Am-rGO.

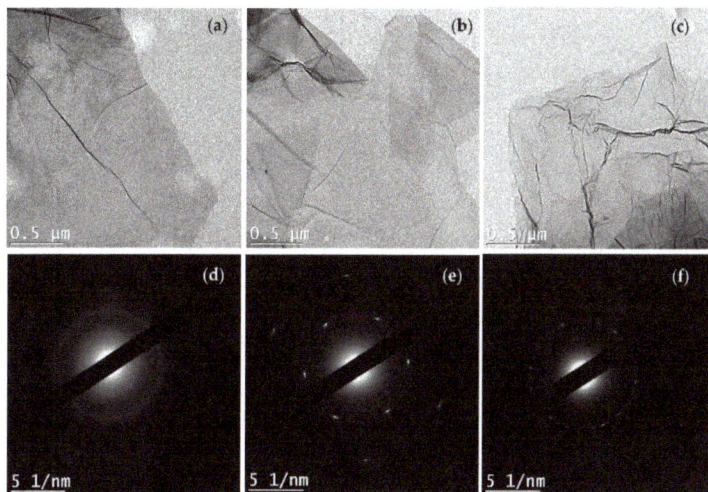

Figure 5. TEM images of (**a**) GO; (**b**) rGO and (**c**) Am-rGO with (**d–f**), respective SAED pattern.

3.2. Nanofibers Morphology

SEM was used to characterize the fabricated PANI/PMMA and PANI/PMMA/Am-rGO nanofibers (see Figure 6a–d). It is observed that the surface of nanofibers are relatively smooth and randomly oriented forming an web-like pattern [18,41]. Yet, owing to the instability of the liquid jet, beads can be observed in the image of PANI/PMMA/Am-rGO. Additionally, the average diameter before adding Am-rGO were in the range 267 ± 55 nm and after adding Am-rGO, the average diameter decreased to the range 133 ± 35 nm. This decrease in the average fiber diameter of PANI/PMMA/Am-rGO in comparison to PANI/PMMA is due to the presence of graphene sheets in the fibers. This could be attributed to the electrical conductivity in the starting solution enhanced by the presence of the graphene where the more conductive the solution, the better the chance of getting thinner fibers [18,41]. Therefore, incorporating graphene into PANI/PMMA solution enhances the conductivity of the solution to be electrospun and as a result of this improved conductivity, the produced fibers become thinner compared with fibers produced from PANI/PMMA solution [18,41].

TEM was conducted to confirm the presence of the graphene filler in the nanofibers. Figure 7a shows that some of the incorporated Am-rGO are randomly embedded in the sidewall of PANI/PMMA nanofibers. Moreover, Figure 7b–c illustrate that along the nanofibers some dark scattering spots could be observed; these aligned dark dots corresponded to graphene flakes in the PANI/PMMA nanofibers. These figures clearly show the individual graphene sheets dispersed in the PANI/PMMA matrix without aggregation, because the lateral size of graphene (a few 100 nm to a few μm) is comparable to the fiber diameter [37–41].

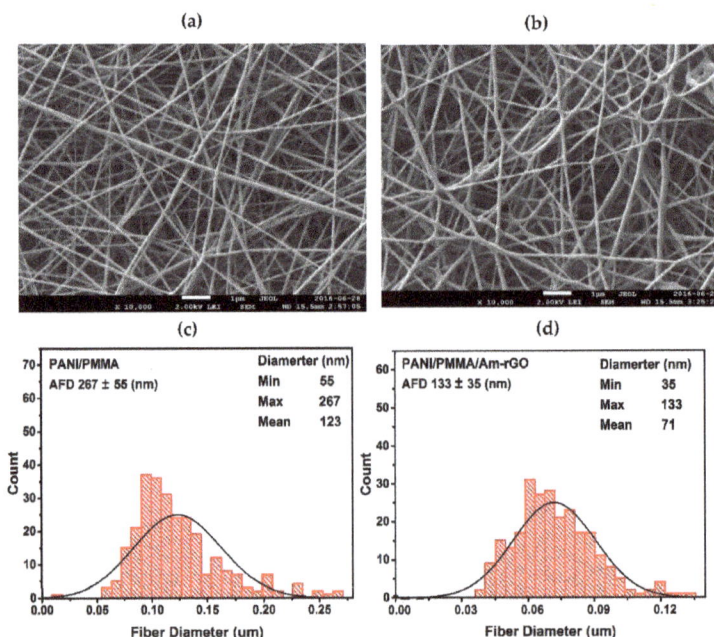

Figure 6. (**a,c**) SEM micrograph of the PANI/PMMA and PANI/PMMA/Am-rGO nanofibers, respectively; (**b,d**) the distribution of dimeters of the PANI/PMMA and PANI/PMMA/Am-rGO nanofibers, respectively.

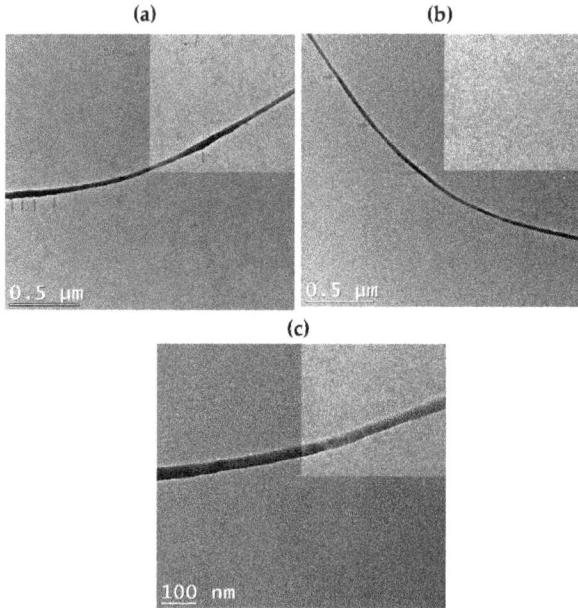

Figure 7. (a–c) High-resolution TEM images of PANI/PMMA/Am-rGO nanofibers at different magnification.

3.3. Thermal Stability

TGA was performed to observe the thermal stability of the rGO, Am-rGO and electrospun PANI/PMMA and PANI/PMMA/Am-rGO nanofiber mat (see Figure 8a,b), respectively, under N_2 atmosphere in temperature range between 20 and 900 °C, with a heating ramp of 10 °C·min^{-1}. GO is thermally unstable resulting in three stages of weight loss. In the first stage, a rapid weight loss occurs at about 173 °C, mostly attributed to the removal of the trapped water molecules and epoxy oxygen functional groups. The second stage occurs at 515 °C which can be attributed to the removal of phenolic groups and decomposition of sp^3 hybridized carbon atoms located at the defect site of GO. These results were congruent with other studies presented in literature [30,37]. On the other hand, the rGO was the most thermally stable, there is almost no weight loss below 600 °C, demonstrating the effective reduction and removal of oxygen functional groups. Comparing with GO, the Am-rGO exhibits good thermal stability. The thermal stability increases and major weight loss starts at temperatures of about 449 °C. This can be ascribed to the decomposition of amino-carbons, which is similar to the previously reported results for functionalization of graphene with amino groups [30,37]. Therefore, the larger thermal stability compared with GO and the greater mass lost compared with rGO indicates the efficient displacement of oxygen moieties by amino groups during the chemical amination of rGO [37].

Incorporating a small amount of Am-rGO in PANI nanofibers improves its thermal stability. As shown in Figure 8b, the thermal degradation temperature of PMMA/PANI/Am-rGO nanofibers increased to ~441 °C, a magnitude higher than that of the PMMA/PANI samples at ~348 °C, accredited to the presence of interfacial bonding [18,41]. In comparison to PMMA/PANI/Am-rGO, the thermal degradation temperature of the PMMA/PANI/rGO was presented around 420 °C, and corroborating the strong interaction exists between the PANI and the Am-rGO. This thermal reinforcement of the electrospun PANI nanofibers by Am-rGO (nano-carbon) fillers is very important in many different technological applications such as those mentioned in the introduction.

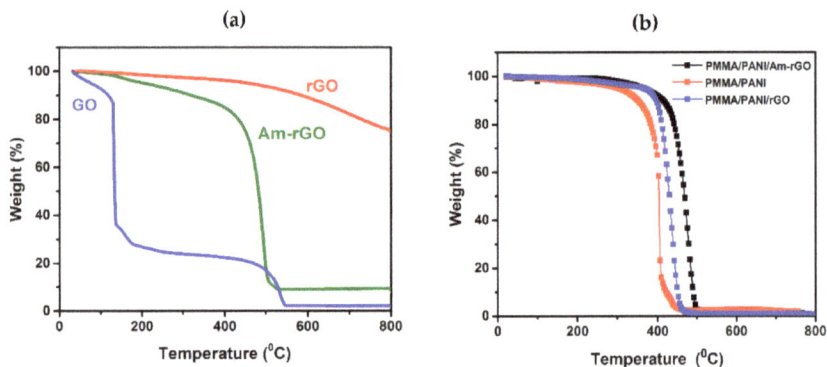

Figure 8. TGA curves (**a**) of GO, rGO and Am-rGO; and (**b**) for electrospun PMMA/PANI/Am-rGO, PMMA/PANI/rGO and PMMA/PANI nanofibers.

4. Conclusions

In this study, Am-rGO was successfully prepared and electrospun nanofibers of PANI/PMMA with amino-functionalized graphene were prepared by a simple electrospinning technique. The FTIR, XPS and Raman spectroscopy analysis confirmed the successful functionalization of the graphene with amino groups, while the TEM observation demonstrated that the addition of amino-functional groups to graphene generated less damage to the graphitic structure of the graphene. SEM micrographs indicated the formation of PANI/PMMA/Am-rGO nanofibers with a diameter ranging between 35 and 133 nm, with a general uniform thickness along the nanofibers. TGA measurements show an improvement of the thermal stability of the PANI in the presence of graphene. In conclusion, the resulting non-woven porous mats with amino-functionalized graphene result in electrospun nanofibers that can be used in future technological applications in various fields.

Acknowledgments: This research was supported by NSERC (National Science and Engineering Research Council of Canada). We are sincerely grateful to the Saudi Ministry of Higher Education for their financial support to Hanan Abdali. Also, we are grateful to Josianne Lefebvre for her indispensable help with the XPS analysis and Gwenaël Chamoulaud at UQAM University for his invaluable help with thermal annealing treatment. We would also like to thank Jean Philippe Masse for his invaluable help with TEM imaging and Samir Elouatik at University of Montreal for his help with the Raman microspectrometer.

Author Contributions: Hanan Abdali designed the experiments, performed the experiments, analyzed the data and drafted the manuscript. Abdellah Ajji designed the experiments, analyzed the data and reviewed and edited the manuscript.

Conflicts of Interest: The authors declare no conflict of interest.

References

1. Huang, Z.; Zhang, Y.; Kotaki, M.; Ramakrishna, S. A review on polymer nanofibers by electrospinning and their applications in nanocomposites. *Compos. Sci. Technol.* **2003**, *63*, 2223–2253. [CrossRef]
2. Gu, B.K.; Shin, M.K.; Sohn, K.W.; Kim, S.I.; Kim, S.J. Direct fabrication of twisted nanofibers by electrospinning. *Appl. Phys. Lett.* **2007**, *90*, 263902. [CrossRef]
3. Doshi, J.; Reneker, D.H. Electrospinning process and application of electrospun fibers. *J. Electrost.* **1995**, *35*, 151–160. [CrossRef]
4. Li, Z.; Wang, C. Effects of working parameters on electrospinning. Chapter 2; In *One-Dimensional Nanostructures*; Springer: Berlin/Heidelberg, Germany, 2013; pp. 15–28.
5. Jabal, J.M.F.; McGarry, L.; Sobczyk, A.; Aston, D.E. Substrate effects on the wettability of electrospun Titania-Poly(vinylpyrrolidone) fiber mats. *Langmuir* **2010**, *26*, 13550–13555. [CrossRef] [PubMed]
6. Jian, F.; HaiTao, N.; Tong, L.; XunGai, W. Applications of electrospun nanofibers. *Chin. Sci. Bull.* **2008**, *53*, 2265–2286.

7. Teo, W.E.; Ramakrishna, S. A review on electrospinning design and nanofiber assemblies. *Nanotechnology* **2006**, *17*, R89–R106. [CrossRef] [PubMed]
8. Nain, A.S.; Wong, J.C.; Amon, C.; Sitti, M. Drawing suspended polymer micro-/nanofibers using glass micropipettes. *Appl. Phys. Lett.* **2006**, *89*, 183105. [CrossRef]
9. Fong, H.; Chun, I.; Reneker, D.H. Beaded nanofibers formed during electrospinning. *Polymer* **1999**, *40*, 4585–4592. [CrossRef]
10. Yang, Q.; Li, Z.; Hong, Y.; Zhao, Y.; Qiu, S.; Wang, C.; Wei, Y. Influence of solvents on the formation of ultrathin uniform poly(vinyl pyrrolidone) nanofibers with electrospinning. *J. Polym. Sci. B* **2004**, *42*, 3721–3726. [CrossRef]
11. Ruiz, J.; Gonzalo, B.; Dios, J.R.; Laza, J.M.; Vilas, J.L.; León, L.M. Improving the process-ability of conductive polymers: The case of polyaniline. *Adv. Polym. Technol.* **2013**, *32*, E180–E188. [CrossRef]
12. Zhang, Y.; Rutledge, G.C. Electrical conductivity of electrospun polyaniline and polyaniline-blend fibers and mats. *Macromolecules* **2012**, *45*, 4238–4246. [CrossRef]
13. Bai, H.; Shi, G. Gas sensors based on conducting polymers. *Sensors* **2007**, *7*, 267–307. [CrossRef]
14. Fratoddia, I.; Vendittia, I.; Camettib, C.; Russoa, M.V. Chemiresistive polyaniline based gas sensors: A mini review. *Sens. Actuators B* **2015**, *220*, 534–548. [CrossRef]
15. Huang, J.; Virji, S.; Weiller, B.H.; Kaner, R.B. Polyaniline nanofibers: Facile synthesis and chemical sensors. *J. Am. Chem. Soc.* **2003**, *125*, 314–315. [CrossRef] [PubMed]
16. Liang, L.; Liu, J.; Windisch, C.F.; Exarhos, G.J.; Lin, Y. Direct assembly of large arrays of oriented conducting polymer nanowires. *Angew. Chem. Int. Ed.* **2002**, *41*, 3665–3668. [CrossRef]
17. Yang, W.; Ratinac, K.R.; Ringer, S.P.; Thordarson, P.; Gooding, J.J.; Braet, F. Carbon nanomaterials in biosensors: Should you use nanotubes or graphene? *Angew. Chem. Int. Ed.* **2010**, *49*, 2114–2138. [CrossRef] [PubMed]
18. Moayeri, A.; Ajji, A. Fabrication of polyaniline/poly(ethylene oxide)/noncovalently functionalized graphene nanofibers via electrospinning. *Synth. Metals* **2015**, *200*, 7–15. [CrossRef]
19. Geim, A.K.; Novoselov, A.K. The rise of graphene. *Nat. Mater.* **2007**, *6*, 183–191. [CrossRef] [PubMed]
20. Potts, J.R.; Dreyer, D.R.; Bielawski, C.W.; Ruoff, R.S. Graphene-based polymer nanocomposites. *Polymer* **2011**, *52*, 5–25. [CrossRef]
21. Du, J.; Cheng, H.M. The fabrication, properties and uses of graphene/polymer composites. *Macromol. Chem. Phys.* **2012**, *213*, 1060–1077. [CrossRef]
22. Klimchitskaya, G.; Mostepanenko, V. Van der Waals and Casimir interactions between two graphene sheets. *Phys. Rev. B* **2013**, *87*, 1–18. [CrossRef]
23. Arbuzov, A.A.; Muradyan, V.E.; Tarasov, B.P.; Sokolov, E.A. Preparation of amino-functionalized graphene sheets and their conductive properties. In Proceedings of the International Conference Nanomaterials: Applications and Properties, the Crimea, Ukraine, 16–21 September 2013; Volume 2.
24. Kim, N.H.; Kuilab, T.; Lee, J.H. Simultaneous reduction, functionalization and stitching of graphene oxide with ethylenediamine for composites application. *J. Mater. Chem. A* **2013**, *1*, 1349–1358. [CrossRef]
25. Zheng, W.; Shen, B.; Zhai, W. Surface Functionalization of Graphene with Polymers for Enhanced Properties. In *New Progress on Graphene Research*; InTech: Janeza Trdine Rijeka, Croatia, 2013.
26. Liu, S.; Liu, X.H.; Li, Z.P.; Yang, S.R.; Wang, J.Q. Fabrication of Free-Standing Graphene/Polyaniline Nanofibers Composite Paper via Electrostatic Adsorption for Electrochemical Supercapacitors. *New J. Chem.* **2011**, *35*, 369–374. [CrossRef]
27. Rodthongkuma, N.; Ruechab, N.; Rangkupana, R.; Vachetd, R.W.; Chailapakule, O. Graphene-loaded nanofiber-modified electrodes for the ultrasensitive determination of dopamine. *Anal. Chim. Acta* **2013**, *804*, 84–91. [CrossRef] [PubMed]
28. Wang, L.; Lu, X.; Leib, S.; Song, Y. Graphene-based polyaniline nanocomposites: Preparation, properties and applications. *J. Mater. Chem. A* **2014**, *2*, 4491–4509. [CrossRef]
29. Song, N.J.; Chen, C.M.; Lu, C.; Liu, Z.; Kongb, Q.Q.; Caib, R. Thermally reduced graphene oxide films as flexible lateral heat spreaders. *J. Mater. Chem. A* **2014**, *2*, 16563–16568. [CrossRef]
30. Tu, Q.; Pang, L.; Chen, Y.; Zhang, Y.; Zhang, R.; Lu, B.; Wang, J. Effects of surface charges of graphene oxide on neuronal outgrowth and branching. *Analyst* **2014**, *139*, 105–115. [CrossRef] [PubMed]

31. Yang, D.; Velamakanni, A.; Bozoklu, G.; Park, S.; Stoller, M.; Piner, R.D.; Stankovich, S.; Jung, I.; Field, D.A.; Ventrice, C.A.; et al. Chemical analysis of graphene oxide films after heat and chemical treatments by X-ray photoelectron and micro-Raman spectroscopy. *Carbon* **2009**, *47*, 145–152. [CrossRef]

32. Ganguly, A.; Sharma, S.; Papakonstantinou, P.; Hamilton, J. Probing the thermal deoxygenation of graphene oxide using high-resolution in situ X-ray-based spectroscopies. *J. Phys. Chem. C* **2011**, *115*, 17009–17019. [CrossRef]

33. Botas, C.; Álvarez, P.; Blanco, C.; Santamaría, R.; Granda, M.; Gutiérrez, M.D.; Reinoso, F.R.; Menéndez, R. Critical temperatures in the synthesis of graphene-like materials by thermal exfoliation–reduction of graphite oxide. *Carbon* **2013**, *52*, 476–485. [CrossRef]

34. Perumbilavil, S.; Sankar, P.; Rose, T.P.; Philip, R. White light Z-scan measurements of ultrafast optical nonlinearity in reduced graphene oxide nanosheets in the 400–700 nm region. *Appl. Phys. Lett.* **2015**, *107*, 051104. [CrossRef]

35. Wang, Y.; Zhou, L.; Wang, S.; Li, J.; Tang, J.; Wang, S.; Wang, Y. Sensitive and selective detection of Hg^{2+} based on an electrochemical platform of PDDA functionalized rGO and glutaraldehyde cross-linked chitosan composite film. *R. Soc. Chem.* **2016**, *6*, 69815–69821. [CrossRef]

36. Wu, N.; She, X.; Yang, D.; Wu, X.; Su, F.; Chen, Y. Synthesis of network reduced graphene oxide in polystyrene matrix by a two-step reduction method for superior conductivity of the composite. *J. Mater. Chem.* **2012**, *22*, 17254–17261. [CrossRef]

37. Navaee, A.; Salimi, A. Efficient amine functionalization of graphene oxide through the Bucherer reaction: An extraordinary metal-free electrocatalyst for the oxygen reduction reaction. *R. Soc. Chem. Adv.* **2015**, *5*, 59874–59880. [CrossRef]

38. Cui, T.; Lv, R.; Huang, Z.H.; Zhu, H.; Jia, Y.; Chen, S.; Wang, K.; Wu, D.; Kang, F. Low-temperature synthesis of multilayer graphene/amorphous carbon hybrid films and their potential application in solar cells. *Nanoscale Res. Lett.* **2012**, *7*, 453. [CrossRef] [PubMed]

39. Song, P.; Zhang, X.; Sun, M.; Cui, X.; Lin, Y. Synthesis of graphene nanosheets via oxalic acid-induced chemical reduction of exfoliated graphite oxide. *R. Soc. Chem.* **2012**, *2*, 1168–1173.

40. Lavanya, J.; Gomathi, N.; Neogi, S. Electrochemical performance of nitrogen and oxygen radio-frequency plasma induced functional groups on tri-layered reduced graphene oxide. *Mater. Res. Express* **2014**, *1*, 1–18. [CrossRef]

41. Barzegar, F.; Bello, A.; Fabiane, M.; Khamlich, S.; Momodu, D.; Taghizadeh, F.; Dangbegnon, J.; Manyala, N. Preparation and characterization of poly(vinyl alcohol)/graphene nanofibers synthesized by electrospinning. *J. Phys. Chem. Solids* **2015**, *77*, 139–145. [CrossRef]

polymers

MDPI

Article

Improving Kinetics of "Click-Crosslinking" for Self-Healing Nanocomposites by Graphene-Supported Cu-Nanoparticles

Neda Kargarfard [1,2], Norman Diedrich [1], Harald Rupp [1], Diana Döhler [1,*]and Wolfgang H. Binder [1,*]

[1] Faculty of Natural Science II, Martin Luther University Halle-Wittenberg, Von-Danckelmann-Platz 4, D-06120 Halle (Saale), Germany; kargarfard@ipfdd.de (N.K.); norman.diedrich@student.uni-halle.de (N.D.); harald.rupp@chemie.uni-halle.de (H.R.)

[2] Leibniz-Institut für Polymerforschung Dresden e. V., Abteilung Reaktive Verarbeitung, Hohe Str. 6, D-01069 Dresden, Germany

* Correspondence: diana.doehler@chemie.uni-halle.de (D.D.); wolfgang.binder@chemie.uni-halle.de (W.H.B.); Tel.: +49-345-55-25907 (D.D.); +49-345-55-25930 (W.H.B.)

Received: 5 December 2017; Accepted: 21 December 2017; Published: 24 December 2017

Abstract: Investigation of the curing kinetics of crosslinking reactions and the development of optimized catalyst systems is of importance for the preparation of self-healing nanocomposites, able to significantly extend their service lifetimes. Here we study different modified low molecular weight multivalent azides for a capsule-based self-healing approach, where self-healing is mediated by graphene-supported copper-nanoparticles, able to trigger "click"-based crosslinking of trivalent azides and alkynes. When monitoring the reaction kinetics of the curing reaction via reactive dynamic scanning calorimetry (DSC), it was found that the "click-crosslinking" reactivity decreased with increasing chain length of the according azide. Additionally, we could show a remarkable "click" reactivity already at 0 °C, highlighting the potential of click-based self-healing approaches. Furthermore, we varied the reaction temperature during the preparation of our tailor-made graphene-based copper(I) catalyst to further optimize its catalytic activity. With the most active catalyst prepared at 700 °C and the optimized set-up of reactants on hand, we prepared capsule-based self-healing epoxy nanocomposites.

Keywords: TRGO; copper nanoparticles; CuAAC crosslinking; self-healing nanocomposite

1. Introduction

Self-healing approaches do have a significant potential in polymeric materials, especially those based on embedded capsule systems [1]. The molecular design of such self-healing materials requires fast and efficient crosslinking processes, which often are afforded by catalytic reactions using homogeneous and heterogeneous chemistry [2]. In the past, a plethora of such processes has been reported, based on e.g., ring-opening metathesis polymerization (ROMP)- (....) [1,3–10], copper(I)-catalyzed azide-alkyne cycloaddition (CuAAC)- (....) [11–27], isocyanate- [28–37], thiol- [38–46] and hydrosilylation-chemistry [47,48], many of them using metal-catalysis.

In the field of ROMP-based self-healing [7,9], the curing behavior of renewable norbornenyl-functionalized isosorbide monomers in the copolymerization with dicyclopentadiene (DCPD) was investigated exhibiting a higher reactivity, consequently facilitating low temperature ROMP. Additionally, higher crosslinking densities were observed, resulting in improved thermal and mechanical properties highlighting the potential of renewable ROMP-monomers towards self-healing applications [9].

In particular, the search for graphene-supported catalysts [12,19,21–23] is ongoing, effecting Ru-based crosslinking within nanocomposites. Thus, graphene oxide (GO)-supported Grubb's catalysts (GO-HG1 and GO-HG2) have been prepared and investigated in the ROMP of 5-ethylidene-2-norbornene at 40 to 60 °C. While showing catalytic activity the amount of catalyst was reduced from 5.0 to 0.5 wt % [7].

We recently reported on a "click"-based crosslinking chemistry useful as a principle for optimized self-healing materials [12]. In this particular system, multivalent azides and alkynes are crosslinked by the use of a Cu(I)-catalyst, acting as the known essential catalytic system for "click"-chemistry [12,49–53]. Of particular importance is the ability to tune the temperature at which crosslinking takes place, thus enabling self-healing at room temperature or even below [11,17]. The central principal of action is the use of encapsulated low molecular weight azides [19,22] or azide-functionalized polymers [11,14,15,17,20,27], and low molecular weight alkynes [15,19,20,22] or alkyne-functionalized polymers [11,14], embedded within capsules sized from 100 nm up to microns [13,15,19,54]. Whereas the uncatalyzed process is conventionally taking place at temperatures close to ~160 °C [55,56], the use of Cu(I) as catalyst can significantly lower the crosslinking temperature, together with an increase in crosslinking efficiency [49–51]. The kinetics of "click-crosslinking" reactions was studied by us [12] and others [24,25] and the huge potential of the CuAAC towards self-healing materials is highlighted by the observation of autocatalytic effects in the melt state. Thus, the formed 1,2,3-triazole rings can act as internal ligands consequently significantly increasing the reaction rate [11], well known before for the addition of external triazole-containing ligands [57]. By additionally designing chelation assisting azides, the self-healing temperature could be tuned, and efficient network formation was even observed below room temperature ($T = 10$ °C) [17].

When probing a large variety of homogeneous and heterogeneous catalysts, the use of finely dispersed Cu_2O on graphene-oxide (TRGO-Cu_2O) displayed high catalytic activity in various "click-crosslinking" reactions in the melt, easy and efficient recyclability in solution experiments as well as high stability against oxygen [21]. Consequently, the catalyst was used to generate graphene-based nanocomposites via the CuAAC without addition of a base or any reducing agents. As the cheap and easy up-scalable catalyst acted as reinforcing filler, the mechanical, thermal and conductive properties of the final resin were improved [22], opening the possibility to generate self-healing capsule-based epoxy nanocomposites showing quantitative healing at room temperature within 36 h [19].

For a capsule-based self-healing approach, the encapsulation of the azides required a careful tuning of their hydrophobicity. It was unclear though, how and whether small changes in the azide-monomer would change its reactivity within the same click-system. Thus, a modification of the azide-monomer to a more hydrophobic surrounding deemed interesting, most of all to facilitate its encapsulation via emulsion-processes. Therefore, in this study we investigate the influence of small substitutions within the trivalent azides on the reaction kinetics investigated via DSC as well as in view of different homogeneous and heterogeneous copper(I) catalysts.

2. Materials and Methods

2.1. Materials

Trimethylol propane (purity > 97%), trimethylol propane triglycidyl ether (technical grade, ^{1}H-NMR spectrum in Supplementary Materials Figure S1), propargyl bromide (80 wt % solution in toluene), sodium sulphate (Na_2SO_4, purity > 99%), sodium azide (NaN_3, purity 99.5%), 4-dimethylaminopyridine (DMAP, purity 99%), graphite (synthetic grade), potassium permanganate ($KMnO_4$, analytical grade), copper(II) acetate hydrate (purity 98%), acetic anhydride (purity > 99%) and decanoyl chloride (purity > 98%) were purchased from Sigma Aldrich (Taufkirchen, Germany) and were used as received. Sodium hydroxide (NaOH, purity > 99%), ammonium chloride (NH_4Cl, purity > 99%), concentrated sulphuric acid (H_2SO_4, 95–98%), hydrochloric acid (HCl, 37%) and *N,N*-dimethylformamide (DMF, purity > 99%) were ordered from Grüssing (Filsum, Germany) and

DMF was freshly distilled from CaH_2 under nitrogen atmosphere before use. Butyryl chloride (purity > 98%) and calcium hydride (CaH_2, purity > 92%) were obtained from Alfa Aesar (Karlsruhe, Germany). Dichloromethane (DCM, purity > 98%) was purchased from Overlack (Mönchengladbach, Germany), chloroform ($CHCl_3$, purity > 98%) was received from VWR (Darmstadt, Germany) and methanol (MeOH, purity > 99.9%) was obtained from Brenntag (Mühlheim an der Ruhr, Germany) and all solvents were distilled prior use. Tetra-*n*-butylammonium bromide (TBAB, purity > 99%) was purchased from TCI (Eschborn, Germany) and sodium chloride (NaCl, purity > 99%), *o*-phosphoric acid (H_3PO_4, 85%), hydroxide peroxide (H_2O_2, 30 wt %) were received from Carl Roth (Karlsruhe, Germany) and were used as received. The synthesis of trimethylolpropane tripropargyl ether (TMPTE, **1**) was done according to literature [22,58] and was further optimized to yield exclusively **1** (see Supplementary Materials, Scheme S1). The synthesis of (azidated trimethylolpropane triglycidyl ether, N_3TMPTE, **2**) was done according to literature [22,59] with slight modifications (see Supplementary Materials Scheme S2).

2.2. Methods

Nuclear magnetic resonance (NMR) spectroscopy: All NMR spectra were recorded on a Varian Gemini 400 spectrometer from Agilent Technologies (Waldbronn, Germany) at 400 MHz at 27 °C. Deuterated chloroform ($CDCl_3$, purity > 99.8%, stabilized with Ag) was used as solvent and was purchased from Chemotrade (Düsseldorf, Germany). The chemical shifts were recorded in ppm and all the coupling constants in Hz. MestRec v.4.9.9.6 (Mestrelab Research, A Coruña, Spain, 2016) was used for data interpretation.

Attenuated total reflection Fourier transformed infrared (ATR-FTIR) spectroscopy: All ATR-FTIR spectra were recorded on a Bruker Tensor VERTEX 70 from Bruker Optics GmbH (Leipzig, Germany) equipped with a heatable Golden Gate Diamond ATR plate from Specac (Orpington, Kent, UK). Opus 6.5 (Bruker Optik GmbH, Leipzig, Germany, 2008) and OriginPro 8G (Version 8.0951, OriginLab Corporation, Northampton, MA, USA, 2008) was used for data interpretation.

Thin layer chromatography (TLC): TLC was performed with TLC aluminum sheets (silica gel 60 F254) obtained from Merck (Darmstadt, Germany). Spots on the TLC plates were visualized by UV light (254 or 366 nm) or by oxidizing agents like "blue stain" consisting of $Ce(SO_4)_2·4H_2O$ (1.0 g, analytical grade, Sigma Aldrich (Taufkirchen, Germany)), $(NH_4)_6Mo_7O_{24}·4H_2O$ (2.5 g, analytical grade, Sigma Aldrich (Taufkirchen, Germany)), dissolved in concentrated H_2SO_4 (6.0 mL, 95–97%, Grüssing (Filsum, Germany)) and distilled water (90.0 mL).

Electrospray ionization time of flight mass spectrometry (ESI-TOF MS): ESI-TOF MS was performed with a micro TOF focus from Bruker Daltonics GmbH (Bremen, Germany) with an electrospray ionization source (ESI source). Samples were dissolved in $CHCl_3$ (HPLC grade, VWR (Darmstadt, Germany)) or MeOH (HPLC grade, VWR (Darmstadt, Germany)) and sodium iodide (purity > 99.9%, Sigma Aldrich (Taufkirchen, Germany)), 20 $mg·mL^{-1}$ acetone (HPLC grade, Sigma Aldrich (Taufkirchen, Germany)) was added. The analyte was injected with a 180 $\mu L·h^{-1}$ flow rate at 180 °C.

Differential scanning calorimetry (DSC): DSC was performed on a differential scanning calorimeter 204F1/ASC Phönix from Netzsch (Selb, Germany). Crucibles and lids made of aluminum were used. Measurements were performed in a temperature range from −20 to 250 °C using heating rates of 5, 10, 15 and 20 $K·min^{-1}$. As purge gas, a flow of dry nitrogen (20 $mL·min^{-1}$) was used. For evaluation of data, the Proteus Thermal Analysis Software (Version 5.2.1, NETZSCH-Geraetebau GmbH, Selb, Germany, 2011) and OriginPro 8G (Version 8.0951, OriginLab Corporation, Northampton, MA, USA, 2008) was used.

Rheology: In situ rheology was performed on an oscillatory plate rheometer MCR 101/SN 80753612 from Anton Paar (Graz, Austria). All measurements were performed within the linear viscoelastic regime (LVE) using a PP08 measuring system. For evaluation of data, the RheoPlus/32 software (V 3.40, Anton Paar Germany GmbH, Ostfildern, Germany, 2008) and OriginPro 8G (Version 8.0951, OriginLab Corporation, Northampton, MA, USA, 2008) were used.

Glass tube furnace: The thermal reduction of **GO-Cu(II)** was performed in a glass tube furnace (Mod. RSR-B120/750/11) from Nabertherm GmbH (Lilienthal, Germany).

Freeze drying: Freeze drying was performed on a LyoQuest freeze dryer from Telstar (Utrecht, The Netherlands) operating at −80 °C and 0.18 mbar.

Ultrasonicator: For the dispersion of the **GO**-species via ultrasonication a sonication tip Vibra Cell VCX500 from Zinsser Analytic (Frankfurt, Germany) was used.

Flame atomic absorption spectroscopy (FAAS): FAAS was performed on a novAA 350 #113A0641 Tech: Flamme spectrometer from Analytik Jena AG (Jena, Germany) using Aspect LS 1.4.1.0 (Analytik Jena AG, Jena, Germany) as software. Therefore, external calibration and calibration via doping were performed. To determine the copper-content within **TRGO-Cu$_2$O** the samples were burned to ash at 800 °C under atmospheric conditions and were dispersed in nitric acid (2 M). This solution was diluted in a 1:1 ratio with a potassium chloride solution (0.2%) and a mixture containing 25 mL of the dispersed nitric sample solution and 25 mL of the potassium chloride solution.

Transmission electron microscopy (TEM): TEM investigations were performed using a EM 900 transmission electron microscope from Carl Zeiss Microscopy GmbH (Oberkochen, Germany) and the images were taken with a SSCCD SM-1k-120 camera from TRS (Moorenweis, Germany). For sample preparation, **TRGO-Cu$_2$O** was dispersed in water and sprayed on a carbon-layered copper grid. After one minute, the excess solution was removed with filter paper and the samples were dried at room temperature.

X-ray diffraction (XRD): XRD-measurements were performed on a D8 X-ray diffractometer from Bruker AXS GmbH (Karlsruhe, Germany). For analysis of the raw data, Diffrac. Suite EVA 3.1 (Bruker AXS GmbH, Karlsruhe, Germany) with an integrated database for the determination of the phases was used as software. For sample preparation, **TRGO-Cu$_2$O** was rubbed in the presence of isopropanol and was put on a glass slide. For evaluation of data, OriginPro 8G (Version 8.0951, OriginLab Corporation, Northampton, MA, USA, 2008) was used.

2.3. General Synthesis Procedure for the Preparation of Trivalent Azides

The synthesis was carried out under a dry atmosphere of nitrogen. A two-necked round-bottom flask equipped with magnetic stir bar, rubber septum and gas tap was heated under vacuum and flushed with nitrogen several times. 4-Dimethylaminopyridine (0.2 eq) was added to **2** (1.0 eq) dissolved in dry DMF and the solution was stirred for ten minutes at room temperature. Afterwards, the desired anhydride or acid chloride (6.0–8.0 eq) was added dropwise to the reaction mixture and the solution was stirred at room temperature. After finishing the reaction, the crude product was either purified by extraction or column chromatography and the obtained product was dried in high vacuo. As determined via NMR-spectroscopy, final products contain up to 25% impurities such as free epichlorohydrine and bivalent residues as trimethylolpropane triglycidyl ether was used in technical grade (see ^{1}H-NMR spectrum in Supplementary Materials Figure S1).

3. Results

3.1. Synthesis of Trivalent Alkyne and Trivalent Azides

In Figure 1, an overview of the synthesized trivalent alkyne **1** and the trivalent azides is given. Trivalent alkyne **1** and trivalent azide **2** have been synthesized according to literature [22,58,59].

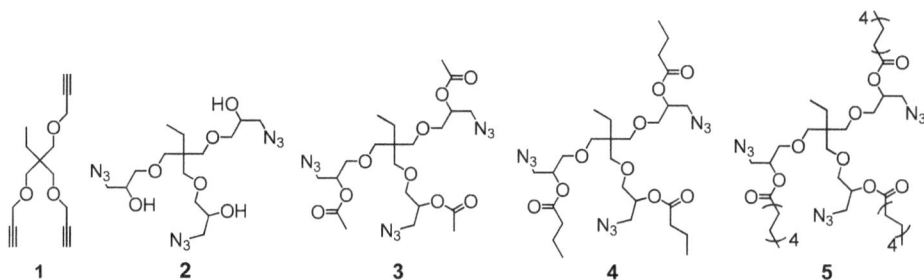

Figure 1. Structures of trivalent alkyne **1** and trivalent azides **2, 3, 4** and **5**.

The synthesis of the trivalent azides **3***, **4**, **4***, **5** and **5*** is presented in Scheme 1, and details are given in the general synthesis procedure in Section 2.3 and in the Supplementary Materials. In brief, the trivalent azide **2** was either converted with acetic anhydride or the desired acid chloride at room temperature in dry DMF and in the presence of DMAP as a nucleophilic basic catalyst.

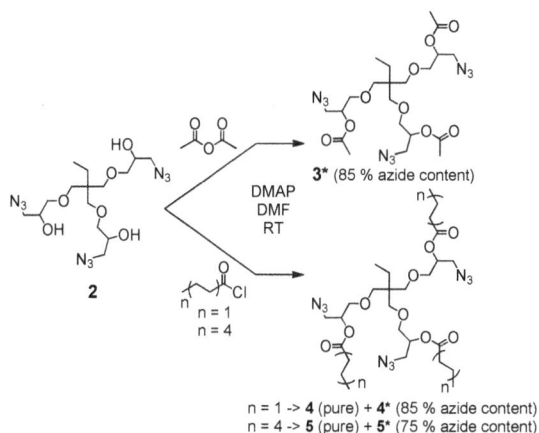

Scheme 1. Synthesis of trivalent azides **3***, **4**, **4***, **5** and **5***.

The trivalent azide **3*** was obtained as a light-yellow, viscous liquid (85% azide content) containing the bivalent azide and free epichlorohydrin present in the starting material (trimethylolpropane triglycidyl ether (technical grade)). This mixture was further studied without purification to investigate its suitability for easy preparable and up-scalable room temperature-based self-healing nanocomposites relying on "click-crosslinking" reactions. For the synthesis of the azides **4** and **5**, a pure trivalent compound (**4** and **5**) and a mixture of bi- and trivalent product (**4*** and **5***, 85 and 75% azide content, respectively) were obtained and further separated via column chromatography. All prepared multivalent azide- and alkyne-functionalized compounds were characterized via NMR- and IR-spectroscopy as well as ESI-TOF mass spectrometry proving their purity and functional group content (for more details see Supplementary Materials Figures S2–S5: ^{1}H- and ^{13}C-NMR spectra of trivalent azides **4** and **5**).

3.2. DSC Investigation of "Click-Crosslinking" Trivalent Alkyne and Trivalent Azides

Thermal analysis by differential scanning calorimetry (DSC) provides useful information about the relationship between the extent of a (crosslinking) reaction and the required time of curing at a certain temperature. Furthermore, information about kinetic parameters can be retained. Thus, DSC analysis is helpful to obtain a wide range of data of the investigated "click-crosslinking" reactions of trivalent alkyne **1** and trivalent azides **3***, **4**, **4***, **5** and **5*** such as the enthalpy of the reaction (ΔH), the onset temperature (T_{onset}), the temperature at the maximum of the DSC curve (T_p) and the apparent activation energy of the reaction ($E_{a, app}$). It should be mentioned, that the experimentally determined activation energies are indicated as apparent activation energies, since it is well known that the physical conditions during crosslinking are at least partially restricted by the mass and heat transport in the solid phase, consequently influencing the internal energy as well as the vibrational states of the investigated reactants [60,61]. Furthermore, it should be emphasized that our prepared **TRGO-Cu₂O** catalyst is composed of several graphene sheets showing a lack in dispersibility with the reactants due to missing functional groups of increasing polarity. Thus, a relatively high apparent activation energy is expected—a phenomenon also observed in graphene oxide nanocomposite epoxy coatings [62].

Via DSC investigations, the "click-crosslinking" reaction conversion can be estimated with respect to a determined reference value, a maximum ΔH value (262 kJ·mol⁻¹) for a reference click reaction between phenylacetylene and benzyl azide when quantitatively forming one triazole unit and therefore being representative of one successful "click" reaction. This reference value is in line with reported literature values for "click" reactions ranging between 210 to 270 kJ·mol⁻¹ [63]. DSC measurements for the "click-crosslinking" reaction of trivalent alkyne **1** with trivalent azide **3*** investigated at a heating rate of 5 K·min⁻¹ in the presence of different homo- and heterogenous copper(I) catalysts (1 mol % per functional group) as well as without catalyst (W/O) are plotted in Figure 2 and the obtained results are summarized in Table 1.

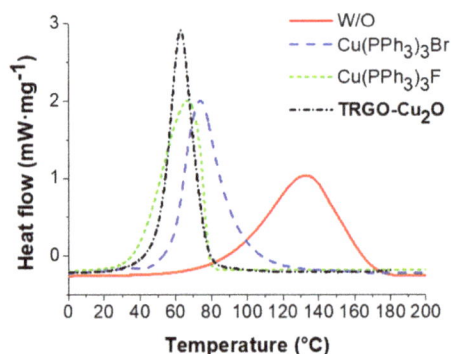

Figure 2. Dynamic scanning calorimetry (DSC) measurements of the "click-crosslinking" reaction of trivalent alkyne **1** and trivalent azide **3*** with Cu(PPh₃)₃Br, Cu(PPh₃)₃F, Cu₂O on graphene-oxide (**TRGO-Cu₂O**, prepared at 600 °C) as well as without catalyst (W/O) at a heating rate of 5 K·min⁻¹.

Table 1. Thermal properties, reaction temperatures (T_{onset} and T_p), reaction enthalpies (ΔH), apparent activation energies ($E_{a, app}$) and conversions of the "click-crosslinking" reaction of trivalent alkyne **1** and trivalent azide **3***, **4**, **4***, **5** or **5*** with different catalysts (1 mol %) as well as without catalyst (W/O) at a heating rate of 5 K·min^{-1}. **TRGO-Cu$_2$O** was prepared at 600 °C.

Entry	Azide	Catalyst	Mass (%)	T_{onset} [1] (°C)	T_p [1] (°C)	ΔH [2] (kJ·mol^{-1})	$E_{a, app}$ (kJ·mol^{-1})	Conversion [3] (%)
1	3*	W/O	-	91	133	205	83	78
2		Cu(PPh$_3$)$_3$Br	3.4	59	74	185	87	71
3		Cu(PPh$_3$)$_3$F	3.2	39	66	191	83	73
4		TRGO-Cu$_2$O	7.4	51	63	177	55	67
5	4	TRGO-Cu$_2$O	6.7	91	92	233	104	89
6	4*	W/O	-	94	125	227	76	87
7		Cu(PPh$_3$)$_3$Br	2.9	54	66	223	70	85
8		Cu(PPh$_3$)$_3$F	3.2	32	50	192	61	73
9		TRGO-Cu$_2$O	6.9	93	94	211	97	81
10	5	TRGO-Cu$_2$O	5.0	98	105	248	133	95
11	5*	W/O	-	102	141	174	100	66
12		Cu(PPh$_3$)$_3$Br	2.2	80	93	149	55	57
13		Cu(PPh$_3$)$_3$F	2.4	38	62	140	103	53
14		TRGO-Cu$_2$O	5.2	32	56	127	85	48

[1] According to our previous publication [22] and our experience, the error is typically $\approx \pm 5$ K. [2] According to our previous publication [22], the error is typically $\approx \pm 6$ kJ·mol^{-1}. [3] Calculated with respect to the enthalpy for 100% click conversion which is $\Delta H = 262$ kJ·mol^{-1} for the reference reaction of phenylacetylene and benzyl azide with 1 mol % of Cu(PPh$_3$)$_3$Br as catalyst.

For the Huisgen cycloaddition of trivalent alkyne **1** and trivalent azide **3*** (Table 1, Entry 1) 78% conversion was achieved corresponding to an enthalpy of 205 kJ·mol^{-1}. Crosslinking took place at high temperatures and T_{onset} and T_p were observed at 91 and 133 °C, respectively. In the case of the homogenous catalysts (Cu(PPh$_3$)$_3$F and Cu(PPh$_3$)$_3$Br, Table 1, Entry 2 and 3), the observed enthalpies were 191 and 185 kJ·mol^{-1}, consequently showing a conversion of 73 and 71%, respectively. By using **TRGO-Cu$_2$O** as a catalyst for "click-crosslinking" **1** and **3*** (Table 1, Entry 4), an enthalpy of 177 kJ·mol^{-1} was observed corresponding to 67% conversion. Moreover, the lowest apparent activation energy (55 kJ·mol^{-1}) and the lowest maximum peak temperature T_p (63 °C) were achieved in the presence of **TRGO-Cu$_2$O** and the lowest T_{onset} (39 °C) by using Cu(PPh$_3$)$_3$F. According to these results, **TRGO-Cu$_2$O** and Cu(PPh$_3$)$_3$F were the best catalysts for the "click-crosslinking" reaction of trivalent alkyne **1** and trivalent azide **3***.

To investigate the activity of the pure trivalent azides **4** and **5** in comparison to the partially functionalized trivalent azides **4*** and **5*** (85 and 75% azide content) together with trivalent alkyne **1** in the CuAAC crosslinking reaction, DSC measurements were run by using **TRGO-Cu$_2$O** as a catalyst (1 mol % per functional group). The DSC thermograms at 5 K·min^{-1} with **TRGO-Cu$_2$O** as a catalyst are plotted in Figure 3a,b and the obtained results are summarized in Table 1.

For "click-crosslinking" trivalent alkyne **1** with trivalent azide **4** (Table 1, Entry 5), the observed enthalpy was 233 kJ·mol^{-1}, corresponding to 89% conversion. In comparison, the observed enthalpy for "click-crosslinking" trivalent alkyne **1** with trivalent azide **4*** (Table 1, Entry 9) showed a slightly lower enthalpy value of 211 kJ·mol^{-1} and a conversion of 81%, related to the presence of the bivalent byproduct consequently lowering the conversion. In contrast, the reaction temperatures (T_p and T_{onset}) were decreased for "click-crosslinking" trivalent alkyne **1** with trivalent azide **5***, mainly attributed to a lower viscosity and therefore, to a faster diffusion. Thus, an enthalpy of 248 kJ·mol^{-1} was observed for "click-crosslinking" trivalent alkyne **1** with trivalent azide **5** (Table 1, Entry 10) at relatively high reaction temperatures corresponding to 95% conversion. In comparison, a lower enthalpy of 127 kJ·mol^{-1} (48% conversion) was measured for the "click-crosslinking" reaction of trivalent alkyne **1** with trivalent azide **5*** (Table 1, Entry 14) while the reaction temperatures were reduced (T_{onset} = 32 °C, T_p = 56 °C). Consequently, further investigations towards easily up-scalable room temperature-based self-healing nanocomposites were continued by using trivalent azides **4*** and **5***.

Polymers **2018**, *10*, 17

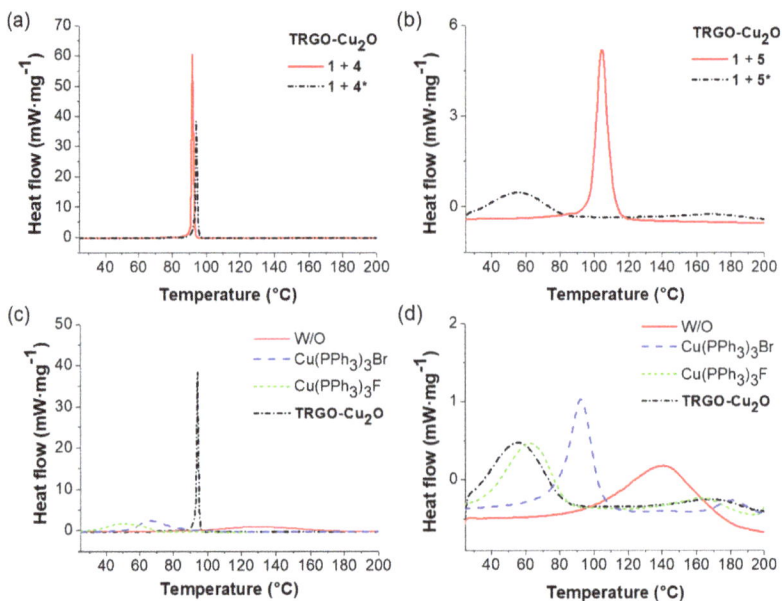

Figure 3. DSC measurements of the "click-crosslinking" reaction of trivalent alkyne **1** and (**a**) trivalent azides **4** (red curve) or **4*** (black curve) with **TRGO-Cu$_2$O**, (**b**) trivalent azides **5** (red curve) or **5*** (black curve) with **TRGO-Cu$_2$O** (prepared at 600 °C), (**c**) trivalent azide **4*** with different catalysts as well as without catalyst (W/O), (**d**) trivalent azide **5*** with different catalysts as well as without catalyst (W/O) (all at a heating rate of 5 K·min^{-1}).

DSC thermograms of "click-crosslinking" trivalent alkyne **1** with trivalent azides **4*** or **5*** without catalyst (Huisgen cycloaddition) as well as with different catalysts (1 mol % per functional group) are illustrated in Figure 3c,d, respectively. For the uncatalyzed crosslinking reaction of trivalent alkyne **1** with trivalent azide **4*** (Table 1, Entry 6), a high reaction enthalpy of 227 kJ·mol^{-1} and high reaction temperatures (T_{onset} = 94 °C, T_p = 125 °C) were observed, corresponding to a conversion of 87%. For the homogenous catalyst Cu(PPh$_3$)$_3$F the lowest enthalpy was observed, while the click reaction happened at relatively low temperatures (Table 1, Entry 8). The determined apparent activation energy for the Huisgen cycloaddition was 75 kJ·mol^{-1}, while an enhanced apparent activation energy was detected for the click reaction using **TRGO-Cu$_2$O** (97 kJ·mol^{-1}). In contrast, a lower apparent activation energy was determined for Cu(PPh$_3$)$_3$F and Cu(PPh$_3$)$_3$Br, showing values of 61 and 70 kJ·mol^{-1} (Table 1, Entry 7 and 8), respectively. When reacting trivalent alkyne **1** with trivalent azide **5***, lower conversions were observed for the (uncatalyzed) Huisgen cycloaddition reaction as well as for all "click-crosslinking" reactions. Thus, a conversion of 66% was achieved for the uncatalyzed reaction (Table 1, Entry 11), corresponding to an enthalpy of 174 kJ·mol^{-1}, while "click-crosslinking" happened at high temperatures (T_{onset} and T_p of 102 and 141 °C). In comparison to the other catalysts, **TRGO-Cu$_2$O** (Table 1, Entry 14) resulted in the lowest enthalpy of 127 kJ·mol^{-1}, which was achieved at T_{onset} and T_p of 32 and 56 °C, respectively. In the case of the homogeneous catalysts (Cu(PPh$_3$)$_3$F and Cu(PPh$_3$)$_3$Br) a similar conversion of 53 to 57% was observed, while lower reaction temperatures were achieved in the presence of Cu(PPh$_3$)$_3$F (Table 1, Entry 13, T_{onset} = 38 °C, T_p = 62 °C). According to the obtained results, Cu(PPh$_3$)$_3$F turned out to be the best catalyst for the "click-crosslinking" reaction of trivalent alkyne **1** and trivalent azides **4*** or **5***.

3.3. DSC Investigation of "Click-Crosslinking" Trivalent Alkyne 1 and Trivalent Azides 3*, 4* and 5* at 0 °C

We were interested in quantifying "click-crosslinking" reactions during preparation and mixing of the components, thus understanding whether "click" reactions at 0 °C play an essential role. Therefore, DSC measurements were applied to investigate the kinetic behavior of the "click-crosslinking" reaction between trivalent alkyne **1** and trivalent azides **3***, **4*** and **5*** with different chain lengths to find a suitable and fast catalytic system for the CuAAC. As the usage of $Cu(PPh_3)_3F$ resulted in high enthalpies at low crosslinking temperatures in the previously performed experiments, $Cu(PPh_3)_3F$ was chosen as catalyst to check the activity of the multivalent azides and alkynes within "click-crosslinking" at 0 °C. Therefore, 1:1 mixtures of trivalent alkyne **1** and different trivalent azides (**3***, **4*** or **5***) together with $Cu(PPh_3)_3F$ (1 mol % per functional group) were prepared. Immediately after preparation of the mixtures, DSC measurements were run at a heating rate of 5 K·min^{-1} to observe the enthalpy of the "click-crosslinking" reaction at time zero (ΔH_0). Afterwards, the mixtures were stored at 0 °C and further DSC investigations were conducted in defined time intervals and the conversion was determined (for more details see Supplementary Materials Figure S4 and Table S1).

Immediately after mixing trivalent alkyne **1** and trivalent azide **3***, the measured enthalpy of the "click-crosslinking" reaction was 186 kJ·mol^{-1}, and T_{onset} and T_p were 39 and 67 °C, respectively (Figure 4, black squares; Supplementary Materials, Table S1, Entry 1). After 48 h storage at 0 °C, the "click-crosslinking" enthalpy decreased around one half of its initial value and reached 82 kJ·mol^{-1}, corresponding to 56% conversion (Supplementary Materials, Table S1, Entry 3). After 312 h, the "click-crosslinking" enthalpy decreased further to 24 kJ·mol^{-1} (13 days, 87% conversion, Supplementary Materials, Table S1, Entry 9). Afterwards, the conversion did not show any significant increase, and the "click-crosslinking" reaction between alkyne **1** and azide **3*** reached its maximum conversion of 93% at 0 °C after 576 h (24 days, Supplementary Materials, Table S1, Entry 12).

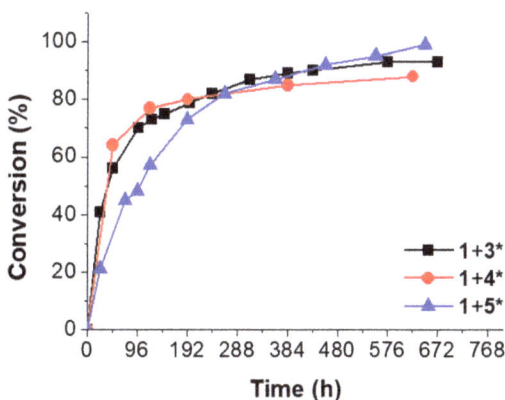

Figure 4. Conversion vs. time of the "click-crosslinking" reaction of trivalent alkyne **1** and trivalent azides **3*** (black squares), **4*** (red circles) or **5*** (blue triangles) at 0 °C with $Cu(PPh_3)_3F$ as catalysts at a heating rate of 5 K·min^{-1}.

The mixture of trivalent alkyne **1** and trivalent azide **4*** showed a high enthalpy of 222 kJ·mol^{-1} immediately after sample preparation, and T_{onset} and T_p were 37 and 54 °C, respectively (Figure 4, red circles; Supplementary Materials, Table S1, Entry 14). After 48 h, the "click-crosslinking" enthalpy decreased to 80 kJ·mol^{-1} and a conversion of 64% was observed (Supplementary Materials, Table S1, Entry 15). After 384 h, the conversion of the "click-crosslinking" reaction at 0 °C increased further to 85%, corresponding to a reaction enthalpy of 33 kJ·mol^{-1} (Supplementary Materials, Table S1, Entry 18). In comparison, the conversion after 624 h did not significantly change and reached a constant value of 88% (27 kJ·mol^{-1}, Supplementary Materials, Table S1, Entry 19). Thus, it was concluded that the maximum "click-crosslinking" conversion obtainable by converting trivalent alkyne **1** and trivalent azide **4*** at 0 °C is below 90%.

Immediately after preparation of the reaction mixture of trivalent alkyne **1** and trivalent azide **5***, a "click-crosslinking" enthalpy of 172 kJ·mol^{-1} was observed, and T_{onset} and T_p were 63 and 86 °C, respectively (Figure 4, blue curve; Supplementary Materials, Table S1, Entry 20). After 96 h, the reaction enthalpy of the "click-crosslinking" reaction decreased to 89 kJ·mol^{-1}, related to a conversion of 48% (Supplementary Materials, Table S1, Entry 23). After 552 h, the enthalpy decreased further to 8 kJ·mol^{-1} (95% conversion), and T_{onset} was not detectable anymore due to the very low reaction enthalpy. After 648 h, a complete "click-crosslinking" conversion was achieved, and no reaction peak was observed.

Moreover, in all crosslinking experiments investigated at 0 °C, the peak temperature T_p and the onset temperature T_{onset} increased in comparison to their initial values. This increase of the crosslinking temperatures is mainly attributed to the early network formation taking place at 0 °C, and is therefore related to a slowed down monomer diffusion.

To sum up the DSC investigations at 0 °C, the "click-crosslinking" reactions of trivalent alkyne **1** with trivalent azides **3*** and **4*** at 0 °C were faster within the first 100 h than the corresponding "click-crosslinking" reaction in the presence of trivalent azide **5***. This phenomenon was mainly attributed to the increasing chain length of the trivalent azide. Thus, under the same conditions, molecules with shorter side chains show faster "click-crosslinking" in comparison to the molecules with the longer chain length, mainly attributed to lower viscosity. Nevertheless, slightly higher conversions of the "click-crosslinking" reaction were observed on long timescales for the trivalent azides with increased chain length.

3.4. Rheology Investigation of "Click-Crosslinking" Trivalent Alkyne 1 and Trivalent Azides 4* and 5*

The viscoelastic and the kinetic behavior of the "click-crosslinking" reaction between trivalent alkyne **1** and trivalent azides **4*** and **5*** (1:1 ratio of azide and alkyne) and the resulting self-healing capability were investigated via in situ rheology. Therefore, the isothermal "click-crosslinking" reaction was directly performed on a rheometer plate at 20 °C using Cu(PPh$_3$)$_3$F (1 mol % per functional group) as a catalyst. The observed crossover times for the "click-crosslinking" reaction of trivalent alkyne **1** and trivalent azides **4*** and **5*** with Cu(PPh$_3$)$_3$F were 190 and 1445 minutes, respectively (see Supplementary Materials, Figure S6). By comparing these times with the crossover time of the "click-crosslinking" reaction of trivalent azide **3*** with the trivalent alkyne **1** which was 35 minutes [22], it was concluded that with increasing chain length of the azide the crossover time increased. This observation was in line with the DSC investigations proving a decreased "click-crosslinking" reactivity with increasing chain length.

3.5. Synthesis and Characerization of TRGO-Cu$_2$O Prepared at Different Temperatures

TRGO-Cu$_2$O was prepared via thermal reduction of copper(II)-modified graphene oxide in a glass tube furnace according to a previously published procedure [21,64,65], while the reduction temperature was varied between 300 to 800 °C, finally obtaining six different batches of the desired heterogeneous copper(I) catalyst (for more details see Scheme S6 in the Supplementary Materials) to optimize the synthesis procedure in terms of the catalytic activity since DSC. Investigations for "click-crosslinking" trivalent alkyne **1** with trivalent azide **3***, **4*** or **5*** revealed slightly higher reaction temperatures (T_{onset} and T_p) in comparison to the homogeneously catalyzed "click-crosslinking" reactions in the presence of Cu(PPh$_3$)$_3$F.

The prepared TRGO-based copper(I) catalysts were investigated via XRD-measurements (see Supplementary Materials Figure S7a) and for all prepared **TRGO-Cu$_2$O** catalysts the characteristic reflex of **GO** at $2\theta = 11°$ has disappeared due to successful reduction. Furthermore, for all prepared catalysts, reflexes at $2\theta = 38°$ and $2\theta = 43°$ were observed related to the formed copper species (pure copper as well as copper(I)). For the **TRGO-Cu$_2$O** samples prepared at 700 and 800 °C, two additional reflexes at $2\theta = 51°$ and $2\theta = 26°$ were detected as characteristic reflexes for pure copper and graphite, respectively. Thus, it could be concluded, that some of the oxidic groups have been eliminated during thermal reduction partially resulting in graphite-like structures. The broad signal at $2\theta = 25°$ observed for all prepared TRGO-based copper(I) catalysts was caused by an interference with the sample holder.

The prepared **TRGO-Cu₂O** samples were further analyzed via TEM investigations (see Supplementary Materials Figure S7b–g) in which the formed nanosized copper(I) particles were visualized. The size of the particles was investigated via Image J. The particle size increased with increasing preparation temperature of **TRGO-Cu₂O**, and average particle-diameters between 25 to 150 nm were determined. Furthermore, it was observed, that TRGO-based catalysts prepared at 700 and 800 °C displayed a more disperse distribution of nanosized copper(I) particles, which may be attributed to the formation of pure copper interacting with graphite-like structures detected in XRD investigations.

3.6. Crosslinking Reactions of Alkynes and Azides in the Presence of *TRGO-Cu₂O* Prepared at Different Temperatures

The catalytic activity of the **TRGO-Cu₂O** catalysts (prepared at different temperatures) towards "click-crosslinking" was investigated via DSC investigations. Therefore, in the first step, a model reaction between phenylacetylene and benzyl azide (1:1 ratio of azide and alkyne) was investigated at a heating rate of $5\ K \cdot min^{-1}$, and the reaction temperatures (T_{onset} and T_p), the reaction enthalpy (ΔH) and the conversion were recorded (for more information see Supplementary Materials Figure S8 and Table S2). While comparing the different catalysts prepared at 300 to 800 °C, it was observed that T_{onset} and T_p decreased with increasing temperature applied during the reduction of copper(II)-modified graphene oxide towards **TRGO-Cu₂O**, in line with the expectation and the increasing size of the formed copper particles. Thereby, one exception was observed, and the catalyst prepared at 500 °C showed the highest peak temperature. Thus, FAAS measurements were performed to determine the loading of **TRGO** with immobilized copper nanoparticles. While most of the prepared catalysts displayed around 8 wt % of copper, a strong decrease was noted for the **TRGO-Cu₂O** synthesized at 500 °C, directly linked to the observed reduced catalytic activity during the click reaction of phenylacetylene and benzyl azide.

In the next step, the catalytic activity was tested in a more complex system suitable for the preparation of room-temperature based self-healing epoxy nanocomposites. "Click-crosslinking" of trivalent alkyne **1** and trivalent azide **3*** (1:1 ratio of azide and alkyne, assuming 66% azide content of **3***) in the presence of different **TRGO-Cu₂O** catalysts prepared from 300 to 800 °C (5 wt %) at a heating rate of $5\ K \cdot min^{-1}$ (see Supplementary Materials Figure S10) was investigated, and the obtained reaction temperatures (T_{onset} and T_p), the reaction enthalpies (ΔH) and the conversions are summarized in Table 2. For comparison, the non-catalyzed reaction between trivalent alkyne **1** and trivalent azide **3*** was repeated, showing similar reaction temperatures (T_{onset} = 96 vs. 91 °C, T_p = 130 vs. 133 °C; Table 2, Entry 1) as described before, but a reduced conversion due to the higher amount of bivalent residues (66 vs. 75% azide content, see Supplementary Materials Figure S7). All DSC measurements were performed in three independent experiments to ensure their reproducibility and to approximate the expected error. Thereby, differences especially in the peak shape were observed to be related to sample preparation and limited blending of the reactants as well as a limited diffusion to the catalyst surface being typical for a reaction directly catalyzed by a solid support [60,61].

Table 2. Thermal properties of the "click-crosslinking" reaction of trivalent alkyne **1** and trivalent azide **3*** with **TRGO-Cu₂O** as a catalyst (prepared at different temperatures) at a heating rate of 5 K·min⁻¹: Reaction temperatures (T_{onset} and T_p), reaction enthalpies (ΔH) and conversions.

Entry	Catalyst	T_{onset} [1] (°C)	T_p [1] (°C)	ΔH [2] (kJ·mol⁻¹)	Conversion [3] (%)
1	W/O	96	130	75	29
2	**TRGO-Cu₂O** (300 °C)	82	97	108	41
3	**TRGO-Cu₂O** (400 °C)	80	100	101	38
4	**TRGO-Cu₂O** (500 °C)	90	105	107	41
5	**TRGO-Cu₂O** (600 °C)	91	100	115	44
6	**TRGO-Cu₂O** (700 °C)	64	80	107	41
7	**TRGO-Cu₂O** (800 °C)	72	91	113	43

[1] Measurements were performed thrice, and the error is ≈ ±5 K. [2] Measurements were performed thrice and the error is ≈ ±6 kJ·mol⁻¹. [3] Calculated with respect to the enthalpy for 100% click conversion, which is $\Delta H = 262$ kJ·mol⁻¹ for the reference reaction of phenylacetylene and benzyl azide with 1 mol % of Cu(PPh₃)₃Br as catalyst.

For "click-crosslinking" trivalent alkyne **1** and trivalent azide **3***, a similar trend was observed as in the model reaction between phenylacetylene and benzyl azide. Thus, the reaction temperatures (T_{onset} and T_p) decreased with increasing reduction temperature applied during the preparation of the different **TRGO-Cu₂O** catalysts. Thereby, the lowest T_{onset} and T_p of 64 and 80 °C, respectively, were observed for "click-crosslinking" trivalent alkyne **1** and trivalent azide **3*** in the presence of **TRGO-Cu₂O** prepared at 700 °C while the catalyst prepared at 500 °C showed the worst result.

The obtained peak temperatures for the click model reaction of phenylacetylene and benzyl azide as well as for the "click-crosslinking" reaction of trivalent alkyne **1** and trivalent azide **3*** were correlated to the preparation temperature of **TRGO-Cu₂O** and the corresponding amount of copper within these catalysts (see Figure 5).

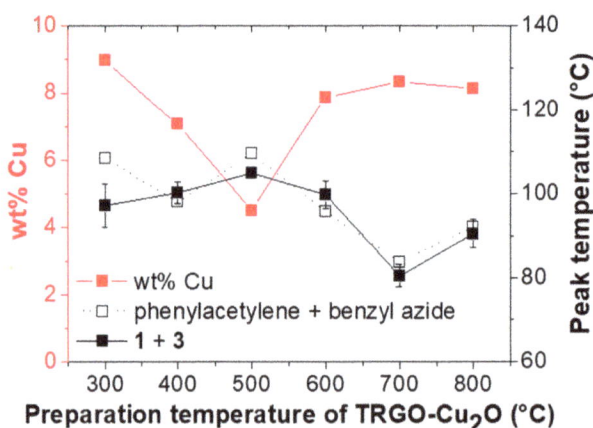

Figure 5. Amount (wt %) of copper within **TRGO-Cu₂O** and peak temperature (T_p) of the click reaction vs. the preparation temperature of **TRGO-Cu₂O** for the click reaction of phenylacetylene and benzyl azide and for the "click-crosslinking" reaction of trivalent alkyne **1** and trivalent azide **3***. Please note that the lines between the measuring points are drawn to guide the eye.

The already mentioned trends within the catalytic activity of **TRGO-Cu₂O** were well highlighted for both click reactions. The catalyst prepared at 500 °C showed a low amount of copper and consequently a high peak temperature during the click reactions, while the best results in terms of copper loading as well as peak temperature were observed for **TRGO-Cu₂O** prepared at 700 °C. The decrease of the copper loading for the catalyst prepared at 500 °C may be related to the decrease of oxygen-functional groups with increasing reduction temperature known to act as reactive sites for the nucleation and growth of metal nanoparticles [66,67]. Following this argumentation, a further decrease in the copper loading together with a decreased catalytic activity would be expected, which was not observed in this particular study. We therefore assume that either the formation of pure copper indicated by XRD investigations is boosting the catalytic activity or that the diffusion of metal atoms, especially favored at higher temperatures, leads to the formation of non-stable metal clusters influencing the catalytic activity [68].

Thus, for the preparation of self-healing nanocomposites according to a previously published procedure [19], the optimized **TRGO-Cu₂O** catalyst prepared at 700 °C was used. Thereby, the trivalent alkyne **1** together with this particular **TRGO-Cu₂O** were directly distributed within the epoxy matrix together with μm-sized capsules filled with trivalent azide **3***. Further self-healing investigations of our optimized healing system as well as the determination of self-healing efficiencies are ongoing in our laboratories and will be part of a future publication.

4. Conclusions

Different low molecular weight, multivalent azides with small structural changes were synthesized and their crosslinking kinetics was investigated in a CuAAC-based curing reaction. Therefore, different homogeneous and heterogeneous copper(I) catalysts were screened and the kinetic parameters such as the reaction temperatures, the enthalpy of the reaction as well as the apparent activation energies were recorded via DSC investigations. We observed, that the "click-crosslinking" reactivity decreased with increasing chain length of the azide. Furthermore, a significant click reactivity of all investigated azides could be proven already at 0 °C.

The reaction conditions for the preparation of our home-made **TRGO-Cu₂O** catalyst were optimized: When increasing the reaction temperature to 700 °C, the resulting copper(I) catalyst displayed the highest catalytic activity as shown in model click reactions as well as in "click-crosslinking" reactions between trivalent alkyne **1** and trivalent azide **3***.

The tuned catalyst was subsequently dispersed in an epoxy matrix together with the trivalent alkyne **1** and the encapsulated trivalent azide **3*** (μm-sized capsules). Further self-healing investigations of the so prepared capsule-based self-healing graphene-supported epoxy nanocomposites are ongoing and will be part of a future publication.

Supplementary Materials: The following are available online at www.mdpi.com/2073-4360/10/1/ 17/s1. Scheme S1: Synthesis of trimethylolpropane tripropargyl ether (TMPTPE, **1**); Scheme S2: Synthesis of azidated trimethylolpropane tripropargyl ether (N₃TMPTPE, **2**); Scheme S3: Synthesis of ((2-((2-acetoxy-3-azidopropoxy)methyl)-2-ethylpropane-1,3-diyl) bis (oxy))bis(3-azidopropane-1,2-diyl) diacetate **3**; Scheme S4: Synthesis of ((2-((3-azido-2-(butyryloxy)propoxy)methyl)-2-ethylpropane-1,3-diyl) bis(oxy))bis (3-azidopropane-1,2-diyl) dibutyrate **4**; Scheme S5: Synthesis of ((2-((3-azido-2-(decanoyloxy)propoxy)methyl)- 2-ethylpropane-1,3-diyl) bis (oxy))bis(3-azidopropane-1,2-diyl) bis(decanoate) **5**; Scheme S6: Synthesis of modified, thermally reduced graphene oxide **TRGO-Cu₂O**; Figure S1: ¹H-NMR spectrum of trimethylolpropane triglycidyl ether; Figure S2: ¹H-NMR spectrum of **4**; Figure S3: ¹³C-NMR spectrum of **4**; Figure S4: ¹H-NMR spectrum of **5**; Figure S5: ¹³C-NMR spectrum of **5**; Figure S6: Rheological behavior of trialkyne **1** and (a) triazide **4*** or (b) triazide **5*** applying Cu(PPh₃)₃F as a catalyst at 20 °C; Figure S7: (a) XRD measurements of **GO-Cu(II)** and **TRGO-Cu₂O** prepared at different temperatures (300 °C–800 °C). Reflexes of Cu, Cu₂O, GO and graphite are shown for comparison. TEM images of **TRGO-Cu₂O** prepared at (b) 300 °C, (c) 400 °C, (d) 500 °C, (e) 600 °C, (f) 700 °C and (g) 800 °C; Figure S8: DSC measurements of the click reaction of phenylacetylene and benzyl azide with **TRGO-Cu₂O** as a catalyst at a heating rate of 5 K·min⁻¹; Figure S9: DSC measurements of the crosslinking reaction of trivalent alkyne **1** and trivalent azide **3*** without catalyst (W/O) at a heating rate of 5 K·min⁻¹; Figure S10: DSC measurements of the "click-crosslinking" reaction of trivalent alkyne **1** and trivalent azide **3*** with **TRGO-Cu₂O** as a catalyst prepared at (a) 300 °C, (b) 400 °C, (c) 500 °C, (d) 600 °C, (e) 700 °C and (f) 800 °C at a heating rate of 5 K·min⁻¹; Table S1: Conversion of the "click-crosslinking" reaction of trivalent alkyne **1** and trivalent azide **3***, **4*** or **5*** at 0 °C with Cu(PPh₃)₃F as a catalyst at a heating rate of 5 K·min⁻¹: Reaction temperatures (T_{onset} and T_p), reaction enthalpies (ΔH) and conversions; Table S2: Thermal properties of the click reaction of phenyl acetylene and benzyl azide with **TRGO-Cu₂O** as a catalyst (prepared at different temperatures) at a heating rate of 5 K·min⁻¹: Reaction temperatures (T_{onset} and T_p), reaction enthalpies (ΔH) and conversions.

Acknowledgments: The authors thank Anette Meister for providing TEM images, Thomas Heymann for providing FAAS measurements and Toni Buttlar for providing XRD measurements. The authors are grateful for Grant DFG BI 1337/8-1 and BI 1137/8-2 of the German Science Foundation within the framework of the SPP 1568 ("Design and Generic Principles of Self-Healing Materials"). Diana Döhler additionally acknowledges the support from the DFG Young Research GRANT SHE-STARS within the SPP 1568. Furthermore, this project has received funding from the European Union's Seventh Framework Program for research, technological development and demonstration under grant agreement no. 313978 which is thankfully acknowledged.

Author Contributions: Neda Kargarfard and Norman Diedrich prepared all described compounds and performed all described characterizations and measurements if not otherwise mentioned. Harald Rupp, Diana Döhler and Wolfgang H. Binder designed the paper and wrote parts of the paper. Diana Döhler and Wolfgang H. Binder designed the work and provided funding.

Conflicts of Interest: The authors declare no conflict of interest.

References

1. White, S.R.; Sottos, N.R.; Geubelle, P.H.; Moore, J.S.; Kessler, M.R.; Sriram, S.R.; Brown, E.N.; Viswanathan, S. Autonomic healing of polymer composites. *Nature* **2001**, *409*, 794–817. [CrossRef] [PubMed]
2. Michael, P.; Döhler, D.; Binder, W.H. Improving autonomous self-healing via combined chemical/physical principles. *Polymer* **2015**, *69*, 216–227. [CrossRef]
3. Huang, G.C.; Lee, J.K.; Kessler, M.R. Evaluation of norbornene-based adhesives to amine-cured epoxy for self-healing applications. *Macromol. Mater. Eng.* **2011**, *296*, 965–972. [CrossRef]
4. Aïssa, B.; Haddad, E.; Jamroz, W.; Hassani, S.; Farahani, R.D.; Merle, P.G.; Therriault, D. Micromechanical characterization of single-walled carbon nanotube reinforced ethylidene norbornene nanocomposites for self-healing applications. *Smart Mater. Struct.* **2012**, *21*, 105028. [CrossRef]
5. Mauldin, T.C.; Leonard, J.; Earl, K.; Lee, J.K.; Kessler, M.R. Modified rheokinetic technique to enhance the understanding of microcapsule-based self-healing polymers. *ACS Appl. Mater. Interfaces* **2012**, *4*, 1831–1837. [CrossRef] [PubMed]
6. Raimondo, M.; Guadagno, L. Healing efficiency of epoxy-based materials for structural applications. *Polym. Compos.* **2013**, *34*, 1525–1532. [CrossRef]
7. Mariconda, A.; Longo, P.; Agovino, A.; Guadagno, L.; Sorrentino, A.; Raimondo, M. Synthesis of ruthenium catalysts functionalized graphene oxide for self-healing applications. *Polymer* **2015**, *69*, 330–342. [CrossRef]
8. Monfared Zanjani, J.S.; Okan, B.S.; Letofsky-Papst, I.; Menceloglu, Y.; Yildiz, M. Repeated self-healing of nano and micro scale cracks in epoxy based composites by tri-axial electrospun fibers including different healing agents. *RSC Adv.* **2015**, *5*, 73133–73145. [CrossRef]
9. Wang, B.; Mireles, K.; Rock, M.; Li, Y.; Thakur, V.K.; Gao, D.; Kessler, M.R. Synthesis and preparation of bio-based ROMP thermosets from functionalized renewable isosorbide derivative. *Macromol. Chem. Phys.* **2016**, *217*, 871–879. [CrossRef]
10. Longo, P.; Mariconda, A.; Calabrese, E.; Raimondo, M.; Naddeo, C.; Vertuccio, L.; Russo, S.; Iannuzzo, G.; Guadagno, L. Development of a new stable ruthenium initiator suitably designed for self-repairing applications in high reactive environments. *J. Ind. Eng. Chem.* **2017**, *54*, 234–251. [CrossRef]
11. Döhler, D.; Michael, P.; Binder, W.H. Autocatalysis in the room temperature copper(I)-catalyzed alkyne-azide "click" cycloaddition of multivalent poly(acrylate)s and poly(isobutylene)s. *Macromolecules* **2012**, *45*, 3335–3345. [CrossRef]
12. Döhler, D.; Michael, P.; Binder, W.H. CuAAC-based click chemistry in self-healing polymers. *Acc. Chem. Res.* **2017**, *50*, 2610–2620. [CrossRef] [PubMed]
13. Döhler, D.; Rana, S.; Rupp, H.; Bergmann, H.; Behzadi, S.; Crespy, D.; Binder, W.H. Qualitative sensing of mechanical damage by a fluorogenic "click" reaction. *Chem. Commun.* **2016**, *52*, 11076–11079. [CrossRef] [PubMed]
14. Döhler, D.; Zare, P.; Binder, W.H. Hyperbranched polyisobutylenes for self-healing polymers. *Polym. Chem.* **2014**, *5*, 992–1000. [CrossRef]
15. Gragert, M.; Schunack, M.; Binder, W.H. Azide/Alkyne-"Click"-Reactions of encapsulated reagents: Toward self-healing materials. *Macromol. Rapid Commun.* **2011**, *32*, 419–425. [CrossRef] [PubMed]
16. Michael, P.; Binder, W.H. A mechanochemically triggered "click" catalyst. *Angew. Chem. Int. Ed.* **2015**, *54*, 13918–13922. [CrossRef] [PubMed]

17. Neumann, S.; Döhler, D.; Ströhl, D.; Binder, W.H. Chelation-assisted CuAAC of star-shaped polymers enables fast self-healing at low temperatures. *Polym. Chem.* **2016**, *7*, 2342–2351. [CrossRef]

18. Raimondo, M.; De Nicola, F.; Volponi, R.; Binder, W.H.; Michael, P.; Russo, S.; Guadagno, L. Self-repairing CFRPs targeted towards structural aerospace applications. *Int. J. Struct. Integr.* **2016**, *7*, 656–670. [CrossRef]

19. Rana, S.; Döhler, D.; Nia, A.S.; Nasir, M.; Beiner, M.; Binder, W.H. "Click"-triggered self-healing graphene nanocomposites. *Macromol. Rapid Commun.* **2016**, *37*, 1715–1722. [CrossRef] [PubMed]

20. Schunack, M.; Gragert, M.; Döhler, D.; Michael, P.; Binder, W.H. Low-temperature Cu(I)-catalyzed "click" reactions for self-healing polymers. *Macromol. Chem. Phys.* **2012**, *213*, 205–214. [CrossRef]

21. Shaygan Nia, A.; Rana, S.; Döhler, D.; Noirfalise, X.; Belfiore, A.; Binder, W.H. Click chemistry promoted by graphene supported copper nanomaterials. *Chem. Commun.* **2014**, *50*, 15374–15377. [CrossRef] [PubMed]

22. Shaygan Nia, A.; Rana, S.; Döhler, D.; Osim, W.; Binder, W.H. Nanocomposites via a direct graphene-promoted "click" reaction. *Polymer* **2015**, *79*, 21–28. [CrossRef]

23. Shaygan Nia, A.; Rana, S.; Döhler, D.; Jirsa, F.; Meister, A.; Guadagno, L.; Koslowski, E.; Bron, M.; Binder, W.H. Carbon-supported copper nanomaterials: Recyclable catalysts for huisgen [3 + 2] cycloaddition reactions. *Chem. Eur. J.* **2015**, *21*, 10763–10770. [CrossRef] [PubMed]

24. Sheng, X.; Mauldin, T.C.; Kessler, M.R. Kinetics of bulk azide/alkyne "click" polymerization. *J. Polym. Sci. Part A Polym. Chem.* **2010**, *48*, 4093–4102. [CrossRef]

25. Sheng, X.; Rock, D.M.; Mauldin, T.C.; Kessler, M.R. Evaluation of different catalyst systems for bulk polymerization through "click" chemistry. *Polymer* **2011**, *52*, 4435–4441. [CrossRef]

26. Vasiliu, S.; Kampe, B.; Theil, F.; Dietzek, B.; Döhler, D.; Michael, P.; Binder, W.H.; Popp, J. Insights into the mechanism of polymer coating self-healing using raman spectroscopy. *Appl. Spectrosc.* **2014**, *68*, 541–548. [CrossRef] [PubMed]

27. Döhler, D.; Peterlik, H.; Binder, W.H. A dual crosslinked self-healing system: Supramolecular and covalent network formation of four-arm star polymers. *Polymer* **2015**, *69*, 264–273. [CrossRef]

28. Cao, S.; Li, S.; Li, M.; Xu, L.; Ding, H.; Xia, J.; Zhang, M.; Huang, K. A thermal self-healing polyurethane thermoset based on phenolic urethane. *Polym. J.* **2017**, *49*, 775–781. [CrossRef]

29. Haghayegh, M.; Mirabedini, S.M.; Yeganeh, H. Preparation of microcapsules containing multi-functional reactive isocyanate-terminated-polyurethane-prepolymer as healing agent, part II: Corrosion performance and mechanical properties of a self-healing coating. *RSC Adv.* **2016**, *6*, 50874–50886. [CrossRef]

30. Haghayegh, M.; Mirabedini, S.M.; Yeganeh, H. Microcapsules containing multi-functional reactive isocyanate-terminated polyurethane prepolymer as a healing agent. Part 1: Synthesis and optimization of reaction conditions. *J. Mater. Sci.* **2016**, *51*, 3056–3068. [CrossRef]

31. Hillewaere, X.K.D.; Teixeira, R.F.A.; Nguyen, L.-T.T.; Ramos, J.A.; Rahier, H.; Du Prez, F.E. Autonomous self-healing of epoxy thermosets with thiol-isocyanate chemistry. *Adv. Funct. Mater.* **2014**, *24*, 5575–5583. [CrossRef]

32. Keller, M.W.; Hampton, K.; McLaury, B. Self-healing of erosion damage in a polymer coating. *Wear* **2013**, *307*, 218–225. [CrossRef]

33. McIlroy, D.A.; Blaiszik, B.J.; Caruso, M.M.; White, S.R.; Moore, J.S.; Sottos, N.R. Microencapsulation of a reactive liquid-phase amine for self-healing epoxy composites. *Macromolecules* **2010**, *43*, 1855–1859. [CrossRef]

34. Park, J.I.; Choe, A.; Kim, M.P.; Ko, H.; Lee, T.H.; Noh, S.M.; Kim, J.C.; Cheong, I.W. Water-adaptive and repeatable self-healing polymers bearing bulky urea bonds. *Polym. Chem.* **2018**, *9*, 11–19. [CrossRef]

35. Schüssele, A.C.; Nübling, F.; Thomann, Y.; Carstensen, O.; Bauer, G.; Speck, T.; Mülhaupt, R. Self-healing rubbers based on NBR blends with hyperbranched polyethylenimines. *Macromol. Mater. Eng.* **2012**, *297*, 411–419. [CrossRef]

36. Ying, H.; Zhang, Y.; Cheng, J. Dynamic urea bond for the design of reversible and self-healing polymers. *Nat. Commun.* **2014**, *5*, 1–9. [CrossRef] [PubMed]

37. Zechel, S.; Geitner, R.; Abend, M.; Siegmann, M.; Enke, M.; Kuhl, N.; Klein, M.; Vitz, J.; Gräfe, S.; Dietzek, B.; et al. Intrinsic self-healing polymers with a high E-modulus based on dynamic reversible urea bonds. *NPG Asia Mater.* **2017**, *9*, 420. [CrossRef]

38. Akiyama, T.; Ushio, A.; Itoh, Y.; Kawaguchi, Y.; Matsumoto, K.; Jikei, M. Synthesis and healing properties of poly(arylether sulfone)-poly(alkylthioether) multiblock copolymers containing disulfide bonds. *J. Polym. Sci. Part A Polym. Chem.* **2017**, *55*, 3545–3553. [CrossRef]

39. Billiet, S.; Van Camp, W.; Hillewaere, X.K.D.; Rahier, H.; Du Prez, F.E. Development of optimized autonomous self-healing systems for epoxy materials based on maleimide chemistry. *Polymer* **2012**, *53*, 2320–2326. [CrossRef]

40. Chakma, P.; Rodrigues Possarle, L.H.; Digby, Z.A.; Zhang, B.; Sparks, J.L.; Konkolewicz, D. Dual stimuli responsive self-healing and malleable materials based on dynamic thiol-Michael chemistry. *Polym. Chem.* **2017**, *8*, 6534–6543. [CrossRef]

41. Kuhl, N.; Geitner, R.; Vitz, J.; Bode, S.; Schmitt, M.; Popp, J.; Schubert, U.S.; Hager, M.D. Increased stability in self-healing polymer networks based on reversible Michael addition reactions. *J. Appl. Polym. Sci.* **2017**, *134*, 44805. [CrossRef]

42. Pepels, M.; Filot, I.; Klumperman, B.; Goossens, H. Self-healing systems based on disulfide-thiol exchange reactions. *Polym. Chem.* **2013**, *4*, 4955–4965. [CrossRef]

43. Yoon, J.A.; Kamada, J.; Koynov, K.; Mohin, J.; Nicolaÿ, R.; Zhang, Y.; Balazs, A.C.; Kowalewski, T.; Matyjaszewski, K. Self-healing polymer films based on thiol–disulfide exchange reactions and self-healing kinetics measured using atomic force microscopy. *Macromolecules* **2011**, *45*, 142–149. [CrossRef]

44. Yue, H.-B.; Fernández-Blázquez, J.P.; Beneito, D.F.; Vilatela, J.J. Real time monitoring of click chemistry self-healing in polymer composites. *J. Mater. Chem. A* **2014**, *2*, 3881–3887. [CrossRef]

45. Zhao, Y.H.; Vuluga, D.; Lecamp, L.; Burel, F. Photoinitiated thiol-epoxy addition for the preparation of photoinduced self-healing fatty coatings. *RSC Adv.* **2016**, *6*, 32098–32105. [CrossRef]

46. Zhu, D.Y.; Cao, G.S.; Qiu, W.L.; Rong, M.Z.; Zhang, M.Q. Self-healing polyvinyl chloride (PVC) based on microencapsulated nucleophilic thiol-click chemistry. *Polymer* **2015**, *69*, 1–9. [CrossRef]

47. Chung, U.S.; Min, J.H.; Lee, P.-C.; Koh, W.-G. Polyurethane matrix incorporating PDMS-based self-healing microcapsules with enhanced mechanical and thermal stability. *Colloids Surf. A* **2017**, *518*, 173–180. [CrossRef]

48. Wei, K.; Gao, Z.; Liu, H.; Wu, X.; Wang, F.; Xu, H. Mechanical activation of platinum—acetylide complex for olefin hydrosilylation. *ACS Macro Lett.* **2017**, *6*, 1146–1150. [CrossRef]

49. Meldal, M.; Tornøe, C.W. *Peptides, the Wave of the Future*; American Peptide Society: San Diego, CA, USA, 2001.

50. Rostovtsev, V.V.; Green, L.G.; Fokin, V.V.; Sharpless, K.B. A stepwise huisgen cycloaddition process: Copper(I)-catalyzed regioselective "ligation" of azides and terminal alkynes. *Angew. Chem. Int. Ed.* **2002**, *41*, 2596–2599. [CrossRef]

51. Tornøe, C.W.; Christensen, C.; Meldal, M. Peptidotriazoles on solid phase: [1,2,3]-triazoles by regiospecific copper(I)-catalyzed 1,3-dipolar cycloadditions of terminal alkynes to azides. *J. Org. Chem.* **2002**, *67*, 3057–3064. [CrossRef] [PubMed]

52. Binder, W.H.; Sachsenhofer, R. 'Click' Chemistry in polymer and materials science. *Macromol. Rapid Commun.* **2007**, *28*, 15–54. [CrossRef]

53. Binder, W.H.; Sachsenhofer, R. "Click"-chemistry in polymer and material science: An update. *Macromol. Rapid Commun.* **2008**, *29*, 952–981. [CrossRef]

54. Zhao, Y.; Döhler, D.; Lv, L.-P.; Binder, W.H.; Landfester, K.; Crespy, D. Facile phase-separation approach to encapsulate functionalized polymers in core–shell nanoparticles. *Macromol. Chem. Phys.* **2014**, *215*, 198–204. [CrossRef]

55. Huisgen, R. 1,3-Dipolar Cycloadditions. Past and Future. *Angew. Chem. Int. Ed.* **1963**, *2*, 565–598. [CrossRef]

56. Huisgen, R. Kinetics and reaction mechanisms: Selected examples from the experience of forty years. *Pure Appl. Chem* **1989**, *61*, 613–628. [CrossRef]

57. Chan, T.R.; Hilgraf, R.; Sharpless, K.B.; Fokin, V.V. Polytriazoles as copper(I)-stabilizing ligands in catalysis. *Org. Lett.* **2004**, *6*, 2853–2855. [CrossRef] [PubMed]

58. Gorman, I.E.; Willer, R.L.; Kemp, L.K.; Storey, R.F. Development of a triazole-cure resin system for composites: Evaluation of alkyne curatives. *Polymer* **2012**, *53*, 2548–2558. [CrossRef]

59. Kantheti, S.; Sarath, P.S.; Narayan, R.; Raju, K.V.S.N. Synthesis and characterization of triazole rich polyether polyols using click chemistry for highly branched polyurethanes. *React. Funct. Polym.* **2013**, *73*, 1597–1605. [CrossRef]

60. Bond, G.C.; Keane, M.A.; Kral, H.; Lercher, J.A. Compensation phenomena in heterogeneous catalysis: General principles and a possible explanation. *Cat. Rev.* **2000**, *42*, 323–383. [CrossRef]

61. Mulokozi, A.M. Kinetic parameters in heterogeneous kinetics. *Thermochim. Acta* **1992**, *197*, 363–372. [CrossRef]

62. Nonahal, M.; Rastin, H.; Saeb, M.R.; Sari, M.G.; Moghadam, M.H.; Zarrintaj, P.; Ramezanzadeh, B. Epoxy/PAMAM dendrimer-modified graphene oxide nanocomposite coatings: Nonisothermal cure kinetics study. *Prog. Org. Coat.* **2018**, *114*, 233–243. [CrossRef]

63. Hein, J.E.; Fokin, V.V. Copper-catalyzed azide-alkyne cycloaddition (CuAAC) and beyond: New reactivity of copper(I) acetylides. *Chem. Soc. Rev.* **2010**, *39*, 1302–1315. [CrossRef] [PubMed]

64. Kyotani, T.; Suzuki, K.-Y.; Yamashita, H.; Tomita, A. Formation of carbon-metal composites from metal ion exchanged graphite oxide. *Tanso* **1993**, *1993*, 255–265. [CrossRef]

65. Hummers, W.S.; Offeman, R.E. Preparation of graphitic oxide. *J. Am. Chem. Soc.* **1958**, *80*, 1339. [CrossRef]

66. Kamat, P.V. Graphene-based nanoarchitectures. anchoring semiconductor and metal nanoparticles on a two-dimensional carbon support. *J. Phys. Chem. Lett.* **2010**, *1*, 520–527. [CrossRef]

67. Goncalves, G.; Marques, P.A.A.P.; Granadeiro, C.M.; Nogueira, H.I.S.; Singh, M.K.; Grácio, J. Surface modification of graphene nanosheets with gold nanoparticles: The role of oxygen moieties at graphene surface on gold nucleation and growth. *Chem. Mater.* **2009**, *21*, 4796–4802. [CrossRef]

68. Gan, Y.; Sun, L.; Banhart, F. One- and two-dimensional diffusion of metal atoms in graphene. *Small* **2008**, *4*, 587–591. [CrossRef] [PubMed]

polymers

MDPI

Article

Thermomechanical Behavior of Polymer Composites Based on Edge-Selectively Functionalized Graphene Nanosheets

Ki-Ho Nam [1,†], Jaehyun Cho [1,†] and Hyeonuk Yeo [2,*]

[1] Institute of Advanced Composite Materials, Korea Institute of Science and Technology (KIST), Jeonbuk 565-902, Korea; khnam@kist.re.kr (K.-H.N.); jaehyun0119@kist.re.kr (J.C.)
[2] Department of Chemistry Education, Chemistry Building, Kyungpook National University, 80, Daehak-ro, Buk-gu, Daegu 41566, Korea
*   Correspondence: yeo@knu.ac.kr; Tel.: +82-53-950-5905
†   K.-H.N. and J.C. contributed equally to this work.

Received: 21 November 2017; Accepted: 23 December 2017; Published: 26 December 2017

Abstract: In this study, we demonstrate an effective approach based on a simple processing method to improve the thermomechanical properties of graphene polymer composites (GPCs). Edge-selectively functionalized graphene (EFG) was successfully obtained through simple ball milling of natural graphite in the presence of dry ice, which acted as the source of carboxyl functional groups that were attached to the peripheral basal plane of graphene. The resultant EFG is highly dispersible in various organic solvents and contributes to improving their physical properties because of its unique characteristics. Pyromellitic dianhydride (PMDA) and 4,4′-oxydianiline (ODA) were used as monomers for constructing the polyimide (PI) backbone, after which PI/EFG composites were prepared by in situ polymerization. A stepwise thermal imidization method was used to prepare the PI films for comparison purposes. The PI/EFG composite films were found to exhibit reinforced thermal and thermo-mechanical properties compared to neat PI owing to the interaction between the EFG and PI matrix.

Keywords: graphene; polyimide; polymer composite; thermo-mechanical properties

1. Introduction

In recent years, polymer composites using graphene derivatives as filler have been studied with the aim of practical application in a wide range of academic and industrial fields [1–5]. In addition to their superior mechanical and electrical properties, because of the advantage of easily granting suitable physical and chemical characteristics for specific purposes, utilizing them as materials with excellent future prospects has been attracting much attention [6–15]. However, graphene as a filler component has a tendency to readily aggregate, and this becomes a major disadvantage in that it is difficult to conjugate its inherently excellent properties. Therefore, dispersing these derivatives uniformly in a polymer matrix rather than maintaining their superior properties has become a priority [16–20]. Accordingly, various methods have been investigated with the aim of achieving a high degree of dispersibility even if this is at the cost of reduced properties [21–26]. One of the most convincing ways is the chemical functionalization of the graphene, which makes it possible to improve the dispersibility to allow production on an industrial scale. However, since the method produces products with a variety of defects, extensive deterioration of the physical properties occurs. This introduces many problems in terms of their utilization as filler such as that they either have a very small reinforcement effect or no effect. In this regard, polymer composites using graphene fillers modified by the traditional chemical methods are problematic in that they are actually unable to provide any reinforcement.

As a result, the key to a successful strategy for developing graphene polymer composites (GPCs) is the enhancement of their dispersibility and the suppression of defects. These two characteristics are almost inversely related.

With this in mind, many researchers have focused on finding a novel method to physically disperse graphene sheets that is sufficiently scalable to enable the mass production of GPCs [27–31]. In particular, one of the promising methods is that of Jeon et al. who reported that chemical functional groups could be effectively introduced at the edge of natural graphite to yield modified graphene through a physical process using ball milling with dry ice [32]. This method holds promise as a very good alternative for a chemical solution process to improve the dispersibility while retaining the unique excellent properties of graphene. This is because the method selectively introduces functional groups at the edge of graphene rather than in the basal plane of graphene. However, the application of edge-selectively functionalized graphene (EFG) to GPCs remains relatively uncommon.

In this paper, we report a series of GPCs using EFG as a reinforcement filler and polyimide (PI), which is one of the outstanding polymer materials, as a matrix polymer [33–35]. Especially, as the EFG filler has a high terminal carboxylic acid ratio, our GPCs would exhibit a strong direct and/or indirect interaction with the amine moieties of polyimides during in situ polymerization [18,19,36]. This effect is expected to facilitate homogeneous mixing as a result of the interactions between the EFG and PI chains, and also to improve the cohesion between the polymer chains, thereby affecting the thermodynamic properties. To grant the interactions to GPCs, there have been many previous studies such as introducing amines moiety in graphene [18,19]. However, the methods created a lot of defects in graphene for the introduction of functional groups, and the chemical reactions at several stages were essential. However, EFG is prepared via a very simple one-step reaction that allows the introduction of specific functional groups from natural graphite with low defects, which is practically usable. In addition, we controlled the level of defects on the filler by modifying the previous report. As a result, we succeeded in obtaining GPCs with greatly enhanced thermal and thermo-mechanical properties. Furthermore, the use of EFG in applications is expected to provide a new perspective for manufacturing GPCs.

2. Materials and Methods

2.1. Materials

Graphite was acquired from Alfa Aesar (Graphite powder, natural, briquetting grade, −100 mesh, 99.9995%) and used without any purification. Dry ice was purchased from Taekyung Chemical Co., Ltd., Seoul, Korea. In addition, pyromellitic dianhydride (PMDA, >98%) and 4,4′-oxydianiline (ODA, >97%) were acquired from Tokyo Chemical Industry Co., Ltd., Tokyo, Japan and used as received. N-Methyl-2-pyrrolidone (NMP) was purified by a two-column solid-state purification system (Glass-contour System, Joerg Meyer, Irvine, CA, USA).

2.2. Experimental Procedure

2.2.1. Preparation of EFG

Edge-selectively functionalized graphene (EFG) was synthesized based on a modification of the procedure reported by Baek et al. [32]. Before the ball-mill process involving pristine graphite, the graphite was vacuum dried at 80 °C for 24 h. Then, in a closed and perfectively sealed stainless steel jar, 5 g of graphite was simply milled with 50 g of dry ice, which is only half the amount recommended in the previous report. Milling was achieved by using a planetary mono ball-mill machine (Pulverisette 6, Fritsch, Germany) at 480 rpm with 50 g of stainless steel balls 5 mm in diameter for 48 h. Because the internal pressure increased to about 100 bar, the carbon dioxide is released from the dry ice in the form of a supercritical fluid; therefore, a smaller amount of dry ice was sufficient. The resulting-EFG was further purified by Soxhlet extraction with an aqueous solution of 1 M HCl to completely acidify the carboxyl derivatives. Then, the products were washed with distilled water and freeze-dried at −120 °C for 72 h.

2.2.2. Preparation of PI/EFG Composites

The PI/EFG composites were fabricated via in situ polymerization and subsequent thermal cyclic dehydration. EFG powder (filler loadings of 0.1–3 wt %) was dissolved in NMP using an ultrasonic bath for 30 min. Next, equivalent molar ratios of ODA (2 mmol) and PMDA (2 mmol) were dissolved in the EFG dispersion with continuous stirring for 24 h in an argon atmosphere. The poly-condensation was identical to that of the viscous polyamic acid (PAA)/EFG solution. After degassing with a vacuum pump, the resulting mixture was spin-coated at 500 rpm onto a fused silica substrate and then pre-baked at 90 °C/2 h and 150 °C/1 h in vacuo. The PAA/EFG films were thermally imidized to PI/EFG composites under specific thermal curing at 200 °C/1 h, 250 °C/30 min, and 300 °C/30 min in a furnace under an argon atmosphere.

2.3. Measurements

The morphologies of the samples were investigated by scanning electron microscopy (SEM, Nova NanoSEM, FEI, Hillsboro, OR, USA) analysis. Fourier transform-infrared (FT-IR) spectra were obtained by using KBr pellets (FT-IR spectrometer, Vertex80v, Bruker, Billerica, MA, USA) in the range 400 to 4000 cm^{-1}. Raman spectroscopy was carried out using a micro-Raman spectrometer (inVia Raman spectrometer, Renishaw, Wotton-under-Edge, UK) with a laser having a 514.5 nm light source with an output of 0.15 mW. Thermogravimetric analysis (TGA) was performed on a Q50 machine (TGA, TA Instruments, New Castle, DE, USA) with an N_2 gas flow at a heating rate of 10 °C/min. Dynamic mechanical analysis (DMA) was conducted by using PI and GPC films (30 mm length, 5 mm wide, and ca. 30 μm thickness) on a Q800 machine (DMA, TA Instruments, New Castle, DE, USA) at a heating rate of 3 °C/min with a load frequency of 1 Hz in air. Electrical conductivity was measured by SM-8311 machine (HIOKI, Nagano, Japan) under 500 V.

3. Results and Discussion

3.1. Synthesis and Characterization of Edge-Selectively Functionalized Graphene (EFG)

3.1.1. Synthesis of EFG

The filler, edge-selectively functionalized graphene (EFG), was prepared by using a slightly modified version of the reported ball-milling process [32]. Similar to the published method, the EFG was functionalized by ball-milling and additional refinement; however, the amount of dry ice was halved to reduce the degree of functionalization as intended. Additionally, the processing time was controlled to prepare larger particles. In particular, ball-milling functionalization using dry ice is known to produce high carboxylic acid derivative content. This derivative can either interact directly with the amine moieties of the co-monomer component by way of covalent bonding during in situ polymerization, or interact indirectly with the imide groups of the resulting PIs by way of hydrogen bonding [18,19,36].

The morphological features of the synthesized EFG and those of pristine natural graphite were compared by SEM observation (Figure 1). The platelets of pristine graphite have an irregular shape of known size, which is the typical natural form. The sheets of synthesized EFG had a similar irregular shape, but the size was much smaller and the surface smoothed. However, as we intended, the size was controlled to be of the order of micrometers even though we first reported the sample to be prepared in ~500 nm size because our experiment was conducted under more mild conditions.

Polymers **2018**, *10*, 29

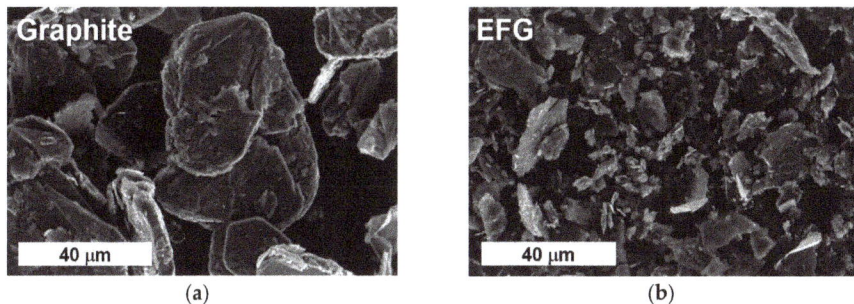

Figure 1. Scanning electron microscopy (SEM) images: (**a**) Natural Graphite; (**b**) Synthesized edge-selectively functionalized graphene (EFG).

3.1.2. Characterization of EFG

The chemical composition of EFG was analyzed by investigating the functionalization results by FT-IR and Raman spectroscopy (Figure 2). The FT-IR spectra (Figure 2a) revealed peaks for the EFG that were not observed in pristine graphite. Specifically, we could identify a peak at 1715 cm^{-1} assigned to C=O stretching and a broad peak around 1250 cm^{-1}, which indicated the coexistence of C–OH (hydroxyl), C–O–C (epoxy), and O=C–OH (carboxyl), respectively [32]. In addition, the peaks at 2920 and 1570 cm^{-1} observed in both EFG and graphite were assigned to sp^2 C–H and aromatic C–C stretching, respectively. These results indicated that the chemo-physical method for functionalizing graphene using a ball-mill and dry ice could efficiently introduce functional groups, especially carboxyl groups, without the use of strong acids such as sulfuric acid or nitric acid and explosive oxidants.

In addition, EFG was further characterized by micro-Raman spectroscopy. The measurements were carried out at both the edge and the basal plane of the pristine graphite and EFG. In the case of the graphite, the I_D/I_G ratio (i.e., the ratio of the intensity of the D-band at 1350 cm^{-1} to that of the G-band at 1585 cm^{-1}) was almost the same for both the edge and the plane. However, interestingly, EFG exhibited a significant difference in the values of the I_D/I_G ratio. The ratio at the edge of EFG was 0.65, which is a significant increase, whereas the value at the center of the sheet plane was almost maintained compared to the ratio for pristine graphite. These results suggest that the composition of the edge was selectively changed into more disordered structures with the sp^2-defects mainly derived from carboxylation [18].

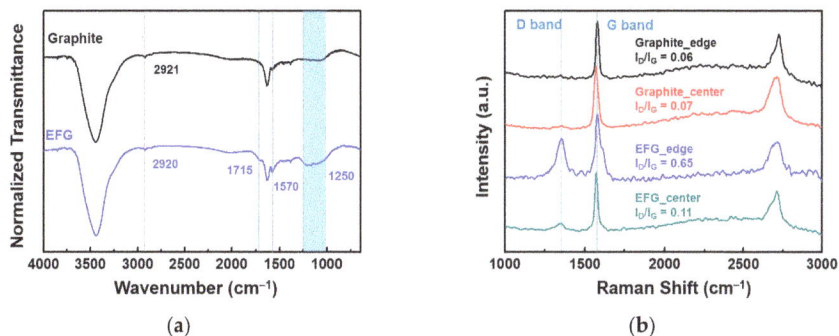

Figure 2. Characterization of EFG and graphite: (**a**) Fourier transform-infrared (FT-IR) spectra; (**b**) Raman spectra and I_D/I_G ratio.

3.2. Fabrication and Thermo-Mechanical Properties of PI/EFG Composites

3.2.1. Fabrication of PI/EFG Composites

The application of EFG was studied by fabricating a series of polymer composites consisting of a typical aromatic polyimide as a matrix polymer and EFG as filler. The detailed procedure is described in Figure 3. First, the EFG was dispersed in NMP with sonication. Subsequently, the poly(amic acid) solution as a pre-polymer for GPC fabrication with various EFG content was prepared by direct in situ polymerization with pyromellitic dianhydride (PMDA) and 4,4′-oxydianiline (ODA), which are the most common monomers for PI synthesis. Lastly, after the GPC films of the PAA states were fabricated, gradual thermal treatment resulted in cured polyimide films. The GPCs were fabricated in four varieties with EFG filler content of 0.1, 0.5, 1.0, and 3.0 wt %, respectively. In addition, as a reference for comparison, the pure PI film was also prepared under the same conditions. All GPC and pure PI films were obtained in the form of a flexible and transparent film that resembled a typical PI film. As the EFG content increased, the GPC became slightly translucent.

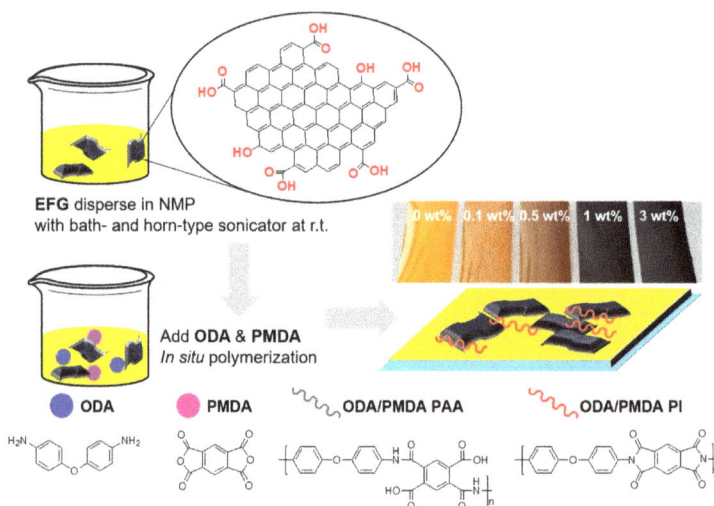

Figure 3. Fabrication process of PI/EFG composites.

3.2.2. Characterization of PI/EFG Composites

The cross-sectional images of the GPC films on the fractured surface were observed by scanning electron microscopy, confirming the morphology of the films and the distribution of the EFG filler (Figure 4). The surface of pure PI appeared smooth without any roughness, but the GPCs had a more coarse morphology. The GPCs with filler loadings of 0.1, 0.5, and 1 wt % exhibited a surface roughness that was almost the same as that of pure PI, although the surface tended to become slightly coarse as the amount of EFG increased, indicating that the EFGs were well dispersed in the matrix polymer. However, the GPC with 3 wt % EFG had an apparently harsh surface morphology in accordance with its high filler loading. In addition, the aggregation of filler platelets was also observed. Even though a little aggregation was observed at a high filler loading, considering that the EFG was not highly functionalized, the GPCs were successfully fabricated with well-dispersed EFGs originating from their direct and/or indirect interfacial interaction. The same trends were also conformed in the optical microscopy images measured to detect the dispersion quality of the fillers in a larger scale (Figure 5). In the 1 and 3 wt % samples, it was observed that the fillers were partially aggregated, but overall, the uniformly dispersed states were confirmed.

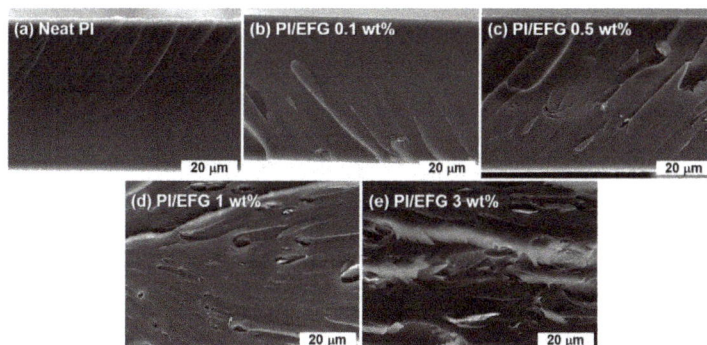

Figure 4. SEM images of GPCs: (**a**) Neat PI; (**b**) PI/EFG 0.1 wt %; (**c**) PI/EFG 0.5 wt %; (**d**) PI/EFG 1 wt %; (**e**) PI/EFG 3.0 wt %.

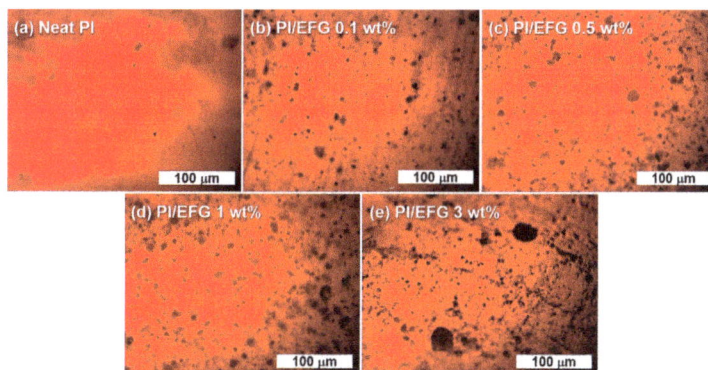

Figure 5. Optical microscope (OM) images of GPCs: (**a**) Neat PI; (**b**) PI/EFG 0.1 wt %; (**c**) PI/EFG 0.5 wt %; (**d**) PI/EFG 1 wt %; (**e**) PI/EFG 3.0 wt %.

3.2.3. Thermal and Thermo-Mechanical Properties of PI/EFG Composites

The thermal and thermo-mechanical properties of the GPCs were examined by thermogravimetric analysis (TGA) and dynamic mechanical analysis (DMA) (Table 1). The thermal stability curves of the GPCs evaluated by TGA appeared almost similar regardless of the filler content because similar thermograms were observed (Figure 6). In fact, the filler loadings of the GPCs were so low that they naturally exhibited similar thermal decomposition behavior. However, upon examining the detailed curves, they tended to display increased decomposition temperatures, i.e., $T_{d,5\%}$ and $T_{d,10\%}$, according to the filler loadings. These results could be interpreted in two ways: EFG is thermally very stable, so the decomposition temperature of the GPCs increases according to their filler content, or, the enhancement is derived from the interaction between EFG and PI chains. As far as the char yields after thermal decomposition up to 800 °C are concerned, the latter effect is considered to be large. Therefore, as we intended, the increased thermal stability could be attributed to the direct/indirect interaction between EFG and PI chains as a consequence of covalent and hydrogen bonding with the carboxylic acid group on the edge of EFG [18,19,36].

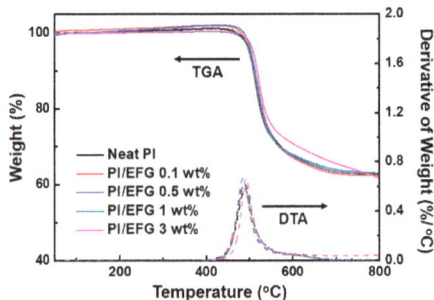

Figure 6. Thermogravimetric analysis (TGA) (solid lines) and derivative weight loss curves (dashed lines) of neat PI and PI/EFG composites.

The DMA measurements of the GPCs were carried out in order to confirm the interaction between the functional groups on EFG and the PI polymer chains as well as to investigate their thermo-mechanical properties (Figure 7). First, the GPCs had a larger storage modulus than pure PI due to the reinforcing effect of the EFG. In addition, the tendency of the modulus to increase was reversed and a decrease was observed owing to the aggregation phenomenon that occurred in the GPC containing 3 wt % EFG. Interestingly, there was a difference in the glass transition temperature (T_g) of the GPCs calculated by the DMA measurements. In general, polymer composites containing heterogeneous fillers have lower T_g values than those of pure polymers due to the effect of the fillers on the inter-chains that causes the interaction between the polymer chains to weaken. In contrast, the GPCs with EFG showed an increasing tendency for T_g as the filler content increased. In particular, the T_g value increased by more than 10 °C for all the GPCs. This significant enhancement in T_g is only possible when specific forces, such as hydrogen bonding or additional chemical bonding, are in effect between the polymer chains [37]. These reinforcement effects would occur additively as a result of the specific interaction between the functional groups at the edge of EFG and the imide groups of the PI chains, and the additional effect of the EFG itself forming a covalent bond with the amine moieties to act as the center in the network branches during in situ polymerization.

Table 1. Thermo-mechanical properties of PI/EFG composites.

Sample (wt %)	$T_{d,5\%}$ Determined by TGA [°C] [1]	$T_{d,10\%}$ Determined by TGA [°C] [2]	Char Yield [%] [3]	T_g Determined by DMA [°C] [4]
Neat PI	500.0	509.7	62.0	393.5
PI/EFG 0.1 wt %	499.9	509.0	62.9	404.6
PI/EFG 0.5 wt %	502.2	512.6	63.3	405.4
PI/EFG 1.0 wt %	505.4	514.3	63.8	405.9
PI/EFG 3.0 wt %	509.6	517.6	61.8	408.6

[1] The decomposition temperatures for 5% weight loss from TGA. [2] The decomposition temperatures for 10% weight loss from TGA. [3] Weight percentage of residues at 800 °C. [4] The values were calculated by the peak points of the Tan δ graph.

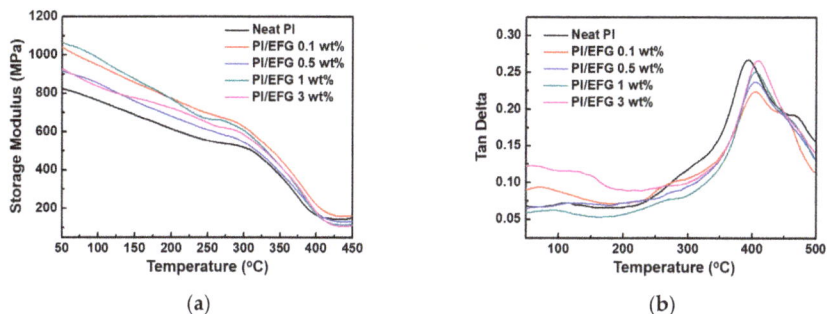

Figure 7. Dynamic mechanical analysis (DMA) curves of neat PI and PI/EFG composites: (**a**) Storage modulus; (**b**) tan δ.

In addition, in order to investigate further strengthening effect by EFG, the electrical conductivity was measured (Figure 8). Since it is well known that graphene shows exceptionally high electron mobility and electrical conductivity, the composites containing EFG would be also expected to exhibit high electrical conductivity. As expected, the electrical conductivity of the GPCs increased significantly as the EFG content increased. Considering that the amount of filler introduced is up to 3 wt %, which is insufficient to form a percolated pathway for electrical conducting, the conductivity increased by about 10^3 orders of magnitude to neat polyimide suggested that introduction of EFG would effectively build partial conducting chain. This reinforced result should have be attributed to both of the interaction between PI chains and EFGs and evenly distributed EFGs in GPCs.

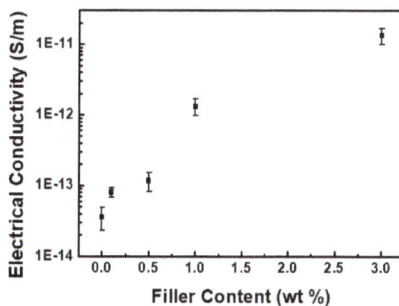

Figure 8. Electrical conductivity of neat PI and PI/EFG composites.

4. Conclusions

In this work, graphene with carboxyl functionalities at the edges of graphene platelets was prepared by the modified ball-milling method in the presence of dry ice. Our method successfully produced a less-functionalized EFG with a larger grain size as confirmed by location-selective micro-Raman spectroscopy.

We utilized the resultant EFG as filler to fabricate GPCs for potential application in such as electronic, automotive, airline, space, defense, and construction industries. A series of GPCs was prepared by varying the EFG contents and using aromatic PI as a matrix polymer, which is one of the representative polymers capable of interacting effectively with the filler. The GPCs were prepared via traditional two-step thermal imidization between PMDA and ODA. During the process, the poly(amic acid) compounds were obtained and used to fabricate the GPC films. The resulting GPC films showed a clear surface morphology except for the 3 wt % sample, which meant that the EFG platelets were well dispersed. In particular, the fact that the EFGs were well dispersed in the

GPCs led to the following three notable enhancements compared with pure PI. First, in terms of their thermal properties, the GPCs exhibited improved thermal stabilities. Furthermore, their mechanical properties observed by DMA were enhanced. Lastly, the T_g values of the GPCs significantly increased. These reinforcements could be attributed to the additional covalent bonds between the EFGs and amine moieties of the PI and the electrostatic interaction between the EFGs and PI chains, as initially designed.

Acknowledgments: This work was supported by the National Research Foundation of Korea (NRF) grant funded by the Korea government (Ministry of Science and ICT) (No. NRF-2017R1C1B5076344). In addition, this research was supported by Kyungpook National University Research Fund, 2017.

Author Contributions: Hyeonuk Yeo designed the entire experimental plan; Jaehyun Cho and Hyeonuk Yeo synthesized and characterized the EFG; Ki-Ho Nam and Hyeonuk Yeo fabricated and characterized the PI/EFG composites; Ki-Ho Nam, Jaehyun Cho, and Hyeonuk Yeo analyzed the data and wrote the manuscript. All authors have approved the final version of the manuscript.

Conflicts of Interest: The authors declare no conflict of interest.

References

1. Stankovich, S.; Dikin, D.A.; Dommett, G.H.; Kohlhaas, K.M.; Zimney, E.J.; Stach, E.A.; Piner, R.D.; Nguyen, S.T.; Ruoff, R.S. Graphene-based composite materials. *Nature* **2006**, *442*, 282–286. [CrossRef] [PubMed]

2. Ramanathan, T.; Abdala, A.A.; Stankovich, S.; Dikin, D.A.; Herrera-Alonso, M.; Piner, R.D.; Adamson, D.H.; Schniepp, H.C.; Chen, X.; Ruoff, R.S.; et al. Functionalized graphene sheets for polymer nanocomposites. *Nat. Nanotechnol.* **2008**, *3*, 327–331. [CrossRef] [PubMed]

3. Huang, X.; Qi, X.; Boey, F.; Zhang, H. Graphene-based composites. *Chem. Soc. Rev.* **2012**, *41*, 666–686. [CrossRef] [PubMed]

4. Sengupta, R.; Bhattacharya, M.; Bandyopadhyay, S.; Bhowmick, A.K. A review on the mechanical and electrical properties of graphite and modified graphite reinforced polymer composites. *Prog. Polym. Sci.* **2011**, *36*, 638–670. [CrossRef]

5. Kuilla, T.; Bhadra, S.; Yao, D.H.; Kim, N.H.; Bose, S.; Lee, J.H. Recent advances in graphene based polymer composites. *Prog. Polym. Sci.* **2010**, *35*, 1350–1375. [CrossRef]

6. Li, A.; Zhang, C.; Zhang, Y.F. Thermal conductivity of graphene-polymer composites: Mechanisms, properties, and applications. *Polymers* **2017**, *9*, 437.

7. Ding, P.; Zhang, J.; Song, N.; Tang, S.F.; Liu, Y.M.; Shi, L.Y. Anisotropic thermal conductive properties of hot-pressed polystyrene/graphene composites in the through-plane and in-plane directions. *Compos. Sci. Technol.* **2015**, *109*, 25–31. [CrossRef]

8. Wang, P.; Zhang, J.J.; Dong, L.; Sun, C.; Zhao, X.L.; Ruan, Y.B.; Lu, H.B. Interlayer polymerization in chemically expanded graphite for preparation of highly conductive, mechanically strong polymer composites. *Chem. Mater.* **2017**, *29*, 3412–3422. [CrossRef]

9. Marra, F.; D'Aloia, A.G.; Tamburrano, A.; Ochando, I.M.; De Bellis, G.; Ellis, G.; Sarto, M.S. Electromagnetic and dynamic mechanical properties of epoxy and vinylester-based composites filled with graphene nanoplatelets. *Polymers* **2016**, *8*, 272. [CrossRef]

10. Wang, X.; Xing, W.Y.; Zhang, P.; Song, L.; Yang, H.Y.; Hu, Y. Covalent functionalization of graphene with organosilane and its use as a reinforcement in epoxy composites. *Compos. Sci. Technol.* **2012**, *72*, 737–743. [CrossRef]

11. Tang, Y.J.; Hu, X.L.; Liu, D.D.; Guo, D.L.; Zhang, J.H. Effect of microwave treatment of graphite on the electrical conductivity and electrochemical properties of polyaniline/graphene oxide composites. *Polymers* **2016**, *8*, 399. [CrossRef]

12. Yan, D.X.; Pang, H.; Li, B.; Vajtai, R.; Xu, L.; Ren, P.G.; Wang, J.H.; Li, Z.M. Structured reduced graphene oxide/polymer composites for ultra-efficient electromagnetic interference shielding. *Adv. Funct. Mater.* **2015**, *25*, 559–566. [CrossRef]

13. Wan, Y.J.; Yang, W.H.; Yu, S.H.; Sun, R.; Wong, C.P.; Liao, W.H. Covalent polymer functionalization of graphene for improved dielectric properties and thermal stability of epoxy composites. *Compos. Sci. Technol.* **2016**, *122*, 27–35. [CrossRef]

14. Bento, J.L.; Brown, E.; Woltornist, S.J.; Adamson, D.H. Thermal and electrical properties of nanocomposites based on self-assembled pristine graphene. *Adv. Funct. Mater.* **2017**, *27*, 1604277. [CrossRef]

15. Shtein, M.; Nadiv, R.; Buzaglo, M.; Kahil, K.; Regev, O. Thermally conductive graphene-polymer composites: Size, percolation, and synergy effects. *Chem. Mater.* **2015**, *27*, 2100–2106. [CrossRef]

16. Shin, D.G.; Yeo, H.; Ku, B.C.; Goh, M.; You, N.H. A facile synthesis method for highly water-dispersible reduced graphene oxide based on covalently linked pyridinium salt. *Carbon* **2017**, *121*, 17–24. [CrossRef]

17. Tang, L.C.; Wan, Y.J.; Yan, D.; Pei, Y.B.; Zhao, L.; Li, Y.B.; Wu, L.B.; Jiang, J.X.; Lai, G.Q. The effect of graphene dispersion on the mechanical properties of graphene/epoxy composites. *Carbon* **2013**, *60*, 16–27. [CrossRef]

18. Lim, J.; Yeo, H.; Goh, M.; Ku, B.C.; Kim, S.G.; Lee, H.S.; Park, B.; You, N.H. Grafting of polyimide onto chemically-functionalized graphene nanosheets for mechanically-strong barrier membranes. *Chem. Mater.* **2015**, *27*, 2040–2047. [CrossRef]

19. Lim, J.; Yeo, H.; Kim, S.G.; Park, O.K.; Yu, J.; Hwang, J.Y.; Goh, M.; Ku, B.C.; Lee, H.S.; You, N.H. Pyridine-functionalized graphene/polyimide nanocomposites; mechanical, gas barrier, and catalytic effects. *Compos. Part B Eng.* **2017**, *114*, 280–288. [CrossRef]

20. Atif, R.; Shyha, I.; Inam, F. Mechanical, thermal, and electrical properties of graphene-epoxy nanocomposites—A review. *Polymers* **2016**, *8*, 281. [CrossRef]

21. Nadiv, R.; Shtein, M.; Buzaglo, M.; Peretz-Damari, S.; Kovalchuk, A.; Wang, T.; Tour, J.M.; Regev, O. Graphene nanoribbon–polymer composites: The critical role of edge functionalization. *Carbon* **2016**, *99*, 444–450. [CrossRef]

22. Chen, D.; Feng, H.; Li, J. Graphene oxide: Preparation, functionalization, and electrochemical applications. *Chem. Rev.* **2012**, *112*, 6027–6053. [CrossRef] [PubMed]

23. Georgakilas, V.; Otyepka, M.; Bourlinos, A.B.; Chandra, V.; Kim, N.; Kemp, K.C.; Hobza, P.; Zboril, R.; Kim, K.S. Functionalization of graphene: Covalent and non-covalent approaches, derivatives and applications. *Chem. Rev.* **2012**, *112*, 6156–6214. [CrossRef] [PubMed]

24. Cho, J.; Jeon, I.; Kim, S.Y.; Jho, J.Y. Improving Dispersion and Barrier Properties of Polyketone/Graphene Nanoplatelet Composites via Noncovalent Functionalization Using Aminopyrene. *ACS Appl. Mater. Interfaces* **2017**, *9*, 27984–27994. [CrossRef] [PubMed]

25. Park, S.; Ruoff, R.S. Chemical methods for the production of graphenes. *Nat. Nanotechnol.* **2009**, *4*, 217–224. [CrossRef] [PubMed]

26. Punetha, V.D.; Rana, S.; Yoo, H.J.; Chaurasia, A.; McLeskey, J.T.; Ramasamy, M.S.; Sahoo, N.G.; Cho, J.W. Functionalization of carbon nanomaterials for advanced polymer nanocomposites: A comparison study between cnt and graphene. *Prog. Polym. Sci.* **2017**, *67*, 1–47. [CrossRef]

27. Parvez, K.; Wu, Z.S.; Li, R.J.; Liu, X.J.; Graf, R.; Feng, X.L.; Mullen, K. Exfoliation of graphite into graphene in aqueous solutions of inorganic salts. *J. Am. Chem. Soc.* **2014**, *136*, 6083–6091. [CrossRef] [PubMed]

28. Ciesielski, A.; Samori, P. Graphene via sonication assisted liquid-phase exfoliation. *Chem. Soc. Rev.* **2014**, *43*, 381–398. [CrossRef] [PubMed]

29. Coleman, J.N. Liquid exfoliation of defect-free graphene. *Acc. Chem. Res.* **2013**, *46*, 14–22. [CrossRef] [PubMed]

30. Paton, K.R.; Varrla, E.; Backes, C.; Smith, R.J.; Khan, U.; O'Neill, A.; Boland, C.; Lotya, M.; Istrate, O.M.; King, P.; et al. Scalable production of large quantities of defect-free few-layer graphene by shear exfoliation in liquids. *Nat. Mater.* **2014**, *13*, 624–630. [CrossRef] [PubMed]

31. Voiry, D.; Yang, J.; Kupferberg, J.; Fullon, R.; Lee, C.; Jeong, H.Y.; Shin, H.S.; Chhowalla, M. High-quality graphene via microwave reduction of solution-exfoliated graphene oxide. *Science* **2016**, *353*, 1413–1416. [CrossRef] [PubMed]

32. Jeon, I.Y.; Shin, Y.R.; Sohn, G.J.; Choi, H.J.; Bae, S.Y.; Mahmood, J.; Jung, S.M.; Seo, J.M.; Kim, M.J.; Wook Chang, D.; et al. Edge-carboxylated graphene nanosheets via ball milling. *Proc. Natl. Acad. Sci. USA* **2012**, *109*, 5588–5593. [CrossRef] [PubMed]

33. Yeo, H.; Lee, J.; Goh, M.; Ku, B.C.; Sohn, H.; Ueda, M.; You, N.H. Synthesis and characterization of high refractive index polyimides derived from 2, 5-bis (4-aminophenylenesulfanyl)-3, 4-ethylenedithiothiophene and aromatic dianhydrides. *J. Polym. Sci. Part A Polym. Chem.* **2015**, *53*, 944–950. [CrossRef]

34. Yeo, H.; Goh, M.; Ku, B.C.; You, N.H. Synthesis and characterization of highly-fluorinated colorless polyimides derived from 4,4'-((perfluoro-[1,1'-biphenyl]-4,4'-diyl)bis(oxy))bis(2,6-dimethylaniline) and aromatic dianhydrides. *Polymer* **2015**, *76*, 280–286. [CrossRef]

35. Nam, K.H.; Kim, H.; Choi, H.K.; Yeo, H.; Goh, M.; Yu, J.; Hahn, J.R.; Han, H.; Ku, B.C.; You, N.H. Thermomechanical and optical properties of molecularly controlled polyimides derived from ester derivatives. *Polymer* **2017**, *108*, 502–512. [CrossRef]
36. Nam, K.H.; Yu, J.; You, N.H.; Han, H.; Ku, B.C. Synergistic toughening of polymer nanocomposites by hydrogen-bond assisted three-dimensional network of functionalized graphene oxide and carbon nanotubes. *Compos. Sci. Technol.* **2017**, *149*, 228–234. [CrossRef]
37. Lewis, C.L.; Stewart, K.; Anthamatten, M. The influence of hydrogen bonding side-groups on viscoelastic behavior of linear and network polymers. *Macromolecules* **2014**, *47*, 729–740. [CrossRef]

polymers

MDPI

Article

Permeability and Selectivity of PPO/Graphene Composites as Mixed Matrix Membranes for CO$_2$ Capture and Gas Separation

Riccardo Rea [1], Simone Ligi [2], Meganne Christian [3], Vittorio Morandi [3], Marco Giacinti Baschetti [1] and Maria Grazia De Angelis [1,*]

[1] Dipartimento di Ingegneria Civile, Chimica, Ambientale e dei Materiali (DICAM), Università di Bologna, Via Terracini 28, 40131 Bologna, Italy; riccardo.rea3@unibo.it (R.R.); marco.giacinti@unibo.it (M.G.B.)
[2] Graphene XT s.r.l., 40131 Bologna, Italy; simone.ligi@graphene-xt.com
[3] CNR-IMM Section of Bologna, via Gobetti, 101-40129 Bologna, Italy; christian@bo.imm.cnr.it (M.C.); morandi@bo.imm.cnr.it (V.M.)
* Correspondence: grazia.deangelis@unibo.it; Tel.: +39-051-209-0410

Received: 5 December 2017; Accepted: 24 January 2018; Published: 29 January 2018

Abstract: We fabricated novel composite (mixed matrix) membranes based on a permeable glassy polymer, Poly(2,6-dimethyl-1,4-phenylene oxide) (PPO), and variable loadings of few-layer graphene, to test their potential in gas separation and CO$_2$ capture applications. The permeability, selectivity and diffusivity of different gases as a function of graphene loading, from 0.3 to 15 wt %, was measured at 35 and 65 °C. Samples with small loadings of graphene show a higher permeability and He/CO$_2$ selectivity than pure PPO, due to a favorable effect of the nanofillers on the polymer morphology. Higher amounts of graphene lower the permeability of the polymer, due to the prevailing effect of increased tortuosity of the gas molecules in the membrane. Graphene also allows dramatically reducing the increase of permeability with temperature, acting as a "stabilizer" for the polymer matrix. Such effect reduces the temperature-induced loss of size-selectivity for He/N$_2$ and CO$_2$/N$_2$, and enhances the temperature-induced increase of selectivity for He/CO$_2$. The study confirms that, as observed in the case of other graphene-based mixed matrix glassy membranes, the optimal concentration of graphene in the polymer is below 1 wt %. Below such threshold, the morphology of the nanoscopic filler added in solution affects positively the glassy chains packing, enhancing permeability and selectivity, and improving the selectivity of the membrane at increasing temperatures. These results suggest that small additions of graphene to polymers can enhance their permselectivity and stabilize their properties.

Keywords: graphene; membranes; gas separation; CO$_2$ capture; permeability; selectivity; PPO

1. Introduction

The removal of CO$_2$ from gaseous streams produced in energy production and energy-intensive industrial processes is one of the most straightforward ways to reduce the global warming effect due to atmospheric CO$_2$ increase [1,2].

Membrane separations are considered suitable, and environmentally friendly, technologies to capture CO$_2$ both in post combustion (removal of CO$_2$ from N$_2$) and pre combustion (removal of CO$_2$ from H$_2$) [3–6]. Mixtures of H$_2$ and CO$_2$ are generated in the production of hydrogen from steam reforming or biomass gasification, and their purification is necessary not only for environmental aims, i.e., for the reduction of the CO$_2$ emitted in the atmosphere, but also for technological purposes, i.e., the purification of H$_2$ for use as a fuel or chemical. Most of those processes are carried out at high temperature, and a high-temperature purification process would allow exploiting the thermal level of the gas stream [4].

In membrane processes, the main way to optimize the separation is to use highly permeable and highly selective membrane materials. However, as pointed out in the literature, it is often not possible to simultaneously increase the permeability and the selectivity, which are governed by a tradeoff mechanism [7].

Poly(2,6-dimethyl-1,4-phenylene oxide) (PPO) is a glassy polymeric material suitable for membrane-based gas separation, due to a high permeability and moderate selectivity [8]. In particular, PPO has a H_2-selective behavior, when exposed to mixtures of H_2 and CO_2, i.e., it exhibits a size-sieving ability, because hydrogen has a smaller kinetic diameter than CO_2 (0.29 vs. 0.33 nm) [9]. The reported ideal H_2/CO_2 selectivity value for PPO is 1.5 at room temperature, while an economically feasible membrane separation process would require higher values [4]. However, this material is stable up to 200 °C and H_2/CO_2 membrane separation performances are known to be significantly increased by increasing temperature [4]. Thus, PPO results to be an interesting candidate for the high temperature separation of this mixture.

A further strategy to enhance the selective behavior of the polymer can be based on the addition of nanosized fillers to the polymer matrix: in particular, nanofillers can modify the chain packing of the polymer, increasing its selectivity. It was reported by many authors that separation performance of polymeric materials can be improved by the addition of nanosized particles of different shapes, such as nanospheres [10–16] and nanotubes [17–19]. Nanometric fillers tend to adjust the chain packing of glassy polymers in a way that affects positively the permselectivity, without creating non selective voids, but rather creating additional selective free volume [20].

Graphene is a nanosized material that has been just recently applied to the field of gas separation membranes, as initially it was mostly considered for improving the barrier effect [21–28]. Indeed, it was noticed that, while a defect-free graphenic layer is virtually impermeable to all molecules, some production techniques, such as chemical vapor deposition (CVD), may introduce a microporosity. Moreover, when the graphene is applied in subsequent layers, permeable channels caused by imperfect adhesion may form. Graphene oxide (GO), on the other hand, naturally contains defects induced by the oxidation process, and is endowed with an intrinsic gas permeability and selectivity. Several studies report interesting results of the application of GO in gas separation [29–31].

Some other works evaluated the combination of graphene, obtained by direct exfoliation of graphite in polymeric solution, to a polymer of intrinsic microporosity, PIM-1. The studies indicate that the optimal concentration of graphene to maximize the CO_2 permeability of PIM-1 is 0.1 wt %, and molecular simulations indicate that the presence of graphene nanolayers modifies the polymer distribution [32–34].

The addition of 1 wt % of monolayers of GO slightly enhances the gas permeability of poly(trimethyl silyl propyne) (PTMSP), and the selectivity for the couple CO_2/He and CH_4/He [35]. The addition of a few layer graphene in the same amount to PTMSP lowers slightly the gas permeability, with factors that increase with decreasing molecule size, and enhances the ideal selectivity for the couples CO_2/He and CH_4/N_2, CH_4/He. The addition of this filler mainly lowers the CO_2 diffusivity, leaving the gas solubility unaltered, while the solubility and diffusivity are both enhanced by addition of GO (lateral dimension 2.0 m, thickness 1.1 nm). Multiple layer graphene (lateral dimension 0.2 m, thickness 2–20 nm) lowers the permeability of PTMSP to a significant extent (up to 30%), with a weight fraction of just 1 wt % [35].

Both GO and few-layer graphene allow to impair the aging process of PTMSP, as tracked with He, N_2, CH_4 and CO_2 permeability. Graphene platelets, due to their high aspect ratio, act as physical barriers to the rearrangement of polymer chains, and the diffusion of free volume domains, which cause the ageing. Such explanation is supported by the fact that ageing is mostly reduced by the fillers that have the higher aspect ratio [35].

In view of the previous findings, in this work, we studied whether the separation performance of PPO with respect to some gas mixtures, can be improved by addition of graphene nanoplatelets. Furthermore, in the aim of assessing the effect of temperature on permselectivity, we tested the membranes performance to temperatures as high as 65 °C.

2. Experimental

2.1. Preparation of Membranes

The solid polymer PPO was purchased by Sigma Aldrich, and chloroform (purity >99.5% Sigma Aldrich, St. Louis, MO, USA) was used as solvent. It is an amorphous material with a rather high T_g (213 °C), that makes it suitable for use in high temperature separations. The melting point is 268 °C and the density at 25 °C is 1.06 g/cm^3. The permeability to hydrogen is rather high, and ranges between 87 and 112 Barrer at 25 °C [36,37]; such property is due to a high fractional free volume of the polymer, which is around 18% [8]. In this work, helium is used instead of H_2 for safety reasons: literature results indicate that the permeability of He is moderately lower than that of H_2 in this polymer, its permeability ranging between 56 and 75 at 25 °C [8,38]. Therefore, the use of He instead of H_2 leads to conservative estimates for the H_2 permeability and H_2/CO_2 selectivity.

Two commercial grades of graphene in powder, produced by Graphene XT, were used: Graphene XT6, with a lateral dimension of 5 m and a thickness between 6.0 and 8.0 nm, and Graphene XT7, with a nominal lateral size of 20 m and a thickness of 2 nm.

The procedure to prepare membranes was optimized considering also a previous work [35].

The polymer was dissolved in chloroform and then graphene powder was added. The resulting suspension was sonicated for at least 15 min, and stirred for one day. In the case of Graphene XT7, the sonication time was increased to 60 min to improve the disaggregation of nanoplatelets, as this materials has a higher aspect ratio than the other filler used. Once the dispersion was complete, the solution was poured on a Petri dish and placed in a clean hood, where the temperature was kept at 50 °C, to ensure fast evaporation of the solvent. The solid film was then removed from the Petri dish, and treated in an oven under vacuum for 1 day at 200 °C. Such treatment not only removes traces of solvent, but also stabilizes the properties of the glassy PPO, with respect to ageing, for a sufficiently long time. It must be noticed, however, that PPO membranes treated at such temperatures show a somewhat lower permeability than untreated ones. Table 1 reports the list of materials produced, the procedure adopted, together with their properties. The photos of the various composite membranes produced are reported in Figure 1: one can notice the increasing darkness of the films with increasing amount of graphene loaded. It must be noticed that a film of pure PPO is completely transparent.

The gases used were N_2 (SIAD, Ozzano Emilia, Italy, purity of 99.999%), CO_2 (SIAD, purity of 99.998%) and He (SIAD, purity of 99.9999%).

Table 1. List of the composite membranes produced and tested.

Graphene Type	Graphene wt % in Polymer (g/100 g of PPO)	Treatment	Sonication Time	Stirring Time	Thickness (m)
XT7	0.3		60 min		41 ± 2.1
XT6 [a]	1	Heating at 200 °C under vacuum for 1 d	15 min	1 d	82 ± 3.9 72 ± 4.2
XT6	5				88 ± 2.6
XT6	15				55 ± 4.6

[a] Two different membranes at 1% of XT6 were produced. The first, 82 μm thickness, was tested at 35 °C, the second, 72 μm thickness, was tested at 65 °C.

Figure 1. Pictures of the various membranes fabricated: (**a**) PPO + 0.3 wt % of Graphene XT7; (**b**) PPO + 1 wt % of Graphene XT6; (**c**) PPO + 5 wt % of Graphene XT6; and (**d**) PPO + 15 wt % of Graphene XT6.

2.2. Permeability Tests

The permeability was measured with a variable pressure apparatus described in previous works [38], by applying an upstream pressure of about 1.4 bar and vacuum on the downstream side. The pure gas permeability at steady state can be calculated from Equation (1), in which $\left.\frac{dp_i}{dt}\right|_{s.s.}$ is the slope of pressure versus time curve at steady state, V_d is the calibrated downstream volume, R is the universal gas constant, T is the system temperature, A is the membrane area, l is the thickness of the sample and $\left(p_i^{up} - p_i^{down}\right)$ is the gas pressure difference across the membrane film.

$$P_i = \left.\frac{dp_i}{dt}\right|_{s.s.} \frac{V_d}{RTA} \frac{l}{\left(p_i^{up} - p_i^{down}\right)} \tag{1}$$

The apparatus is placed in a thermostatic chamber which can reach temperatures in the order of 80 °C. Tests were conducted at 35 °C; and for most samples we also carried out tests at 65 °C to investigate temperature effect on permeability of the different gases investigated, i.e., He, N_2 and CO_2.

The permeability value is also affected by the uncertainty on the membrane thickness, which depends on the sample considered. The error on ideal selectivity of a single sample however is negligible, because such value is unaffected by downstream volume, permeation area and membrane thickness values as it is clear from Equation (1).

Ideal selectivity for the different gas couple of interest is calculated using Equation (2), which is valid in the case of negligible downstream pressure, as in the tests considered here:

$$\alpha_{ij} = \frac{P_i}{P_j} \tag{2}$$

where the selectivity is defined as $\alpha_{ij} \equiv \frac{y_i^{downstream}/y_i^{upstream}}{y_j^{downstream}/y_j^{upstream}}$. In general, real selectivity can differ from the ideal selectivity, which is estimated using the permeability values performed on the pure gases, but the latter still represents a good indication of the material performance with respect to other membranes as most of the data reported in literature refer to this kind of selectivity [7].

In many cases, we were able also to determine the characteristic time of permeation, i.e., the time required to reach a stable flux across the membrane, the so-called time-lag, t_L, that is related to the gas diffusivity in the membrane by the following relation [39]:

$$D = \frac{1}{6}\frac{l^2}{t_L} \tag{3}$$

The time lag is actually estimated as the intercept, on the time axis, of the asymptotic straight line representing the downstream gas pressure versus time. It can also be noticed that, due to the phase equilibrium between the gas and the polymer phase, according to which the gas is absorbed onto the polymer surface proportionally to its solubility coefficient S_i, a straight forward relation holds true between the gas permeability, and its diffusivity and solubility coefficient in the polymer:

$$P_i = D_i S_i \tag{4}$$

According to which, the selectivity can be decomposed into a diffusivity-based, and a solubility-based contribution:

$$\alpha_{ij} = \left(\frac{D_i}{D_j}\right)\left(\frac{S_i}{S_j}\right) \tag{5}$$

Usually, for a same gas couple, the two contributions are, respectively, lower than unity and higher than unity. In general, the diffusivity-selectivity prevails, so that smaller gases are generally more permeable than larger ones in most polymeric membranes. In some cases, the solubility prevails, as happens for instance when CO_2 is concerned. For this reason, not all polymeric membranes have a H_2/CO_2 selectivity higher than 1, but some of them, due to a very high CO_2 solubility, exhibit a CO_2-selective behavior [40].

The temperature dependence of permeability, diffusivity and solubility is governed by Arrhenius-like laws, as follows:

$$P_i = P_i(T_0)\exp\left(-\frac{E_{P,i}}{RT}\right) \tag{6}$$

$$D_i = D_i(T_0)\exp\left(-\frac{E_{D,i}}{RT}\right) \tag{7}$$

$$S_i(T) = S_i(T_0)\exp\left(-\frac{\Delta H_{S,i}}{RT}\right) \tag{8}$$

where E_P and E_D are the permeation and diffusion activation energies, respectively, which are usually positive for gases in glassy polymers, and ΔH_S is the sorption enthalpy, which is usually negative. Accordingly, the permeability and diffusivity increase with temperature, while the solubility decreases with it.

Due to Equations (4)–(8), the permeation activation energy has a diffusivity and solubility contribution as follows:

$$E_{P,i} = E_{D,i} + \Delta H_{S,i} \tag{9}$$

For the selectivity dependence on temperature is concerned, one can combine Equations (2) and (6) to obtain:

$$\alpha_{i,j}(T) = \alpha_{i,j}(T_0) \exp\left(-\frac{E_{P,i} - E_{P,j}}{RT}\right) \tag{10}$$

which indicates that the selectivity increases with temperature if the more permeable gas has a higher activation energy than the less permeable one, while it decreases with temperature if the opposite is true. In general, the second situation is more frequent, as less permeable gases have higher activation energies. The couples H_2/CO_2 and He/CO_2 in PPO form an exception, because H_2 and He are more permeable at room temperature than CO_2, but also have a higher activation energy of permeation. The activation energy of CO_2 is low, due to a high contribution of sorption, an exothermic process, on the permeation. Therefore, for H_2-selective membranes, as the ones of the present paper, increasing the temperature enhances the H_2 permeability and H_2/CO_2 selectivity.

3. Results and Discussion

3.1. SEM Analysis

First, neat graphene in powder was characterized. The Graphene XT7 was chosen for this analysis and two images are reported in Figure 2a,b that show some wrinkles. The effect of dispersing the platelets in water is that of swelling and opening the structure, as shown in Figure 2c,d. Then, the effect of sonication on the sample morphology was analyzed. Pictures of the sample dispersed in water and sonicated for only 10 min exhibit lateral sizes, from the picture analyzed and reported in Figure S1 in the Supplementary Materials, larger than what is declared by the supplier, between 26 and 63 micrometers (based on seven platelets). In Figure S2 in the Supplementary Materials, we report similar pictures taken on a sample of Graphene XT7 (in water suspension) sonicated for 15 h. As can be seen, the platelets analyzed have a lateral size ranging between 34 and 43 micrometers, also based on seven platelets. Therefore, it seems that the sonication produces just a slight decrease of the average lateral size after 15 h, but the size distribution of the platelets becomes narrower.

Pictures of the cross section of PPO + graphene membranes are reported in Figure 3. It is apparent that, by increasing the concentration of graphene, the cross section becomes filled with graphenic domains. Those structures are aligned perpendicular to the membrane cross section, thereby increasing the tortuosity of diffusing molecules. The images also show that some voids form at the interface of PPO and graphene. Images with higher magnification were also taken (Figure 4), and show that the surface of graphene platelets remains similar to the one observed in the neat powder (Figure 2).

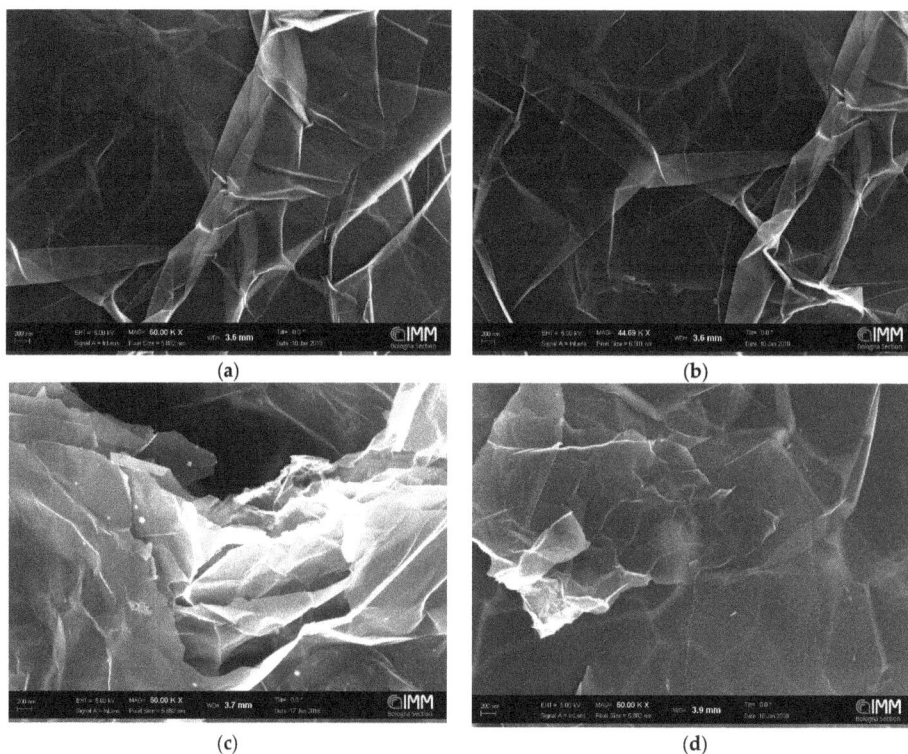

Figure 2. SEM images of: dry Graphene XT7 powder (**a**,**b**); and of the Graphene XT7 dispersed in water: before sonication (**c**); and after a 15 h sonication (**d**).

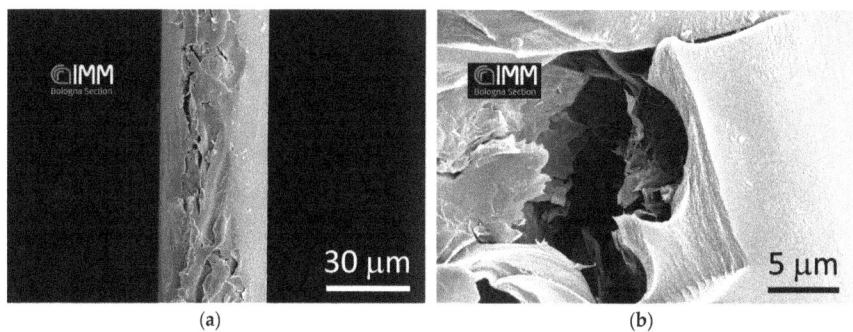

Figure 3. *Cont.*

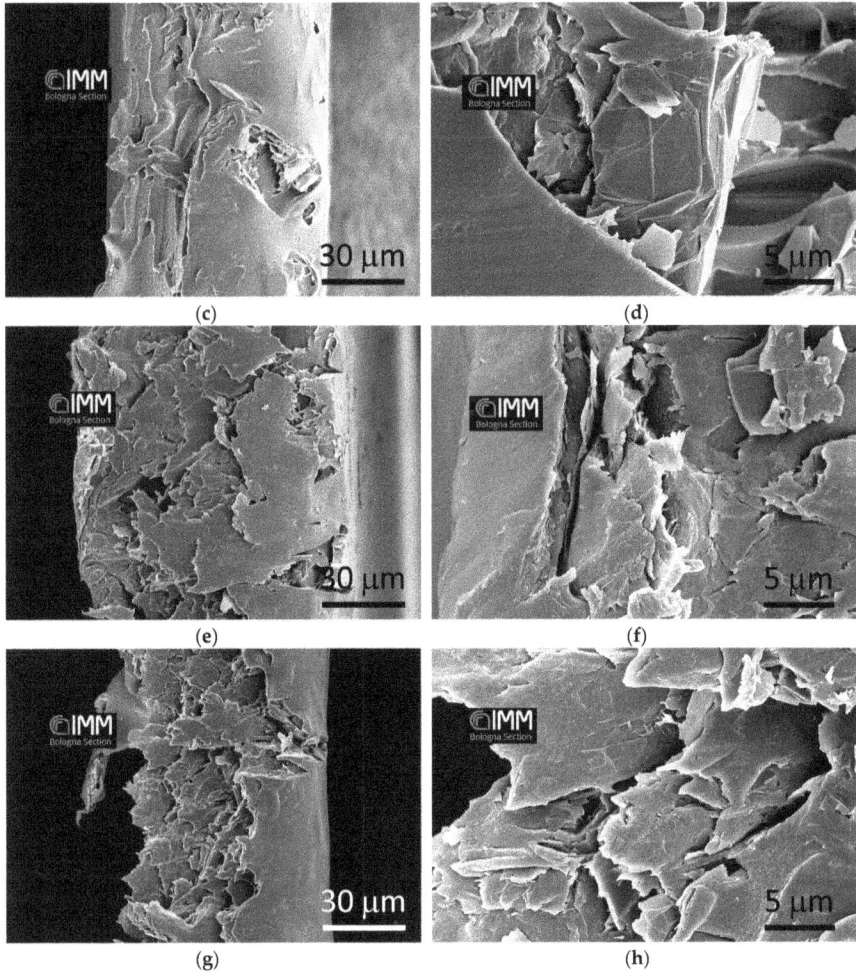

Figure 3. SEM images of membranes of: (**a**,**b**) PPO/0.3 wt % of Graphene XT 7; (**c**,**d**) PPO/1 wt % of Graphene XT 6; (**e**,**f**) PPO/5 wt % of Graphene XT 6; and (**g**,**h**) PPO/15 wt % of Graphene XT 6.

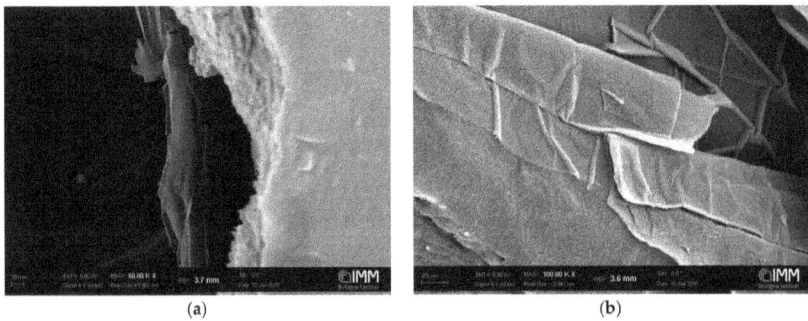

Figure 4. *Cont.*

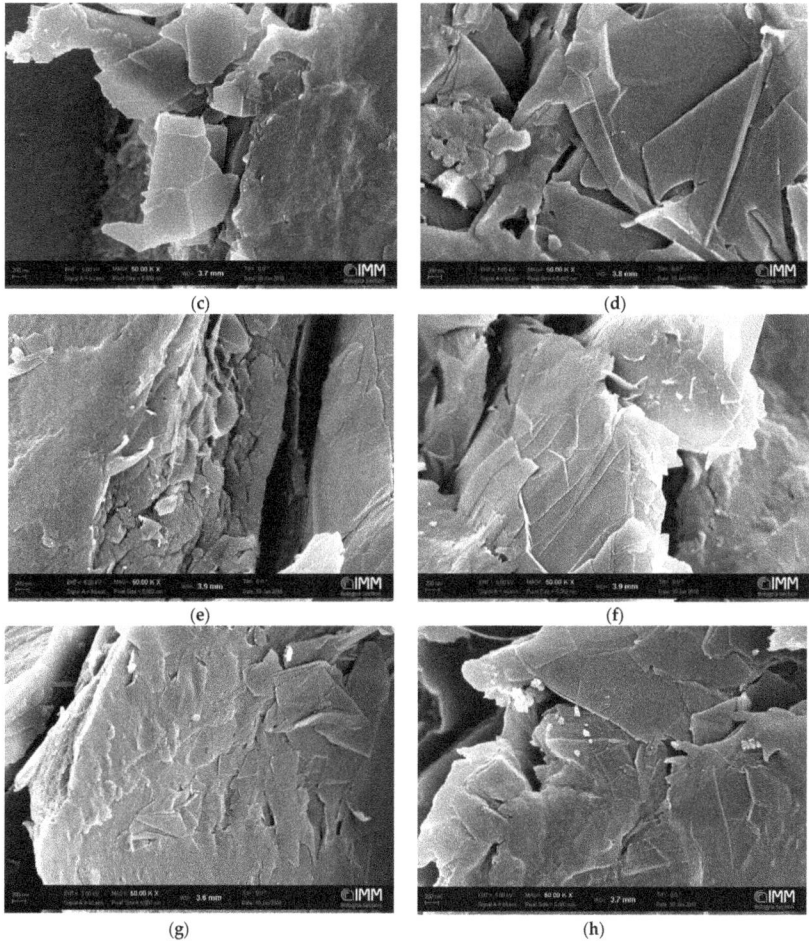

Figure 4. SEM images of membranes of: (**a**,**b**) PPO/0.3 wt % of Graphene XT 7; (**c**,**d**) PPO/1 wt % of Graphene XT 6; (**e**,**f**) PPO/5 wt % of Graphene XT 6; and (**g**,**h**) PPO/15 wt % of Graphene XT 6.

3.2. Permeability

In Table 2 and Figure 5, we report the permeability values measured at 35 and 65 °C in membranes with increasing loadings of graphene.

Table 2. Permeability of the various gases in PPO and composite membranes.

Permeability at 35 °C, Barrer	PPO	PPO/0.3 wt % Graphene XT7	PPO/1 wt % Graphene XT6	PPO/5 wt % Graphene XT6	PPO/15 wt % Graphene XT6
He	78 ± 3.8	86 ± 4.2	86 ± 4.1	68 ± 2.0	38 ± 3.2
N_2	3.0 ± 0.2	3.5 ± 0.2	3.6 ± 0.2	2.8 ± 0.1	1.8 ± 0.2
CO_2	61 ± 2.0	62 ± 2.9	60 ± 2.9	51 ± 1.5	27 ± 2.3
Permeability at 65 °C, Barrer					
He	114 ± 5.0	-	116 ± 6.7	81.0 ± 2.4	51.6 ± 4.4
N_2	5.00 ± 0.4	-	4.64 ± 0.3	3.31 ± 0.1	-
CO_2	69.3 ± 2	-	61.9 ± 3.6	42.3 ± 1.2	27.6 ± 2.4

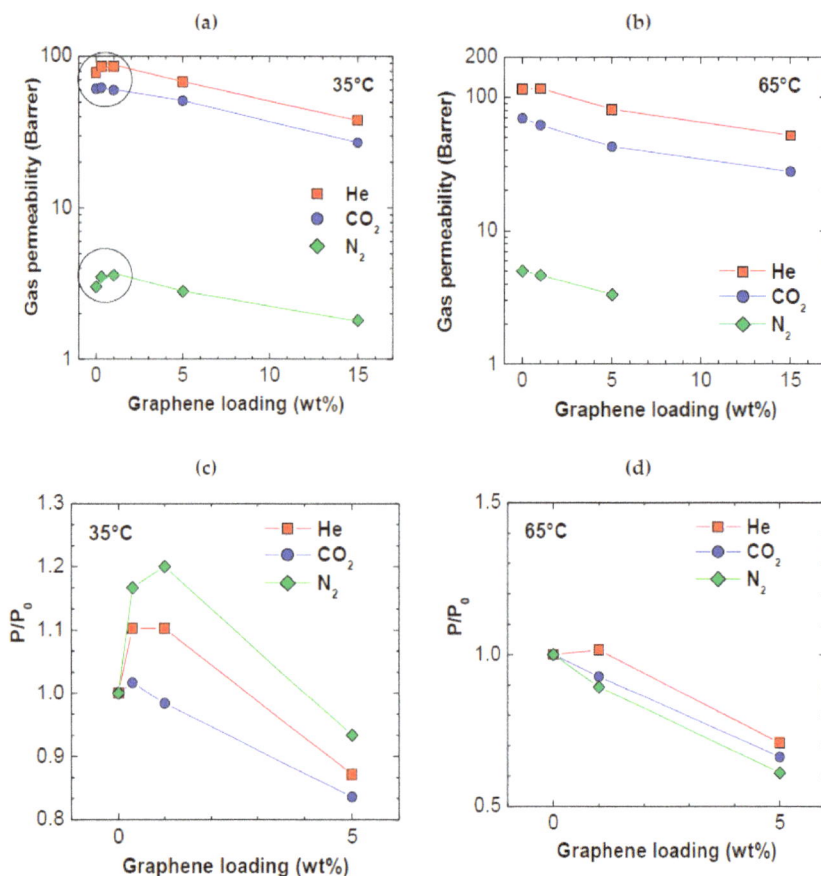

Figure 5. Gas permeability at: (**a**) 35 °C; and (**b**) at 65 °C, as a function of graphene loading in PPO. Variation of PPO permeability after addition of graphene vs. graphene loading at: (**c**) 35 °C; and (**d**) 65 °C.

It can be seen that, while the addition of small amounts (below 1 wt %) of both types of graphene enhance the He permeability, by about 10%, increasing the filler loading to higher amounts (5 and 15 wt %) reduces the permeability, for all gases considered. This is because the wide graphene platelets, in this range of concentration, significantly increase the tortuosity of gases diffusive path in the membrane. The permeability of CO_2 seems unaffected by the presence of graphene up to a loading of 1 wt %. The permeability of N_2, on the other hand, follows a same qualitative behavior as that of helium, increasing by about 20% for graphene loadings lower than 1 wt %, and decreasing for higher values. The permeability enhancement observed could be due to various factors, such as the formation of additional free volume at the interphase between polymer and filler, and the permeation of the gas through graphene layers.

At 65 °C, no increase of permeability could be detected after addition of graphene: we believe that, at this temperature, the transition between the regime where the free volume increase prevails and the one in which the tortuosity increase dominates occurs at very small graphene loadings, or the first regime does not occur at all. At higher temperatures, indeed, the mobility of both diffusing gases and polymeric chains is higher, and the diffusion and permeation process relies less on the presence of intrinsic free volume in the solid matrix.

3.3. Selectivity

The ideal selectivity values calculated with Equation (2) at 35 °C are reported in Table 3 and Figure 6. Three types of separations are considered: He/CO_2, He/N_2 and CO_2/N_2. The first separation (He vs. CO_2) is favored by high temperatures, i.e., both permeability and selectivity increase with increasing temperature, due to higher activation energy of permeation for He than for CO_2. The other separations considered, namely He/N_2 and CO_2/N_2 are, in terms of selectivity, not favored by temperature, as the selectivity of polymers for such couples decreases with temperature, due to the unfavorable difference between the permeation activation energies of the two gases (see Equation (10)).

Table 3. Ideal selectivity in PPO and composite membranes.

Ideal Selectivity at 35 °C	PPO	PPO/0.3 wt % Graphene XT7	PPO/1 wt % Graphene XT6	PPO/5 wt % Graphene XT6	PPO/15 wt % Graphene XT6
He/CO_2	1.28	1.39	1.43	1.33	1.41
He/N_2	26.0	24.6	23.9	24.3	21.1
CO_2/N_2	20.3	17.7	16.7	18.2	15.0
Ideal Selectivity at 65 °C	**PPO**				
He/CO_2	1.65	-	1.88	1.92	1.87
He/N_2	22.9	-	25.0	24.5	-
CO_2/N_2	13.9	-	13.3	12.8	-

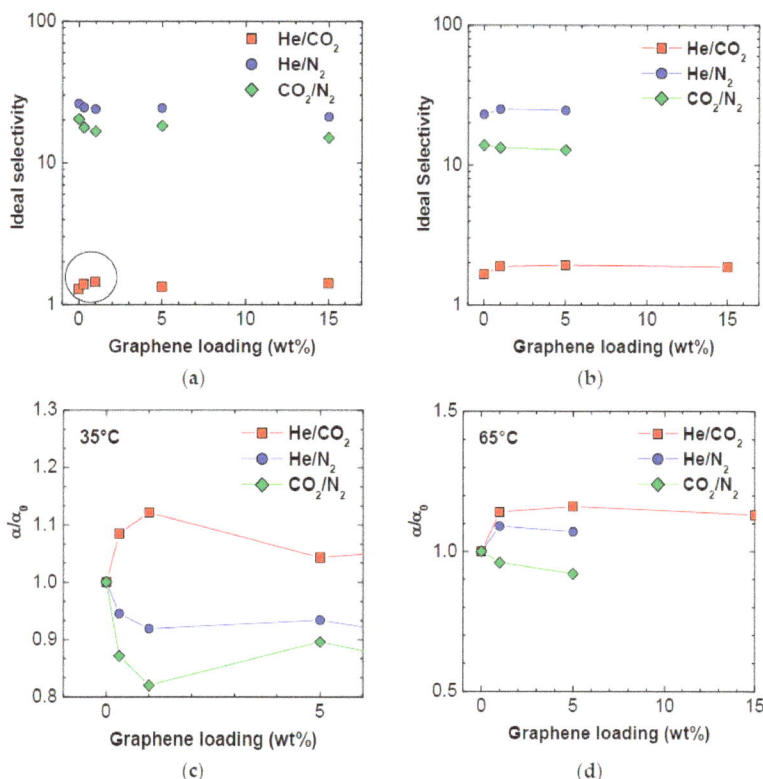

Figure 6. Gas ideal selectivity at: (**a**) 35 °C; and (**b**) at 65 °C, as a function of graphene loading in PPO. Variation of PPO selectivity after addition of graphene vs. graphene loading at (**c**) 35 °C and (**d**) 65 °C.

The addition of graphene in all proportions produces an increase of He/CO_2 selectivity of the polymer at 35 °C, of about 10%. For the He/N_2 separation, the selectivity values decrease monotonically with increasing graphene content at 35 °C, going from 26.0 to 21.1 at the largest loading. For the CO_2/N_2 mixture, the selectivity also decreases monotonically with increasing graphene loading, from 20.3 to 15.0. Such trends are due to the gas-dependent variations of permeability induced by addition of graphene. The gas that experiences the highest increase of permeability after addition of small loadings of graphene is the largest one, namely nitrogen. For samples containing larger amounts of graphene, the smallest reduction of permeability recorded is again that of nitrogen. Thus, both He/N_2 and CO_2/N_2 selectivity decrease. Such trend is attributable mostly to the kinetic term of permeability, the diffusivity, which is strongly affected by molecule size, as will be seen in the following. When the He/CO_2 selectivity is considered, on the other hand, the size of the molecule plays a less important role, as CO_2 permeability is also strongly affected by solubility, which depends on the molecule condensability rather than on its size. Therefore, for this couple of gases, the size-sieving capacity of the membrane slightly increases with the addition of graphene, rather than decreasing. When the temperature is increased to 65 °C, the situation slightly changes, in particular for the case of the He/N_2 mixture, for which graphene addition positively affects the selectivity. The incorporation of filler is also beneficial for the He/CO_2 separation, similar to the lower temperature case. The CO_2/N_2 selectivity, on the other hand, decreases with graphene addition also at 65 °C. The different trends of selectivity at different temperatures are due to the effect that graphene has on the thermal dependence of the gas permeability in the polymer, which we will discuss below.

3.4. Diffusivity

The diffusivity is a kinetic quantity that contributes to the permeability, according to Equation (4). In particular, diffusivity is strongly affected by variations of the polymer free volume, and by the chain packing of the membrane. Therefore, it is expected to be the quantity that is most strongly affected by the introduction of filler with a peculiar morphology, such as the high aspect ratio graphene platelets with one nanometric dimension. Furthermore, in size-selective polymers such as the one that we are considering, diffusivity dictates the selective behavior, rather than solubility, as happens for instance in the case of higher free volume glassy polymers such as PTMSP and PIM-1. In Table 4, we report the variations observed for the diffusion coefficients as a function of graphene loading. It can be seen that, when high filler loadings are considered, a marked decrease of diffusivity is measured, which is because the graphenic nanoplatelets act as physical barriers and increase tortuosity. At lower loadings, the behavior shows a large scattering. The diffusivity data obtained at 35 and 65 °C are also reported in Figure 7a,b.

Table 4. Diffusivity of the various gases in PPO and composite membranes.

Diffusivity at 35 °C, cm²/s	PPO	PPO/0.3 wt % Graphene XT7	PPO/1 wt % Graphene XT6	PPO/5 wt % Graphene XT6	PPO/15 wt % Graphene XT6
He	$(5.0 \pm 0.5) \times 10^{-6}$	$(7.8 \pm 0.8) \times 10^{-7}$	$(3.6 \pm 0.3) \times 10^{-6}$	$(10 \pm 0.6) \times 10^{-7}$	$(4.4 \pm 0.8) \times 10^{-7}$
N_2	$(4.3 \pm 0.4) \times 10^{-8}$	$(5.5 \pm 0.5) \times 10^{-8}$	$(4.6 \pm 0.4) \times 10^{-8}$	$(3.7 \pm 0.2) \times 10^{-8}$	$(1.2 \pm 0.2) \times 10^{-8}$
CO_2	$(6.3 \pm 0.6) \times 10^{-8}$	$(5.5 \pm 0.5) \times 10^{-8}$	$(6.2 \pm 0.6) \times 10^{-8}$	$(5.1 \pm 0.3) \times 10^{-8}$	$(3.0 \pm 0.5) \times 10^{-8}$
Diffusivity at 65 °C, cm²/s					
He	$(1.5 \pm 0.2) \times 10^{-5}$	-	$(2.2 \pm 0.3) \times 10^{-6}$	$(6.9 \pm 0.4) \times 10^{-6}$	$(1.5 \pm 0.7) \times 10^{-7}$
N_2	$(1.1 \pm 0.1) \times 10^{-7}$	-	$(7.1 \pm 0.8) \times 10^{-8}$	$(7.2 \pm 0.4) \times 10^{-8}$	-
CO_2	$(1.4 \pm 0.1) \times 10^{-7}$	-	$(1.4 \pm 0.2) \times 10^{-7}$	$(9.5 \pm 0.6) \times 10^{-8}$	$(6.7 \pm 1.1) \times 10^{-8}$

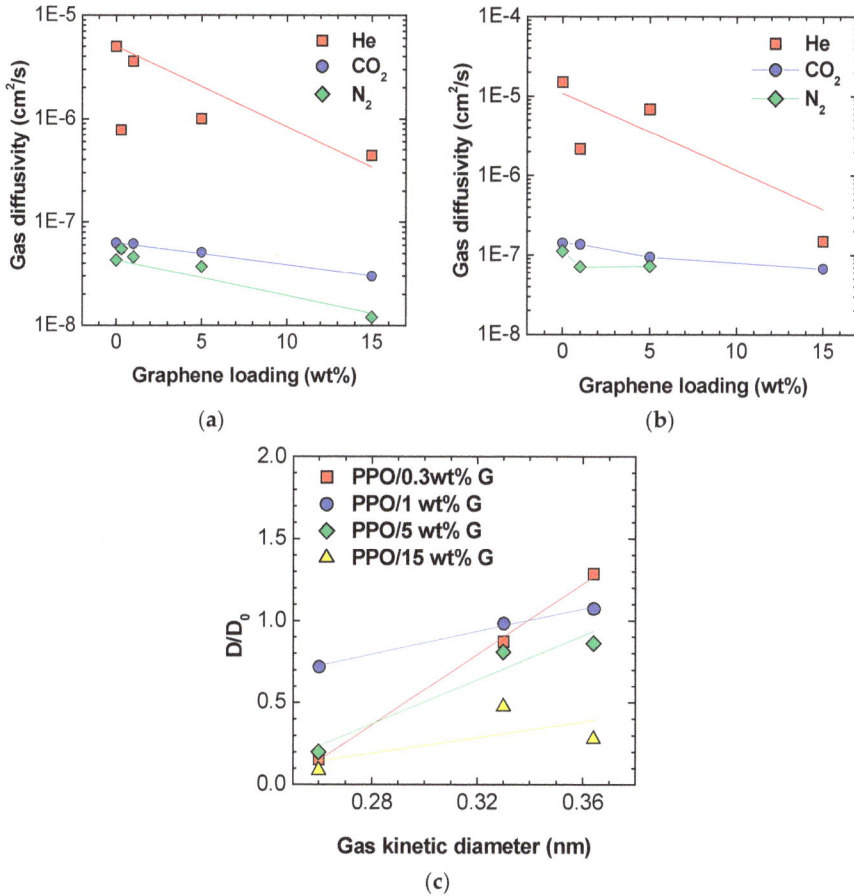

Figure 7. Diffusivity of the various gases in the PPO composite membranes at various values of graphene loadings at: (**a**) 35 °C; and (**b**) 65 °C; (**c**) Gas-dependent effect of graphene on membrane diffusivity at 35 °C: the chart shows the relative increase of membrane gas diffusivity after graphene addition versus the size of the gas molecule for various loadings of graphene.

It is often seen, in mixed matrix membranes comprising nanometric fillers, that the variations of permeability induced by filler are gas size-dependent [15,16]. Usually, porous fillers with a size selective ability increase more effectively the permeability towards smaller gases, while the opposite can be observed in the case of nanoscopic impermeable fillers such as fumed silica [15,16]. In our previous work, we have found that the addition of graphene reduces more strongly the permeation of small gases, such as Helium, and to a lower extent the permeation of larger molecules, such as Nitrogen [35]. In this work, we observe a similar trend: we reported diffusivity variations after graphene addition versus the kinetic diameter of penetrating molecule in Figure 7c. We see values of D/D_0 close to 1 for the larger molecule, namely N_2, and much smaller than unity for the smallest molecule, Helium.

To effectively correlate the observed variations in permeability to variations of diffusivity, we plotted a parity curve, reporting the observed variations in permeability after addition of graphene, P/P_0, to the ones observed for diffusivity, D/D_0, in Figure 8. It can be seen that, while for CO_2 and, less markedly, for N_2, the permeability trend is strongly associated to the diffusivity one, confirming that diffusion plays a strong role in such membranes, for He, the behavior is different. For this gas, the permeability of composite membranes is reduced less than the diffusivity by addition of graphene. Although the error associated to the evaluation of the diffusivity from the time lag measurement is larger in the case of Helium, which can cause some significant random scattering, such trend is systematic, i.e., all the data fall above the bisector. If the deviations were due merely to scattering associated to experimental error, they should have lied both above and below the bisector. According to the solution–diffusion model expressed by Equation (4), one should have:

$$\frac{P}{P_0} = \frac{D}{D_0}\frac{S}{S_0} \qquad (11)$$

so that, if $P/P_0 = D/D_0$, as it approximately happens in the case of N_2 and CO_2, one concludes that $S/S_0 = 1$, i.e., that the solubility of such gases is not affected by the presence of graphene. On the other hand, the same equation applied to the case of helium indicates that $S/S_0 > 1$, i.e., that graphene enhances the solubility of helium in the membrane. Although the solubility of helium in solids is generally small in absolute value, we are considering relative variations here, which can be significant. Furthermore, the helium molecule, due to its small size, could be adsorbed onto the graphene layers interface but also between the layers of graphene platelets. This solubility behavior counterbalances the strong reduction of diffusivity observed in the case of helium.

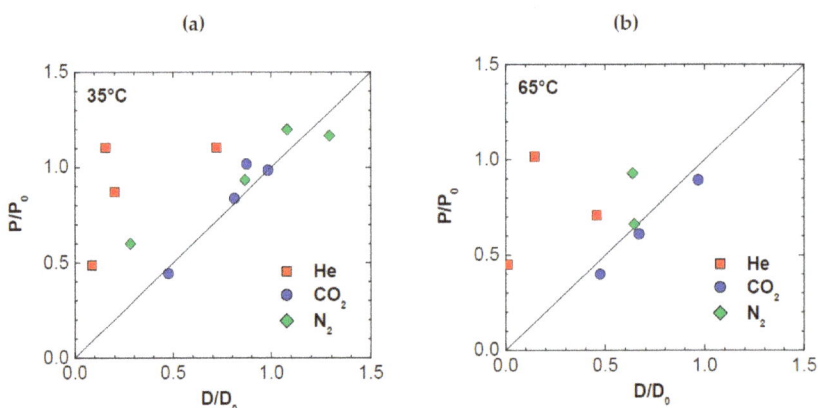

Figure 8. Parity plot showing the correlation between the variation of diffusivity induced by addition of graphene and the corresponding variation of permeability at: (**a**) 35 °C; and (**b**) 65 °C.

3.5. Analysis of Temperature Effect

It is very interesting to analyze the effect of temperature on those composite membranes. Normally, the transport properties of polymers, especially the permeability, are strongly dependent on temperature, and, to our knowledge, this is the first analysis of the temperature effect on transport properties of polymer/graphene composites for separation.

Figure 9a reports the temperature effect on permeability, as a function of graphene loading, and explains the observed trend of selectivity with temperature. It is clear from the graph that the role of graphene is to reduce strongly the positive dependence of gas permeability on temperature, and even cause an inversion of trend, in the case of CO_2. These data appear for the first time and indicate that graphene nanoplatelets hinder the polymer mobility and flexibility. Indeed, it is known that the strong increase of gas diffusivity and permeability observed in polymers at increasing temperature is mostly due to the thermally-enhanced polymer chain flexibility and mobility, which makes the diffusive jumps of gas molecules more frequent. When increasing amounts of graphene are added, up to 5 wt %, to PPO, the thermal activation of diffusivity and permeability are strongly inhibited. In particular, adding 5 wt % of graphene to PPO reduces the relative increase of permeability (corresponding to a 30 °C increase) from 0.7 to 0.2 for N_2, and from 0.47 to 0.36 for He. The relative dependence of CO_2 permeability in pure PPO is much weaker, for the reasons discussed above, and numerically is expressed by a value of +0.13, which becomes −0.17 when the sample contains 5 wt % graphene. At graphene loadings higher than 5 wt %, the effect discussed above is partially lost, possibly because the graphene nanoplatelets start to form aggregates and lose part of their ability to immobilize the polymer.

(a)   (b)

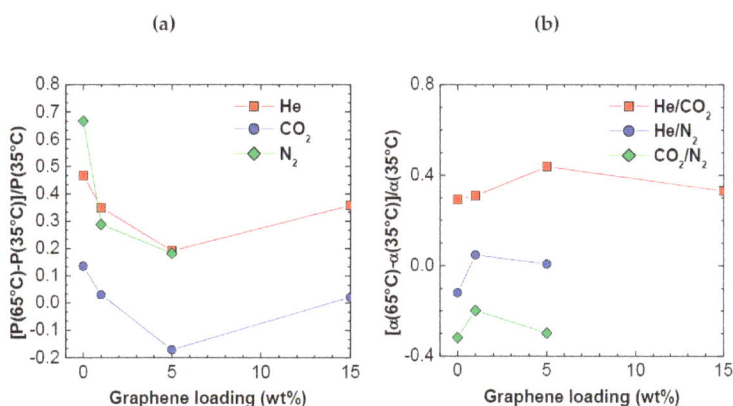

Figure 9. Relative variation of: (**a**) permeability; and (**b**) selectivity with temperature versus graphene loading of different PPO composite membranes.

In Figure 9b, we reported the relative selectivity variation in the temperature range inspected. The behavior is extremely clear and indicates that, for all the gases considered, the addition of even small amounts of graphene affects positively the selectivity, in all types of separation. For He/CO_2 selectivity, favored by temperature, graphene further enhances this trend. For CO_2/N_2 separations, unfavored by temperature, the negative effect of a temperature increase seems mitigated by graphene addition. For He/N_2 the selectivity starts to increase, rather than decrease, with temperature, after graphene addition. Such trends are a natural consequence of the permeability trend previously discussed, and indicate that graphene can have a marked, and beneficial, effect on the polymer mobility, which reduces the negative effects of temperature on selectivity. Indeed, at higher temperatures, the polymer becomes more flexible and loses part of its size selectivity. The addition of graphene in small amounts seems to mitigate such trend, due to the peculiar, thin and long, shape of graphene platelets which hinder polymer mobility and the loss of discriminating ability. Such effect is totally consistent with what observed when adding graphene to other glassy polymers prone to ageing: it was shown that graphene reduces ageing via the same mechanism, i.e., by acting as a physical constraint, or stabilizer, of the polymer matrix [35].

3.6. Comparison with Other Polymers

The permeability variations observed with graphene have comparable order of magnitude as the ones obtained by adding 1 wt % of graphene into other glassy polymers, namely PTMSP [35] and PIM-1 [33]. In the case of PTMSP/graphene, the nominal amount of graphene in the composite membrane was 1 wt %, and the procedure used to fabricate the membrane was exactly the same used for PPO in this work, as well as the initial features of the graphene added [35]. In the case of PIM-1, the graphene content was varied between 0.1% and 2.43%, and the graphene was exfoliated in situ from graphite in the polymer solution, after a long sonication of 84 h [33]. Due to this process, it is expected that the graphene particles in such case have a smaller aspect ratio than the ones used in this work. The data of permeability variation in the various polymers inspected, after addition of 1 wt % of graphene in the case of PPO and PTMSP, and of 0.71 wt %, in the case of PIM-1, are reported in Figure 10a. Data were obtained at different temperatures in the range 25–35 °C. It can be noticed that the order of magnitude of permeability variation is the same in all polymers inspected, and the trend is increasing with increasing kinetic dimeter, indicating that the permeability of larger gases is enhanced more (or reduced less) than that of smaller gases. This could be because, for large gases, even a small adjustment of the internal free volume associated to graphene addition can make a big difference in the permeability value. It must be noted that both PIM-1 and PTMSP have much higher permeability values than PPO, and that PTMSP has a higher free volume than PPO. Plus, both such polymers, as far as the He/CO_2 separation is concerned, are CO_2-selective, rather than He-selective: this is due to their high free volume, and to a strong impact of the solubility on the separation. In Figure 10b,c, we report, for comparison, the trend of permeability variation for two gases, He and CO_2, in the PPO of this work, and in PIM-1. We can notice that, in the case of helium, no significant differences between the two polymers are observed.

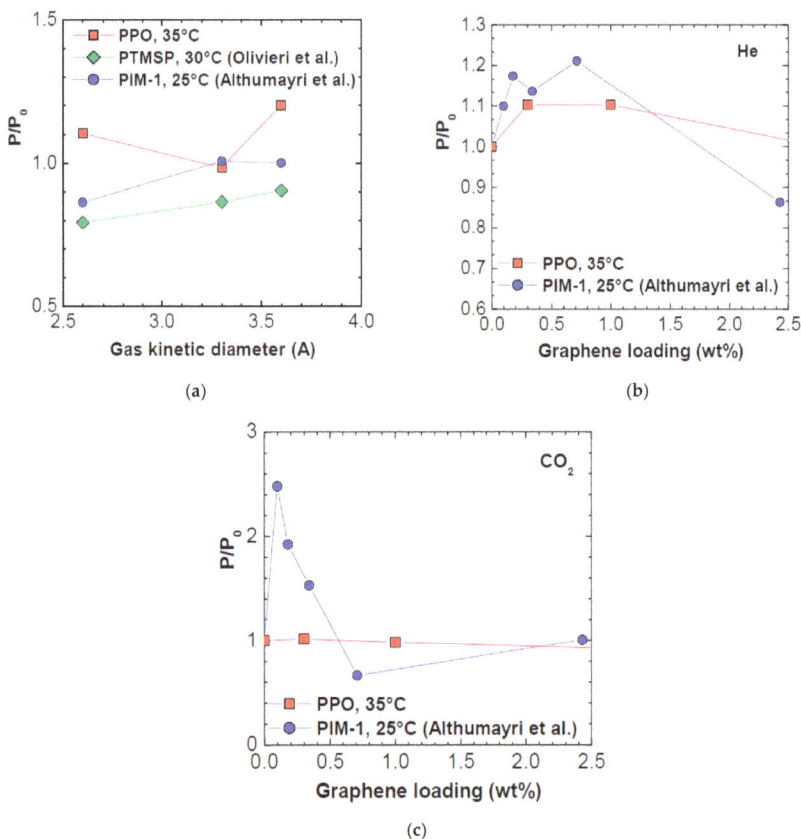

(a)

(b)

(c)

Figure 10. (**a**) Permeability variation induced by graphene addition, as a function of the gas kinetic diameter, in three glassy polymers: PPO (this work), PTMSP [35] and PIM-1 [33]. The weight fraction of graphene added was 1 wt % in the case of PPO and PTMSP; 0.71 wt % in the case of PIM-1. He permeability variation (**b**); and CO_2 permeability variation (**c**) after addition of graphene in PPO (this work) and PIM-1 versus graphene loading.

When it comes to CO_2 (and similar results are noticed for N_2), that is a slightly larger molecule, the situation changes strongly: while in PPO the CO_2 permeability is not much affected by graphene, in the loading range inspected, marked increases of permeability are observed at very low loadings (0.1 wt %) in PIM-1. One can say that such difference can be due to the, probably, smaller aspect ratio of the graphene used in the paper by Althumayri et al., [33]. Furthermore, a significant difference may also arise from the fact that PPO is a size selective polymer, with He/CO_2 selectivity higher than 1, while PIM-1 has a He/CO_2 selectivity much smaller than unity.

The comparison with other polymers case indicates that, as far as graphene addition to gas separation polymers is concerned, it seems that is advisable to work in the low loading range to retain the beneficial effects. The quantitative enhancement obtained are still limited, but there is large room for improvement by working on the morphological aspects, i.e., the aspect ratio of graphene, which could be optimized by varying the preparation method. Furthermore, the effect of a surface chemistry modification of graphene platelets on the final composite properties has not been studied yet: chemistry affects the solubility contribution to permeability, but also the adhesion and morphology of the composite material. Both aspects will be investigated in future works.

4. Conclusions

We fabricated mixed matrix membranes based on PPO and increasing amounts of few layer graphene, from 0.3 to 15 wt %, and tested them for the permeability of He, N_2 and CO_2 at 35 and 65 °C.

In general, the best effect of graphene addition is observed when the loadings are small, below 1%, and this was attributed to the fact that, at high filler loading, the effect of increasing the tortuosity prevails on the effect of enhancing the polymer chain distribution. Such aspect is in agreement with previous findings on other glassy polymer filled with graphene. The selectivity for He/CO_2 and He/N_2 is also positively affected by addition of graphene, and the effect is observed at both 35 and 65 °C. Furthermore, the graphene addition allows adjusting the thermal dependence of permeability and selectivity, generally improving it, by acting as a physical constraint to the relaxation of polymer chains which compromises size selectivity at high temperatures.

The permeability trend follows closely the diffusivity one, confirming the validity of the solution-diffusion mechanism, and that solubility of gases is less strongly affected than diffusivity by graphene addition, for all gases except helium, which might be adsorbed onto graphene insertions.

These preliminary results suggest that small additions of graphene to PPO can enhance the permselectivity, as it does in other glassy polymers. The effect could be quantitatively improved by optimizing the aspect ratio of particles, and testing chemical modification of graphene.

Supplementary Materials: The supplementary materials are available online at www.mdpi.com/2073-4360/10/ 2/129/s1.

Acknowledgments: This work has been performed in the framework of the European Project H2020 NANOMEMC2 "NanoMaterials Enhanced Membranes for Carbon Capture", GA No. 727734.

Author Contributions: Riccardo Rea and Meganne Christian performed the experiments; Riccardo Rea and Meganne Christian analyzed the data; Simone Ligi and Vittorio Morandi contributed materials and analysis tools; Maria Grazia De Angelis and Marco Giacinti Baschetti wrote the paper

Conflicts of Interest: The authors declare no conflict of interest.

References

1. Aaron, D.; Tsouris, C. Separation of CO_2 from Flue Gas: A Review. *Sep. Sci. Technol.* **2005**, *40*, 321–348. [CrossRef]
2. Bains, P.; Psarras, P.; Wilcox, J. CO_2 capture from the industry sector. *Prog. Energy Combust. Sci.* **2017**, *63*, 146–172. [CrossRef]
3. George, G.; Bhoria, N.; AlHallaq, S.; Abdala, A.; Mittal, V. Polymer membranes for acid gas removal from natural gas. *Sep. Purif. Technol.* **2016**, *158*, 333–356. [CrossRef]

4. Merkel, T.C.; Zhou, M.; Baker, R.W. Carbon dioxide capture with membranes at an IGCC power plant. *J. Membr. Sci.* **2012**, *389*, 441–450. [CrossRef]

5. Marano, J.J.; Ciferino, J.P. Integration of Gas Separation Membranes with IGCC: Identifying the right membrane for the right job. *Energy Procedia* **2009**, *1*, 361–368. [CrossRef]

6. Powell, C.E.; Qiao, G.G. Polymeric CO_2/N_2 gas separation membranes for the capture of carbon dioxide from power plant flue gases. *J. Membr. Sci.* **2006**, *279*, 1–49. [CrossRef]

7. Robeson, L.M. The upper bound revisited. *J. Membr. Sci.* **2008**, *320*, 390–400. [CrossRef]

8. Huang, Y.; Paul, D.R. Effect of molecular weight and temperature on physical aging of thin glassy poly(2,6-dimethyl-1,4-phenylene oxide) films. *J. Polym. Sci. B* **2007**, *45*, 1390–1398. [CrossRef]

9. Mehio, N.; Dai, S.; Jiang, D. Quantum Mechanical Basis for Kinetic Diameters of Small Gaseous Molecules. *J. Phys. Chem. A* **2014**, *118*, 1150–1154. [CrossRef] [PubMed]

10. Merkel, T.C.; He, Z.; Pinnau, I.; Freeman, B.D.; Meakin, P.; Hill, A.J. Effect of Nanoparticles on Gas Sorption and Transport in Poly(1-trimethylsilyl-1-propyne). *Macromolecules* **2003**, *36*, 6844–6855. [CrossRef]

11. Merkel, T.C.; He, Z.; Pinnau, I.; Freeman, B.D.; Meakin, P.; Hill, A.J. Sorption and Transport in Poly(2,2-bis(trifluoromethyl)-4,5-difluoro-1,3-dioxole-*co*-tetrafluoroethylene) Containing Nanoscale Fumed Silica. *Macromolecules* **2003**, *36*, 8406–8414. [CrossRef]

12. Merkel, T.C.; Freeman, B.D.; Spontak, R.J.; He, Z.; Pinnau, I.; Meakin, P.; Hill, A.J. Ultrapermeable, Reverse-Selective Nanocomposite Membranes. *Science* **2002**, *296*, 519–522. [CrossRef] [PubMed]

13. Ferrari, M.C.; Galizia, M.; De Angelis, M.G.; Sarti, G.C. Gas and Vapor Transport in Mixed Matrix Membranes Based on Amorphous Teflon AF1600 and AF2400 and Fumed Silica. *Ind. Eng. Chem. Res.* **2010**, *49*, 11920–11935. [CrossRef]

14. Galizia, M.; De Angelis, M.G.; Messori, M.; Sarti, G.C. Mass transport in hybrid PTMSP/Silica membranes. *Ind. Eng. Chem. Res.* **2014**, *53*, 9243–9255. [CrossRef]

15. De Angelis, M.G.; Gaddoni, R.; Sarti, G.C. Gas Solubility, Diffusivity, Permeability, and Selectivity in Mixed Matrix Membranes Based on PIM-1 and Fumed Silica. *Ind. Eng. Chem. Res.* **2013**, *52*, 10506–10520. [CrossRef]

16. De Angelis, M.G.; Sarti, G.C. Gas sorption and permeation in mixed matrix membranes based on glassy polymers and silica nanoparticles. *Curr. Opin. Chem. Eng.* **2012**, *1*, 148–155. [CrossRef]

17. Kim, S.; Jinschek, J.R.; Chen, H.; Sholl, D.S.; Marand, E. Scalable Fabrication of Carbon Nanotube/Polymer Nanocomposite Membranes for High Flux Gas Transport. *Nano Lett.* **2007**, *7*, 2806–2811. [CrossRef] [PubMed]

18. Khan Muntazim, M.; Filiz, V.; Bengtson, G.; Shishatskiy, S.; Rahman, M.; Abetz, V. Functionalized carbon nanotubes mixed matrix membranes of polymers of intrinsic, microporosity for gas separation. *Nanoscale Res. Lett.* **2012**, *7*, 504. [CrossRef] [PubMed]

19. Kim, S.; Pechar, T.W.; Marand, E. Poly(imide siloxane) and carbon nanotube mixed matrix membranes for gas separation. *Desalination* **2006**, *192*, 330–339. [CrossRef]

20. De Angelis, M.G.; Sarti, G.C. Solubility and Diffusivity of Gases in Mixed Matrix Membranes Containing Hydrophobic Fumed Silica: Correlations and Predictions Based on the NELF Model. *Ind. Eng. Chem. Res.* **2008**, *47*, 5214–5226. [CrossRef]

21. Zhu, J.; Lim, J.; Lee, C.-H.; Joh, H.-I.; Kim, H.C.; Park, B.; You, N.-H.; Lee, S. Multifunctional polyimide/graphene oxide composites via in situ polymerization. *J. Appl. Polym. Sci.* **2014**, *131*, 40177. [CrossRef]

22. Kim, H.W.; Yoon, H.W.; Yoon, S.-M.; Yoo, B.M.; Ahn, B.K.; Cho, Y.H.; Shin, H.J.; Yang, H.; Paik, U.; Kwon, S.; et al. Selective Gas Transport Through Few-Layered Graphene and Graphene Oxide Membranes. *Science* **2013**, *342*, 91. [CrossRef] [PubMed]

23. Li, H.; Song, Z.; Zhang, X.; Huang, Y.; Li, S.; Mao, Y.; Ploehn, H.J.; Bao, Y.; Yu, M. Ultrathin, Molecular-Sieving Graphene Oxide Membranes for Selective Hydrogen Separation. *Science* **2013**, *342*, 95. [CrossRef] [PubMed]

24. Yoo, B.M.; Shin, J.E.; Lee, H.D.; Park, H.B. Graphene and graphene oxide membranes for gas separation applications. *Curr. Opin. Chem. Eng.* **2017**, *16*, 39–47. [CrossRef]

25. Sun, C.; Wen, B.; Bai, B. Recent advances in nanoporous graphene membrane for gas separation and water purification. *Sci. Bull.* **2015**, *60*, 1807–1823. [CrossRef]

26. Huang, L.; Zhang, M.; Li, C.; Shi, G. Graphene-Based Membranes for Molecular Separation. *J. Phys. Chem. Lett.* **2015**, *6*, 2806–2815. [CrossRef] [PubMed]

27. Yoon, H.W.; Cho, Y.H.; Park, H.B. Graphene-based membranes: Status and prospects. *Philos. Trans. R. Soc. A Math. Phys. Eng. Sci.* **2016**, *374*, 20150024. [CrossRef] [PubMed]

28. Yoo, B.M.; Shin, H.J.; Yoon, H.W.; Park, H.B. Graphene and graphene oxide and their uses in barrier polymers. *J. Appl. Polym. Sci.* **2013**, *131*, 39628. [CrossRef]

29. Kim, H.; Yoon, H.; Yoo, B.; Park, J.; Gleason, K.; Freeman, B.D.; Park, H.B. High performance CO_2-phylic graphene oxide membranes under wet conditions. *Chem. Commun.* **2014**, *50*, 13563–13566. [CrossRef] [PubMed]

30. Shen, J.; Liu, G.; Huang, K.; Jin, W.; Lee, K.-R.; Xu, R. Membranes with Fast and Selective Gas-Transport Channels of Laminar Graphene Oxide for Efficient CO_2 Capture. *Angew. Chem.* **2015**, *127*, 588–592. [CrossRef]

31. Zhao, L.; Cheng, C.; Chen, Y.-F.; Wang, T.; Du, C.-H.; Wu, L.-G. Enhancement on the permeation performance of polyimide mixed matrix membranes by incorporation of graphene oxide with different oxidation degrees. *Polym. Adv. Technol.* **2015**, *26*, 330–337. [CrossRef]

32. Gonciaruk, A.; Althumayri, K.; Harrison, W.J.; Budd, P.M.; Siperstein, F.R. PIM-1/graphene composite: A combined experimental and molecular simulation study. *Microporous Mesoporous Mater.* **2015**, *209*, 126–134. [CrossRef]

33. Althumayri, K.; Harrison, W.J.; Shin, Y.; Gardiner, J.M.; Casiraghi, C.; Budd, P.M.; Bernardo, P.; Clarizia, G.; Jansen, J.C. The influence of few-layer graphene on the gas permeability of the high-free-volume polymer PIM-1. *Philos. Trans. R. Soc. A* **2016**, *374*, 20150031. [CrossRef] [PubMed]

34. Shin, Y.; Prestat, E.; Zhou, K.-G.; Gorgojo, P.; Althumayri, K.; Harrison, W.; Budd, P.M.; Haigh, S.J.; Casiraghi, C. Synthesis and characterization of composite membranes made of graphene and polymers of intrinsic microporosity. *Carbon* **2016**, *102*, 357–366. [CrossRef]

35. Olivieri, L.; Ligi, S.; De Angelis, M.G.; Cucca, G.; Pettinau, A. Effect of Graphene and Graphene Oxide Nanoplatelets on the Gas Permselectivity and Aging Behavior of Poly(trimethylsilyl propyne) (PTMSP). *Ind. Eng. Chem. Res.* **2015**, *54*, 11199–11211. [CrossRef]

36. Aguilar-Vega, M.; Paul, D.R. Gas transport properties of phenylene ethers. *J. Polym. Sci. B* **1993**, *31*, 1577–1589. [CrossRef]

37. Le Roux, J.D.; Paul, D.R.; Kampaby, J.; Lagow, R.J. Surface fluorination of poly(phenylene oxide) composite membranes; Part I. Transport properties. *J. Membr. Sci.* **1994**, *90*, 1–35. [CrossRef]

38. Minelli, M.; De Angelis, M.G.; Doghieri, F.; Marini, M.; Toselli, M.; Pilati, F. Oxygen permeability of novel organic–inorganic coatings: I. Effects of organic–inorganic ratio and molecular weight of the organic component. *Eur. Polym. J.* **2008**, *44*, 2581–2588. [CrossRef]

39. Crank, J. *The Mathematics of Diffusion*; Oxford Press: London, UK, 1956.

40. Luo, S.; Stevens, K.A.; Park, J.S.; Moon, J.D.; Liu, Q.; Freeman, B.D.; Guo, R. Highly CO_2-selective gas separation membranes based on segmented copolymers of poly(ethylene oxide) reinforced with pentiptycene-containing polyimide hard segments. *ACS Appl. Mater. Inter.* **2016**, *8*, 2306–2317. [CrossRef] [PubMed]

polymers

MDPI

Article

Effects of Carbon Nanotubes/Graphene Nanoplatelets Hybrid Systems on the Structure and Properties of Polyetherimide-Based Foams

Hooman Abbasi *, Marcelo Antunesand José Ignacio Velasco

Departament de Ciència dels Materials i Enginyeria Metal·lúrgica, Centre Català del Plàstic, Universitat Politècnica de Catalunya (UPC·BarcelonaTech), C/Colom 114, E-08222 Terrassa, Barcelona, Spain; marcelo.antunes@upc.edu (M.A.); jose.ignacio.velasco@upc.edu (J.I.V.)
* Correspondence: hooman.abbasi@upc.edu; Tel.: +34-937-837-022; Fax: +34-937-841-827

Received: 31 January 2018; Accepted: 19 March 2018; Published: 21 March 2018

Abstract: Foams based on polyetherimide (PEI) with carbon nanotubes (CNT) and PEI with graphene nanoplatelets (GnP) combined with CNT were prepared by water vapor induced phase separation. Prior to foaming, variable amounts of only CNT (0.1–2.0 wt %) or a combination of GnP (0.0–2.0 wt %) and CNT (0.0–2.0 wt %) for a total amount of CNT-GnP of 2.0 wt %, were dispersed in a solvent using high power sonication, added to the PEI solution, and intensively mixed. While the addition of increasingly higher amounts of only CNT led to foams with more heterogeneous cellular structures, the incorporation of GnP resulted in foams with finer and more homogeneous cellular structures. GnP in combination with CNT effectively enhanced the thermal stability of foams by delaying thermal decomposition and mechanically-reinforced PEI. The addition of 1.0 wt % GnP in combination with 1.0 wt % CNT resulted in foams with extremely high electrical conductivity, which was related to the formation of an optimum conductive network by physical contact between GnP layers and CNT, enabling their use in electrostatic discharge (ESD) and electromagnetic interference (EMI) shielding applications. The experimental electrical conductivity values of foams containing only CNT fitted well to a percolative conduction model, with a percolation threshold of 0.06 vol % (0.1 wt %) CNT.

Keywords: nanocomposites; graphene; carbon nanotubes; hybrid nanoparticles; polyetherimide foams; electrical conductivity; percolation; ultrasonication

1. Introduction

Polyetherimide (PEI) has recently become popular for use in advanced applications, due to its outstanding combination of high mechanical properties, flame and chemical resistance, and high thermal and dimensional stability. The preparation of PEI-based foams reinforced with carbon-based nanoparticles using water vapor induced phase separation (WVIPS) has shown promising results in terms of homogeneity and filler dispersion [1–4]. The addition of carbon-based nanofillers to PEI has created a suitable candidate for various advanced applications, such as fuel cells and electromagnetic interference (EMI) shielding [5,6]. Additionally, foaming could facilitate desirable features such as density reduction, damping properties, high thermal insulation, and the potential improvement of electrical conductivity and electromagnetic absorption by promoting wave scattering [7–9]. The combination of functional nanoparticles and foaming has a high potential to generate new lightweight composites with high specific strength and multifunctionality [10]. Simultaneous enhancements in electrical and mechanical properties, with the addition of carbon-based nanosized fillers such as graphene nanoplatelets (GnP) or carbon nanotubes (CNT), owing to their high aspect ratio (AR) and exceptional mechanical and electrical properties, have brought important advantages over non-carbon-based nanofillers [11–13].

Polymers **2018**, *10*, 348

Various attempts to prepare hybrid CNT/graphene materials have been carried out to create transparent conductors [14–17], electrodes [18], electron field emitters [19], field effect transistors [17], supercapacitors [20], and Li-ion batteries [21,22]. Maxian et al. [23] numerically investigated the electrical percolation behavior of porous systems incorporating carbon 1D- and 2D-fillers assuming perfect random filler distribution and realistically modeling 'foaming' by displacing the fillers. Their model was able to successfully capture experimental trends [24] showing the increased conductivity and lower percolation threshold of porous compared to non-porous systems. Sensitivity analysis demonstrated that the electrical percolation behavior of the porous polymer systems incorporating 1D- and 2D-nanofillers was significantly affected by four main factors: (i) porosity level, (ii) type of filler, (iii) filler alignment, and (iv) filler aspect ratio (AR). The type of filler and its AR played the most important role in establishing the percolation threshold.

The present work aimed to extend the applicability of PEI composites by considering the combination of two strategies: foaming of the composites by means of WVIPS and the use of a hybrid nanofiller system based on GnP and CNT. In terms of the first strategy, we have already shown in previous works that WVIPS foaming is an effective method to obtain medium-density PEI foams with homogeneous structures, and that the addition of GnP to PEI and foaming can lead to components with enhanced electrical conductivity. This is crucial in applications requiring high EMI shielding, and especially those where EM absorption mechanisms play a key role, such as in stealth technology [25,26]. On the other hand, it has been shown that hybrid nanofillers based on platelet-like GnP and other conductive nanoparticles such as CNT may promote the formation of an efficient conductive network [27–29].

This work considered the preparation of composites based on PEI and different proportions of dispersed CNT (from 0.0 to 2.0 wt %) and GnP (from 0.0 to 2.0 wt %), to give a total nanofillers amount of 2.0 wt %, their foaming by WVIPS, and their characterization in terms of microstructure, cellular structure, thermal stability, viscoelastic behavior and electrical conductivity. We predominantly focused on analyzing how the addition of GnP affects the electrical conduction behavior of the resulting composite foams.

2. Experimental

2.1. Materials

Thermoplastic polyetherimide (PEI), with the commercial name Ultem 1000, manufactured by Sabic (Riyadh, Saudi Arabia), was used. PEI Ultem 1000 has a density of 1.27 g/cm^3 and a glass transition temperature (T_g) of 217 °C.

Graphene nanoplatelets, known as GnP (commercial name xGnP M-15 and density of 2.2 g/cm^3), were supplied by XG Sciences (Lansing, MI, USA). These nanofillers are formed using stacks of graphene nanoplatelets with an average thickness of 6–8 nm and a lateral size of 15 μm, with an approximate surface area of 120–150 m^2/g and an electrical conductivity of 10^7 and 10^2 S/m, respectively, measured parallel and perpendicular to their surface, as reported by the manufacturer.

Multi-wall carbon nanotubes (MWCNT), from now on referred to as CNT, with a carbon content >95%, density of 2.1 g/cm^3 and characteristic dimensions of 6–9 nm × 5 μm, were purchased from Sigma Aldrich (Saint Louis, MO, USA). These multi-walled carbon nanotubes were prepared by chemical vapor deposition, using cobalt and molybdenum as catalysts.

N-methyl pyrrolidone (NMP) was acquired from Panreac Química SA (Barcelona, Spain) with a purity of 99%, a boiling point of 202 °C, and a flash point of 95 °C.

2.2. Foam Preparation

Two sets of foams were prepared by WVIPS: a first series of foams containing only CNT ("CNT series"), particularly 0.1, 0.5, 1.0, and 2.0 wt % of CNT; and a second series containing the hybrid nanofillers system through the combination of different amounts of GnP (from 0.0 to 2.0 wt %) and CNT (from 0.0 to 2.0 wt %), for a total nanofiller amount of 2.0 wt % ("Hybrid series").

A detailed explanation of the WVIPS process is given in previous works [25,26]. In this study, the preparation of foams began with the dispersion of 0.5 g of CNT into 200 mL of NMP, which is known to be a proper solvent for carbon-based suspension at room temperature [30]. High power probe sonication was applied for 60 min using a Fisher Scientific FB-705 ultrasonic processor with a 12 mm solid tip probe at 20% amplitude and 20 kHz output, applying a total amount of energy of 380 kJ (30–60 W), kept at a constant temperature of 50 °C by placing the suspension inside an ice-bath. In the case of the Hybrid series, the corresponding amount of GnP was initially sonicated at 100% amplitude for 30 min, followed by the sonication of CNT in the GnP-NMP suspension. In the following step, PEI was dissolved in the suspension containing the sonicated particles (15.0 wt % PEI solution) at 75 °C and kept stirring at 450 rpm for a period of 24 h. Afterwards, the filler-rich solution was diluted with PEI-NMP (15.0 wt % PEI in NMP) to obtain the 0.1, 0.5, 1.0, and 2.0 wt % CNT-filled composites alongside with the hybrid compositions of 1.5–0.5, 1.0–1.0, and 0.5–1.5 CNT-GnP. The composites corresponding to the Hybrid series (total CNT-GnP amount of 2.0 wt %) are specifically referred to as 2/0, 1.5/0.5, 1/1, 0.5/1.5 and 0/2, with the first number corresponding to the wt % of CNT and the second one to the wt % of GnP.

Subsequently, each solution was poured on a flat glass exposed to air, with an average measured humidity of 75% at room temperature for 4 days, which promoted foaming of the polymer by means of WVIPS. The resulting foams were then washed with a 50/50 mixture of ethanol and water followed by extraction of the remaining solvent, utilizing hot water, stirring at 90 °C for 7 days and intensively drying under vacuum at 140 °C for 7 additional days to fully extract the residual NMP. The typical density of the prepared foams was 0.3–0.5 g/cm³, with a final thickness of around 5 mm. Samples were later cut directly from the prepared foams and used in the several characterizations.

2.3. Testing Procedure

The density of the foams was measured according to the ISO-845 standard procedure. The porosity of the foam, understood as its void percentage, could be directly obtained from the density values of the foam and respective unfoamed material according to the following expression:

$$Porosity\ (\%) = \left(1 - \frac{\rho}{\rho_s}\right) \times 100 \tag{1}$$

where ρ and ρ_s is the density of the foam and density of the solid unfoamed material, respectively.

The morphology of the foams was analyzed using a JEOL (Tokyo, Japan) JSM-5610 scanning electron microscope (SEM). Samples were fractured using liquid nitrogen and a thin layer of gold was sputter deposited onto their surface with a BAL-TEC (Los Angeles, CA, USA) SCD005 Sputter Coater (Ar atmosphere). The values of the average cell size (Φ), cell nucleation density, and cell density (N_0 and N_f, respectively, both in cells/cm³) were measured and calculated, respectively, from the analysis of ×300 magnification SEM images using the intercept counting method, a procedure presented in detail in [31]. Five ×300 magnification SEM images were analyzed for each foam. Particularly, N_0 and N_f were determined assuming an isotropic distribution of spherical cells according to:

$$N_0 = \left(\frac{n}{A}\right)^{3/2} \left(\frac{\rho_s}{\rho}\right) \tag{2}$$

$$N_f = \frac{6}{\pi\Phi^3}\left(1 - \frac{\rho}{\rho_s}\right) \tag{3}$$

where n is the number of cells counted in each SEM image and A is the area of the SEM image in cm². In Equations (2) and (3), N_0 represents the number of cells per volume of unfoamed material and N_f represents the number of cells per volume of foamed material. For foams that displayed a dual cell size distribution, the proportion of area occupied by each cell population was taken into account when calculating N_0 and N_f.

The analysis of the characteristic (002) diffraction planes of GnP and CNT, as well as the possible crystalline characteristics of PEI in the foams, was carried out by means of wide-angle X-ray diffraction (XRD) using a PANalytical diffractometer (Almelo, The Netherlands) running with CuKα ($\lambda = 0.154$ nm) at 40 kV and 40 mA. Scans were performed from 2° to 60° using a scan step of 0.033°.

A TGA/DSC 1 Mettler Toledo (Columbus, OH, USA) STAR System analyzer was used to study the thermal stability of foams, using samples of around 10.0 mg, heating from 30 to 1000 °C at 10 °C/min under a nitrogen atmosphere (constant flow of 30 mL/min) and analyzing the weight loss evolution with temperature.

Thermomechanical analysis was used to study the viscoelastic behavior of foams, particularly their storage and loss moduli (E' and E'', respectively), as well as PEI's glass transition temperature (T_g). A DMA Q800 from TA Instruments (New Castle, DE, USA) was used in a single cantilever configuration. Samples were analyzed from 30 to 300 °C at a heating rate of 2 °C/min, applying a dynamic strain of 0.02% and frequency of 1 Hz. Rectangular shape specimens were prepared with a length of 35.5 ± 1.0 mm, width of 12.5 ± 1.0 mm, and thickness of 3.0 ± 0.5 mm. Three different measurements were performed for each sample (error < 5%).

Samples of 20 mm × 20 mm × 1 mm were prepared to measure electrical conductivity using a 4140B model HP pA meter/dc voltage source with a two-probe set. The surfaces of the samples in contact with the copper electrode pads were covered with a thin layer of colloidal silver conductive paint, with an electrical resistance between 0.01 and 0.1 Ω/cm^2 to guarantee perfect electrical contact. A direct current voltage was applied with a range of 0–20 V, voltage step of 0.05 V, hold time of 10 s, and step delay time of 5 s.

The electrical conductivity (σ, in S/m) was calculated using:

$$\sigma = 1/\rho_v \tag{4}$$

and

$$\rho_v = RA_{E.C}/d \tag{5}$$

where ρ_v ($\Omega \cdot$m) is the electrical volume resistivity, R is the electrical resistance of the sample (in Ω), $A_{E.C}$ is the area of the surface in contact with the electrode (in m^2), and d is the distance between the electrodes (in m).

Considering that porosity could affect the surface area in contact with the electrode, the cell size and the cell density of foams were used to apply a correction to the values of electrical conductivity (σ_{corr}) by taking into account variations in effective surface area as follows:

$$\sigma_{corr} = \frac{d}{R(A_{non\text{-}cell} + A_{cell\text{-}hemisphere})} \tag{6}$$

where $A_{non\text{-}cell}$ is the $A_{E.C}$ with the cell section area excluded and

$$A_{cell\text{-}hemisphere} = \left(\frac{n}{A}\right)A_{E.C}\left(2\pi\frac{\Phi^2}{4}\right) \tag{7}$$

Therefore:

$$A_{non\text{-}cell} + A_{cell\text{-}hemisphere} = A_{E.C} + \left(\left(\frac{n}{A}\right)A_{E.C}\left(\pi\frac{\Phi^2}{4}\right)\right) \tag{8}$$

the values of n, A, and Φ were obtained by analyzing SEM micrographs, and represent the number of cells, the corresponding area of the micrograph, and average cell size, respectively.

3. Results and Discussion

3.1. Cellular Structure

The composition of both series of foams and respective densities are presented in Table 1. As can be seen, foam density was affected by the addition of CNT, as density grew higher with increasing CNT concentration.

Table 1. Composition of the foams and their respective densities.

Series	Sample code	Density	CNT (wt %)	GnP (wt %)	Total filler (vol %)
Carbon nanotubes (CNT)	0.1 CNT	0.32	0.1	0.0	0.06
	0.5 CNT	0.31	0.5	0.0	0.30
	1 CNT	0.40	1.0	0.0	0.61
	2 CNT	0.57	2.0	0.0	1.22
Hybrid	1.5/0.5	0.43	1.5	0.5	1.18
	1/1	0.41	1.0	1.0	1.19
	0.5/1.5	0.31	0.5	1.5	1.21
	0/2	0.33	0.0	2.0	1.16

The characteristic micrographs showing the general cellular structure of both the CNT and Hybrid series are, respectively, presented in Figures 1 and 2. The morphological characteristics of all the samples are compiled in Table 2.

Table 2. Cellular structure characteristics, average cell sizes, and cell densities of the CNT and Hybrid series foams.

Sample code	Homogeneity	Cell type	Porosity (%)	Cell size * (μm)	Cell density (Cells/cm^3)	
					N_f	N_0
0.1 CNT	Homogeneous (unimodal)	Closed	74.8	Low 23.4 (6.7)	1.1×10^8	4.4×10^8
0.5 CNT	Homogeneous (quasi-unimodal)	Slightly inter-connected	75.7	Medium 55.3 (19.9)	8.5×10^6	3.2×10^7
1 CNT	Heterogeneous (dual)	Partially inter-connected	69.0	High 194.6 (41.1) Low 17.7 (8.3)	1.2×10^7	5.2×10^6
2 CNT	Homogeneous (unimodal)	Inter-connected	55.7	Low 26.9 (19.3)	5.4×10^7	5.1×10^7
1.5/0.5	Heterogeneous (dual)	Partially inter-connected	66.6	High 81.0 (29.1) Low 14.2 (10.1)	5.3×10^7	1.5×10^7
1/1	Heterogeneous (dual)	Partially inter-connected	68.3	High 84.6 (37.0) Low 15.6 (9.1)	9.5×10^7	1.7×10^7
0.5/1.5	Homogeneous (unimodal)	Closed	76.1	Medium 71.2 (24.0)	4.0×10^6	8.8×10^6
0/2	Homogeneous (unimodal)	Closed	74.3	Low 33.2 (10.4)	3.9×10^7	1.2×10^8

* Cell size standard deviation is presented between parentheses.

The cellular structure of the CNT series foams changed from homogeneous closed-cell (0.1 CNT) to homogeneous porous inter-connected (2 CNT). Foams with intermediate CNT concentrations (0.5 CNT and 1 CNT) showed complex cellular structures. Sample 0.5 CNT displayed a quasi-unimodal structure, with slightly inter-connected cells, bigger than those of the 0.1 CNT foam. Sample 1 CNT developed a dual cell size population of partially inter-connected cells with very different cell sizes.

Significant differences were observed in terms of the cellular structure of the foams with varying CNT content, which could be the consequence of two main simultaneous effects of CNT during cell formation and growth. On the one hand, CNT interfered with solvent/non-solvent exchange, slowing it down. As a result, the cell nucleation rate decreased, what leads to increased cell sizes, as observed up until 1 wt % CNT. On the other hand, due to the strong surface interaction between CNT and PEI, as the amount of CNT increased the mobility of the polymer decreased in the solution, that is, the CNT surface strongly limited polymer mobility, favoring the formation of inter-connected cells/porous structure. Due to this effect, the inter-connected cell structure of the foams increased with the CNT content up until 2 wt % CNT.

In the 1 CNT foam, a dual cell structure developed as a consequence of the two different cell nucleation stages. As the first formed cells grew, the remaining solution got richer in polymer due to the extraction of the solvent, leading to the nucleation of a second cell population. It has recently been reported how polymer concentration plays a key role in the resultant cell size of foams prepared by WVIPS [32].

In the particular case of the 2 CNT foam, the high restriction to polymer mobility prevented the formation of cells, resulting in a generalized inter-connected porous structure. In this case, the solvent exchange with water was not confined within the cells and, rather, expelled from the foam through the formed porous structure.

The foam porosity decreased with increasing CNT and, as mentioned before, the foam density increased with increasing CNT content. The cause behind the increased density and decreased porosity of the foams containing the higher CNT concentrations (1 wt % CNT and 2 wt % CNT) seems to be due to a collapse of the inter-connected cell structure.

The Hybrid series foams displayed increased cellular structure homogeneity with a decreasing concentration of CNT (see Figure 2), which coincides with the results obtained for the CNT series. As can be seen in Figure 2a,b, Hybrid series foams with a higher concentration of CNT (1.5/0.5 and 1/1 samples) showed a dual cell size distribution because of two different cell nucleation steps. Additionally, comparing the SEM images of the 0.5/1.5 and 1/1 hybrids (Figure 2b,c) with the images of 0.5 CNT and 1 CNT (Figure 1b,c), it can be seen that the addition of GnP provided further cellular structure homogeneity.

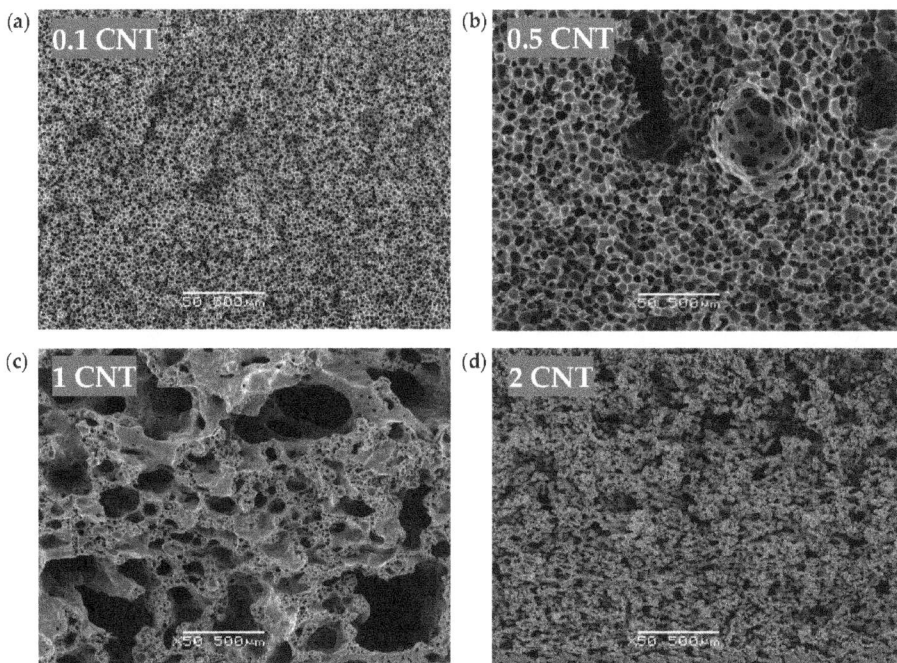

Figure 1. Typical SEM images at ×50 magnification showing the general cellular structure of the CNT series foams: (**a**) 0.1 wt %, (**b**) 0.5 wt %, (**c**) 1.0 wt %, and (**d**) 2.0 wt % of CNT.

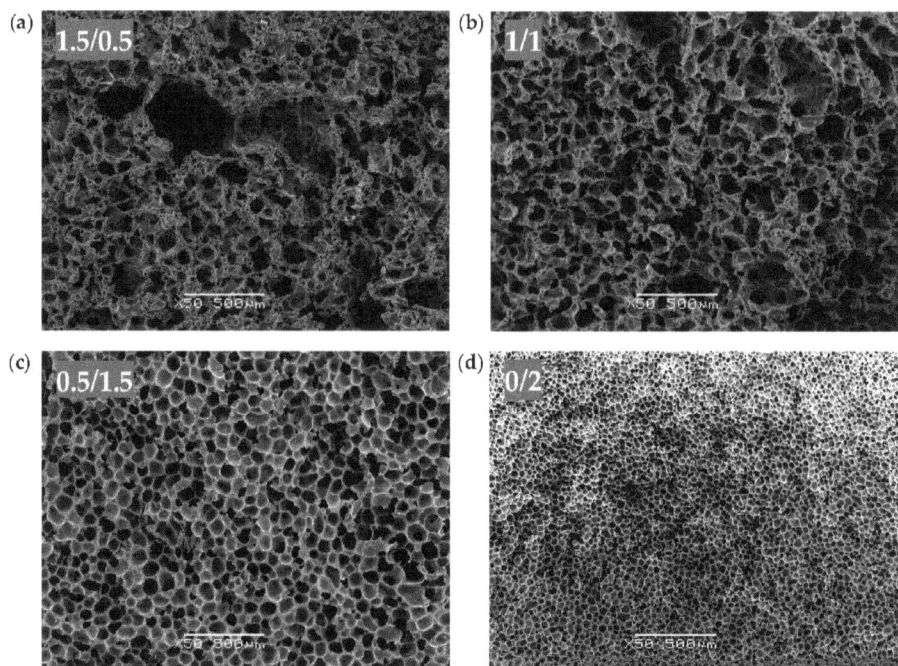

Figure 2. Typical SEM images at ×50 magnification showing the general cellular structure of the Hybrid series foams: (**a**) 1.5 wt % CNT and 0.5 wt % GnP, (**b**) 1.0 wt % CNT and 1.0 wt % GnP, (**c**) 0.5 wt % CNT and 1.5 wt % GnP, and (**d**) 0.0 wt % CNT and 2.0 wt % GnP.

As in the CNT series, the average cell size in the Hybrid series displayed a general positive trend along with the increasing content of CNT, which was the direct result of CNT slowing down the solvent exchange process.

The porosity of the Hybrid series also showed a general decreasing trend with increasing CNT content, as well as an increasing trend in density. As can be seen, the influence of GnP on the morphology of the Hybrid series foams was less relevant than CNT.

As can be seen in Figure 3a, the X-ray spectra of the CNT series foams demonstrated the proper dispersion and possible exfoliation of the nanotubes throughout PEI's matrix, as foams with 0.1, 0.5, and 1.0 wt % of CNT showed an absence of the (002) diffraction plane found at $2\theta = 25.8°$, characteristic of CNT. The disappearance of this peak was related to the dispersion and partial exfoliation of the nanotubes promoted by the combined effects of high power sonication and later foaming. On the other hand, as seen in Figure 3b, some of the Hybrid series foams presented two peaks at 25.8° and 26.5°, respectively corresponding to the characteristic (002) crystal plane of CNT and GnP. This dual peak was observed for 1.5/0.5 and 1/1 foams, which could indicate the absence of total exfoliation of nanoparticles throughout PEI's matrix. These results reflect one of the potential causes of the enhancement of electrical conductivity for these composites, due to the formation of an effective conductive network by physical contact between GnP and CNT.

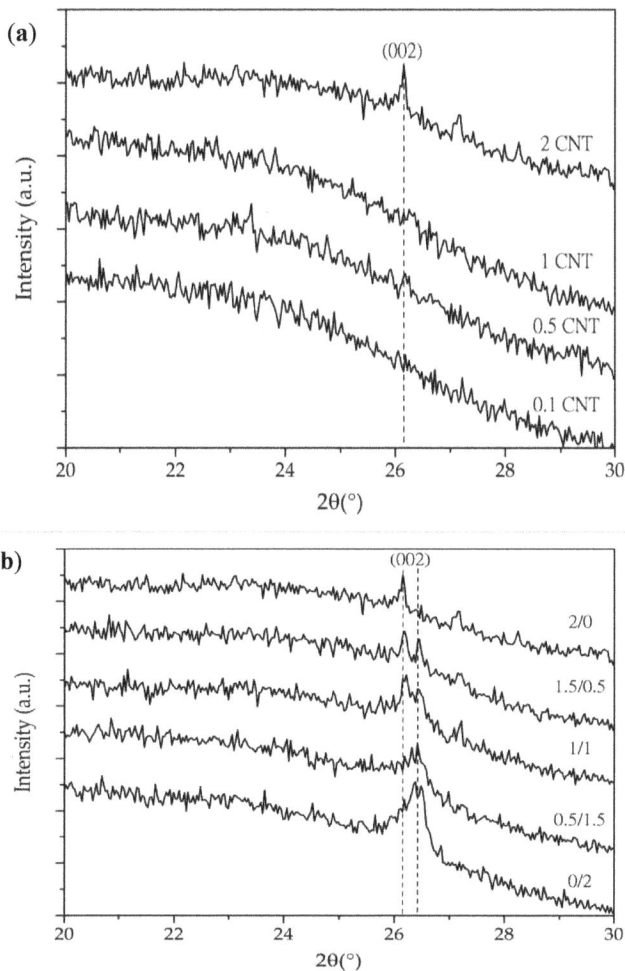

Figure 3. X-ray spectra of (**a**) CNT series foams and (**b**) Hybrid series foams showing the characteristic (002) diffraction plane of CNT and the (002) diffraction plane of CNT and GnP, respectively.

3.2. Thermal Stability

Thermogravimetric analysis of the foams demonstrated a decomposition retardancy that was induced by increasing the concentration of CNT (see the thermograms shown in Figure 4 and the results presented in Table 3). This could be attributed to the fact that the carbon nanotubes seemed to favor a better physical barrier in the prepared foams or a higher thermal conductivity, facilitating heat dissipation and, therefore, avoiding the accumulation of heat at a certain point [33,34]. Additionally, the combination of GnP and CNT resulted in a more complex decomposition behavior. GnP nanoparticles seemed to have more influence on delaying foam degradation, which could be expected from the barrier effect of GnP induced by its layered-shape hindering the escape of volatile gases. As a consequence, by increasing the concentration of CNT in the Hybrid series, foams showed faster degradation. It has been reported [35,36] that layered-shaped particles could promote a barrier effect by increasing tortuosity and delaying the discharge of volatile products, therefore slowing down the decomposition process.

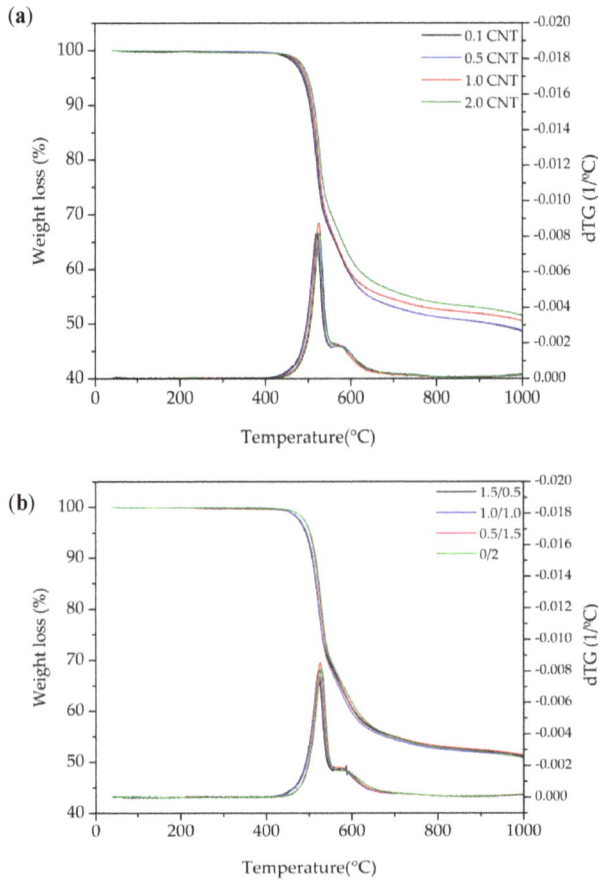

Figure 4. TGA and DTG thermograms of (**a**) CNT series foams, and (**b**) Hybrid series foams.

Table 3. Thermogravimetric results for CNT series foams and Hybrid series foams.

Series	CNT (wt %)	GnP (wt %)	Decomposition temperature (°C)			Residue at 1000 °C (wt %)
			Onset	T_{max}	40 wt % Loss	
	0.1	0.0	495.6	520.6	591.0	48.6
	0.5	0.0	497.5	521.0	592.2	48.8
CNT	1.0	0.0	500.7	523.7	594.3	50.6
	2.0	0.0	505.0	526.3	617.5	51.5
	1.5	0.5	499.4	522.2	604.7	51.1
Hybrid	1.0	1.0	495.2	523.4	597.7	50.9
	0.5	1.5	505.3	526.7	606.4	51.3
	0.0	2.0	506.9	529.8	612.6	50.8

The influence of the additional CNT on the morphology of the foams (see Section 3.1 Cellular Structure) could have also had an effect in altering the velocity of decomposition by increasing the surface area of cells and their interconnectivity.

As can be seen in Table 3, a delay of around 10 °C was observed when the concentration of CNT increased from 0.1 to 2.0 wt %. Later, the decomposition followed this behavior with a steeper trend, reaching around 25 °C of delay for a 40 wt % loss (maximum temperature value around 617 °C for the sample with 2.0 wt % CNT). The Hybrid series did not follow the same trend. As mentioned before, foams with a high amount of GnP could have benefited from the layered-shape of the particles by delaying decomposition. Particularly, the 0.5/1.5 Hybrid series foam seemed to exceed the sample with 2.0 wt % of CNT at the onset temperature of decomposition by a close margin, slowly falling behind at more advanced stages of thermal decomposition. Increasing the concentration of CNT in the Hybrid series foams did not seem to enhance thermal stability when compared to the initial sample with only 0.5 wt % of CNT.

3.3. Viscoelastic Behavior

Thermomechanical analysis of the foams showed that two main factors affected their viscoelastic behavior: the density and cellular structure of the foams, and the concentration of CNT. Both factors are closely related, as the amount of CNT had a clear direct effect on the final cellular structure of the foams (see Section 3.1 Cellular Structure). As can be observed in Figure 5a, the highest measured value of E' at 30 °C, directly related to the behavior of the elastic portion of the material, corresponded to the foam with the highest amount of CNT in both series (2.0 wt % in CNT series and 1.5 wt % in Hybrid series). This was related to a higher foam density induced by the presence of CNT.

The foam's structural influence on mechanical behavior could be extracted from the changes observed in the specific storage modulus of the foams (Figure 5b). The results showed that the morphological differences caused by increasing the amount of CNT could result in a counter effect, reducing the specific storage modulus of CNT series foams. However, the addition of GnP and its mentioned effect on the cellular structure in the Hybrid series foams resulted in higher values of the specific storage modulus, opening up a possible strategy to exploit the mechanical reinforcing effects of these carbon nanoparticles for these type of foams.

With respect to the viscous response, results presented in Table 4 suggest that nanofillers could have opposing effects on the viscous behavior of the foams. On the one hand, a lubricating effect facilitates the mobility of PEI molecules surrounding the nanofillers and, on the other, there is a restrictive element to molecular mobility due to an enhanced surface interaction with the polymer molecules. CNT series foams showed a 1–3 °C decrease in the maximum temperatures of both tan δ and E'' when increasing the amount of CNT from 0.1 to 2.0 wt %. Considering that these values reflect PEI's T_g, the decreases seem to be related to higher molecular mobility when increasing the concentration of CNT. There are several studies suggesting an increase in molecular mobility is a consequence of CNT addition [34,37]. Liu et al. [37] suggested that MWCNT have self-lubricating properties, as they are formed by sp^2 bonded cylindrical layers that can easily slide and move upon each other as inter-layer interaction is controlled by weak Van der Waals forces.

Table 4. Glass transition temperatures for CNT series foams and Hybrid series foams obtained from the maximum of tan δ and loss modulus (E'') and their corresponding intensity and full width at half maximum (FWHM).

Series	CNT (wt %)	GnP (wt %)	T_g Max tan δ (°C)	T_g Max E'' (°C)	tan δ Intensity	FWHM in tan δ	E'' Intensity (MPa)	FWHM in E''
CNT	0.1	0.0	229.1	223.5	1.56	10.6	28.1	9.0
	0.5	0.0	227.3	221.0	1.86	11.0	29.9	7.8
	1.0	0.0	227.4	220.9	1.88	11.1	43.5	7.7
	2.0	0.0	226.4	221.5	1.30	11.4	73.6	7.5
Hybrid	1.5	0.5	227.9	218.7	1.67	14.4	60.3	15.7
	1.0	1.0	228.7	220.9	1.73	12.3	54.7	10.6
	0.5	1.5	226.0	218.8	1.73	12.0	44.6	7.9
	0.0	2.0	228.6	223.3	2.03	9.5	68.4	6.1

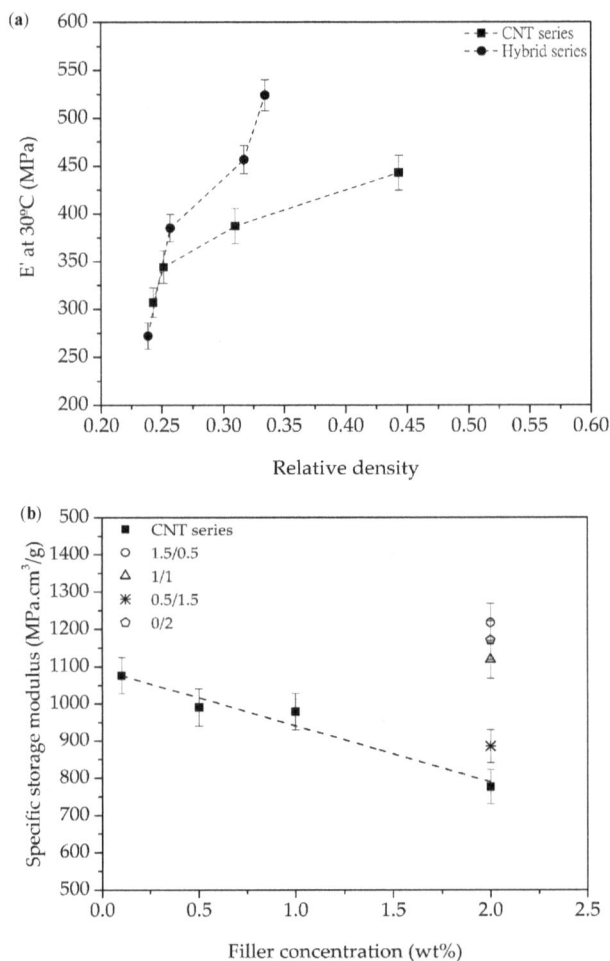

Figure 5. (a) Evolution of the storage modulus (E') measured at 30 °C with increasing relative density and (b) evolution of the specific storage modulus at 30 °C with increasing filler concentration for CNT series foams and Hybrid series foams.

Other works have shown that nanofillers can act to restrict molecular mobility due to the high and strong interfacial interaction that can be established between nanometric-sized particles, such as carbon nanotubes or graphene, and polymer molecules [38,39]. As can be seen in Table 4, Hybrid series foams did not show any clear tendency related to the two previously mentioned factors.

The intensity of the characteristic peak of the T_g in the E'' curve showed clear increases when augmenting the concentration of CNT in the CNT series foams from 0.1 to 2.0 wt %, indicating an increase in the amount of the viscous response of the foams. On the other hand, Hybrid series foams showed that the combination of CNT and GnP could induce a reduction in the viscous response when compared to foams having the same amount of either nanofiller alone. The increase in the intensity of tan δ could be associated with the energy damping response of the foams, as the dual structure in the CNT series seemed to play a role in dissipating energy alongside the presence of a higher amount of CNT. However, the 2.0 wt % CNT foam presented a low value of tan δ, which could be related to its particular cellular structure, formed by inter-connected open pores. As Hybrid series foams displayed more homogeneous cellular structures when increasing the concentration of GnP, they showed an increase in tan δ.

Variations in the width of the loss modulus and tan δ peaks corresponding to the T_g were quantified by measuring their FWHM (see values presented in Table 4). A slight decrease was observed in terms of the FWHM measured in the loss modulus (E'') for CNT series when increasing the amount of CNT, while for Hybrid series the FWHM decreased when the GnP concentration increased. It has been demonstrated by several authors that a lower FWHM value of the loss modulus peak is indicative of higher molecular relaxation [40,41]. Comparatively, as the decrease in the FWHM was lower for the CNT series when increasing the amount of CNT, these foams displayed a lower restriction to molecular relaxation when compared to the Hybrid series. The addition of CNT-GnP hybrids seemed to restrict chain segment relaxation, as Hybrid series foams presented globally higher FWHM values measured in the loss modulus when compared to CNT series (besides globally higher peak intensities). In terms of tan δ, both CNT and Hybrid series presented significant FWHM increases with rising CNT concentration, related to efficient PEI restriction by the nanotubes, which acted by avoiding localized strains [42]. This could be another factor assisting energy dissipation in these foams. Additionally, Hybrid series foams showed more pronounced energy dissipation as a result of a synergic effect between CNT and GnP, which disappeared in the sample containing only GnP.

3.4. Electrical Conductivity

The porosity of both CNT series foams and Hybrid series foams was taken into account when calculating their effective surface area and was used to determine a corrected value of electrical conductivity (σ_{corr}). Calculations showed that the corrected values could drop to around half when considering the presence of a porous surface (see values presented in Table 5). These corrected values have been used in further discussions.

Table 5. Electrical conductivity and corrected electrical conductivity values for CNT series foams and Hybrid series foams.

Series	Sample code	σ (S/m)	σ_{corr} (S/m) *
CNT	0.1 CNT	8.7×10^{-12}	4.5×10^{-12} (1.2×10^{-12})
	0.5 CNT	1.2×10^{-3}	6.4×10^{-4} (2.1×10^{-4})
	1 CNT	5.8×10^{-3}	3.3×10^{-3} (8.0×10^{-4})
	2 CNT	9.2×10^{-3}	6.4×10^{-3} (2.5×10^{-3})
Hybrid	1.5/0.5	6.9×10^{-3}	3.7×10^{-3} (1.4×10^{-3})
	1/1	1.6×10^{-2}	8.8×10^{-3} (3.5×10^{-3})
	0.5/1.5	9.6×10^{-4}	5.9×10^{-4} (1.5×10^{-4})
	0/2	3.9×10^{-12}	2.2×10^{-12} (6.1×10^{-13})

* Standard deviation of the corrected electrical conductivity is presented between parentheses.

The electrical conductivity results presented in Figure 6 illustrate the significant influence of the CNT-GnP hybrid network in inducing higher electrical conductivity when compared to foams containing only CNT. Among Hybrid series foams, the 1/1 hybrid foam was the one that displayed the highest value of electrical conductivity, which was related to the formation of an optimum conductive network within the cell struts of PEI for electrical conduction.

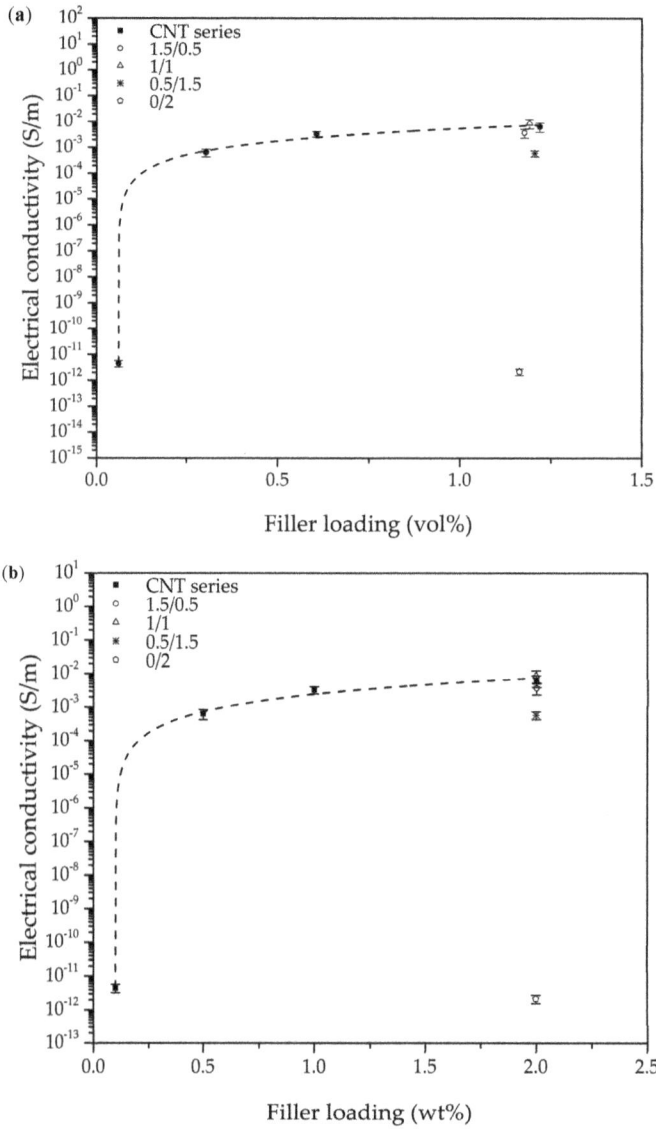

Figure 6. *Cont.*

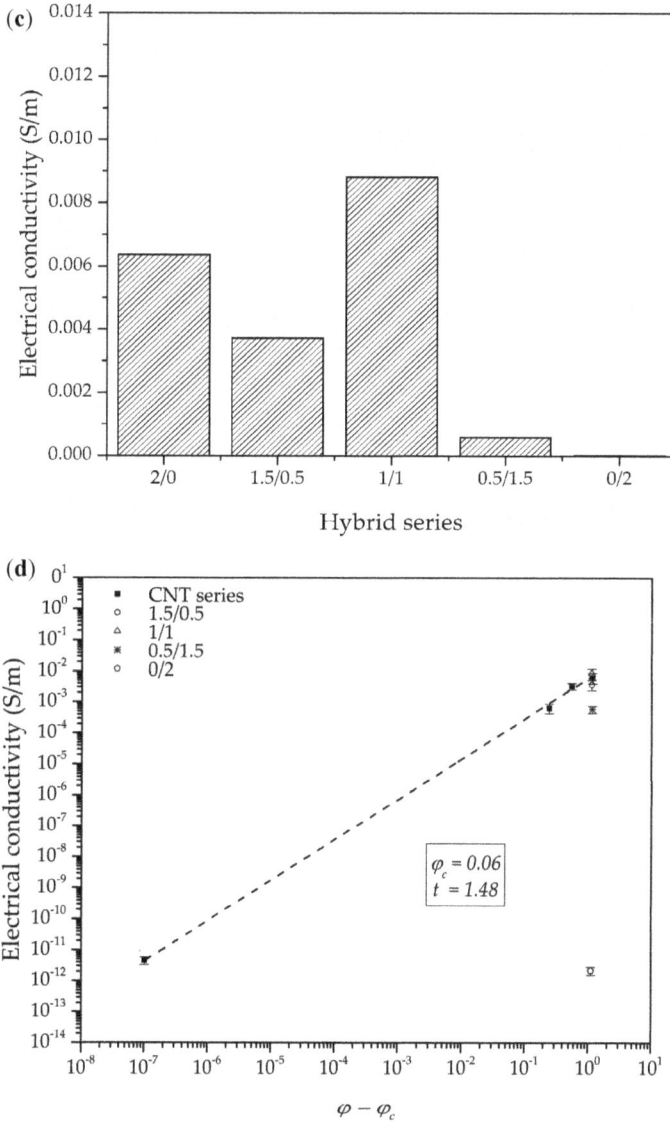

Figure 6. Electrical conductivity evolution with filler loading ((**a**) vol % and (**b**) wt %) for CNT series foams and Hybrid series foams; (**c**) comparison of the electrical conductivity of Hybrid series foams; and (**d**) evolution of electrical conductivity with reduced filler loading ($\phi - \phi_c$) assuming a percolative conduction model.

As previously mentioned, the X-ray spectra of the Hybrid series foams (Figure 3b) illustrated the appearance of two peaks corresponding to the (002) diffraction plane of CNT and GnP that could indicate the incomplete exfoliation of nanofillers. However, a good distribution of the nanoparticles resulted in the formation of a proper conductive network. Moreover, as seen in the high magnification micrographs presented in Figures 7 and 8, a certain level of physical contact between CNTs was obtained within the cell walls, which induced electrical conduction through the formation of an effective percolative network. This physical contact between nanofillers was more evident in the Hybrid series foams, with the 1/1 hybrid foam apparently displaying an ideal distribution of nanofillers in the cell struts in terms of forming an effective conductive pathway (Figure 8).

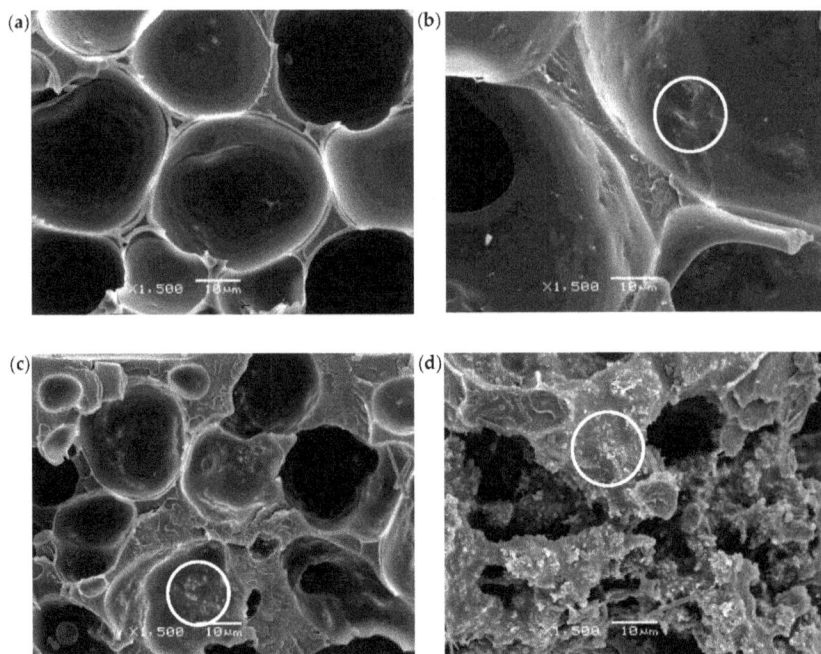

Figure 7. High magnification SEM images of CNT series foams: (a) 0.1% CNT; (b) 0.5% CNT; (c) 1.0% CNT; and (d) 2.0% CNT. White circles show physical contact between CNT.

As can be seen in Figure 6a, CNT series foams displayed increasingly higher values of electrical conductivity when increasing the amount of CNT up to 2.0 wt % (equivalent to 1.22 vol %). As shown, when increasing the amount of CNT from 0.1 to 0.5 wt % (0.06 to 0.30 vol %), electrical conductivity significantly improved from 4.5×10^{-12} to 6.4×10^{-4} S/m. The 1/1 hybrid foam displayed even greater electrical conductivity of 8.8×10^{-3} S/m, positioning itself as one of the highest registered electrical conductivity measurements for polymer-based foams, with only 2.0 wt % of conductive fillers (see Figure 6c). This could be explained by two causes assisting the formation of an effective conductive network: firstly, high power sonication has been proven to have a large influence on enhancing the dispersion level of carbon-based nanofillers in liquid suspensions; and secondly, the combination of GnP and CNT exhibited a synergic effect due to high individual AR levels and their impact on the cellular structure of the resulting foams. A number of works have achieved relatively low percolation thresholds for unfoamed composites with the addition of hybrid carbon-based conductive nanoparticles [27–29]. Additionally, Wu et al. [43] conducted a study using CNT and carbon black (CB) as hybrid fillers for a biodegradable polylactide composite, showing the synergic effect between both fillers in controlling cell size and forming an effective network, significantly enhancing the electrical conductivity of the foamed composites, especially when compared to similar foams containing only CNT. Maxian et al. [23] also studied the combination of GnP and CNT using a new numerical model considering nanofiller random distribution in a porous polymeric matrix in order to predict the electrical percolation behavior of polymer-based composites. In their simulations, the hybrid system exhibited significantly lower percolation values in porous systems when compared to the prediction given by the rule of mixtures, demonstrating the synergic effects of combining CNT and GnP.

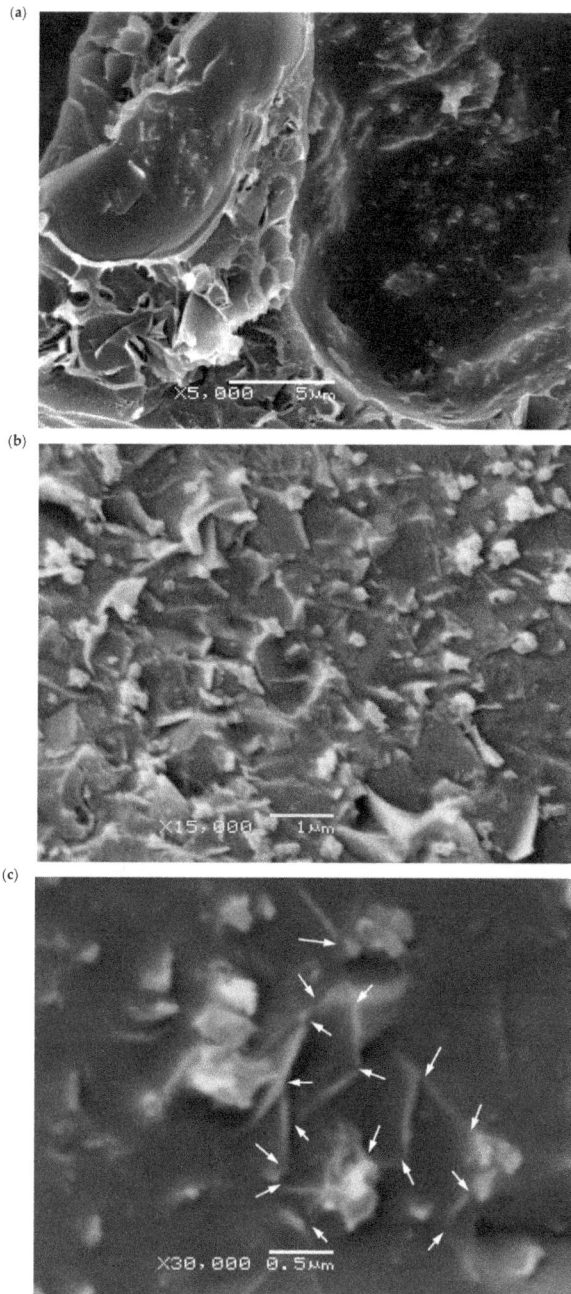

Figure 8. (a–c) Characteristic high magnification SEM images showing nanoparticle dispersion in Hybrid series foams. White arrows in (c) show physical contact between nanoparticles.

A tunneling conduction mechanism has been considered as a suitable model for various polymer foams containing carbon-based nanoparticles, with a range of nanofiller concentrations before the formation of a continuous conductive path [25,40,41]. Initially, a tunnel-like conduction mechanism was considered for both CNT and Hybrid series foams, but no linear trend was observed due to the high electrical conductivity of foams with a nanofiller concentration above 0.1 wt %. Therefore, a percolative conduction model was used, as the nanofiller content was enough to establish an electrical percolation network. The percolative model has been used extensively for polymeric composites containing CNT as a conductive nanofiller [23,43–48].

In a percolative model, the electrical conductivity (σ) above a certain critical concentration (ϕ_c), commonly called the percolation threshold, is given by:

$$\sigma \propto (\phi - \phi_c)^t \tag{9}$$

where ϕ is the volume fraction of particles in the material and t is the percolation exponent [2,49]. As the filler content increases above the threshold value, conductivity rises drastically, indicating the formation of a conductive path. As presented in various cases, the critical exponent t is commonly assumed to depend on particle dimensionality, with calculated values of around $t \approx 1.3$ and $t \approx 2.0$ corresponding to two and three dimensional systems, respectively [48,50–52]. The results of electrical conductivity as a function of nanofiller loading are presented in Figure 6a,b for all foams, with a dash line illustrating the fit to Equation (9). The fitting curve gave a value of the percolation threshold of $\phi_c = 0.061$ vol % (0.1 wt %) and an exponent of $t = 1.48$ (Figure 6d). As can be seen, the percolation threshold almost overlaps with the CNT concentration corresponding to the CNT series foam having the lowest CNT amount (0.1 wt %). This could indicate that the precise calculation of the percolation threshold requires increasing the number of foams with compositions near this value. Additionally, for a distribution of particles, the excluded volume concept gives the following relation between the percolation threshold and the AR of the nanofiller(s) [53]:

$$\phi_c \approx 1/\eta \tag{10}$$

where the analysis indicates that most solid composites seem to provide experimental values similar to those that were theoretically predicted when the particles are homogeneously distributed in the composite. For randomly oriented tubular-like particles such as carbon nanotubes:

$$\eta = L/W \tag{11}$$

with L and W representing the length and diameter of the particle, respectively. The calculated theoretical value of the threshold was based on the average size of CNT obtained from high magnification micrographs (1 μm $\geq L \geq$ 0.2 μm) and the assumption that the diameter of the nanotubes remained constant after sonication ($W = 0.0075$ μm, as indicated by the manufacturer). The obtained value of ϕ_c was estimated to be $\leq$0.1 wt % (0.01–0.04 vol %), slightly behind the experimentally calculated value. Assuming the presence of well-dispersed particles after ultrasonication, the theoretical calculation seems to show that the percolation threshold could be even lower than estimated for this particular system.

While a value of $t = 1.48$ seems to suggest the existence of a two dimensional charge transport system, various studies with obtained t values around 1.3 have claimed otherwise [52,54]. Bauhofer and colleagues [48] conducted a review of over 147 experimental studies on the electrical percolation of polymer composites with CNT, concluding that the value of t could not be related to any dimensional parameter and, therefore, no well-founded conclusion about CNT network geometry could be provided from most of the experimentally achieved values of t.

The electrical conductivity values showed that these foams could fulfil the requirements necessary for applications such as electrostatic discharge (ESD) and electromagnetic interference (EMI) shielding with conductive filler concentrations as low as 2.0 wt %. Various studies have shown that EMI shielding efficiency depends on many factors, including the electrical conductivity of the material [55–57]. EMI shielding materials are required to have an electrical resistance below 10^5 Ω, while the range for materials for ESD applications falls between 10^{12} and 10^5 Ω [58]. Our previous studies have shown that foaming can also enhance EMI shielding efficiency in cases referring to PC-based nanocomposites [8,9]. Therefore, the high conductivity of these foams could make them suitable candidates for EMI shielding and ESD applications.

4. Conclusions

In terms of cellular structure, the addition of CNT resulted in PEI foams with distinctive structures depending on the amount of CNT: a homogeneous unimodal distribution of closed cells (0.1 wt % CNT), a heterogeneous dual distribution with both closed as well as inter-connected pores (1.0 wt % CNT), and a homogeneous unimodal distribution of inter-connected open pores (2.0 wt % CNT). A similar tendency was observed for CNT-GnP Hybrid series foams and a similar dual structure was formed with both smaller pores and closed cells by adding 0.5 wt % GnP, while keeping 1.5 wt % CNT. Altering the composition to a 1/1 ratio of CNT and GnP resulted in an increase in the homogeneity of the structure, with both bigger closed cells and inter-connected pores, showing that the addition of GnP favored the formation of a finer and more homogeneous cellular structure. These different cellular structures resulted as a consequence of at least two combined effects: (a) the effect of CNT on the kinetics of solvent/non-solvent exchange, and (b) the effect of CNT on the mobility of the polymer. The studied systems were sufficiently complex and demand further investigation in order to elucidate the responsible mechanisms behind the formation and evolution of the cellular structure and the influence of each type of nanoparticle.

CNT series foams displayed an increasingly higher decomposition retardancy when increasing the amount of CNT, while the addition of GnP in combination with CNT seemed to have a high influence in delaying foam degradation, as expected from the more effective barrier effect of layered graphene in hindering the escape of volatile gases during combustion.

Two main factors were found to affect the viscoelastic behavior of the foams: their density and cellular structure and the amount of CNT, given that, indirectly, the amount of CNT also had an important effect in setting the final cellular structure characteristics of the foams. As expected, foams that displayed the highest values of storage modulus were the ones with the highest amount of CNT, related to a higher reinforcing effect of CNT when compared to GnP. Meanwhile, the heterogeneity and existence of open pores, as a consequence of CNT incorporation, resulted in lower values of the specific storage modulus when compared to foams with a lower amount or without CNTs. The incorporation of GnP in the Hybrid series foams and its effect on the formation of a more homogeneous cellular structure resulted in a rise in the specific storage modulus.

Regarding the viscous response, while the addition of only CNT led to foams with lower glass transition temperatures, related to a higher mobility of PEI molecules due to a lubricating effect of CNT, the combination of GnP and CNT did not significantly affect the viscous response of PEI.

All values of measured electrical conductivity were corrected taking into account the porosity of foams and their respective effective surface area. Although foams containing only CNT already displayed high electrical conductivity values, comparatively, the combination of 1.0 wt % GnP and 1.0 wt % CNT resulted in the foam with the highest electrical conductivity (8.8×10^{-3} S/m). This was related to the formation of an optimum conductive network within the cell struts of PEI for electrical conduction by physical contact between particles, as assessed by X-ray diffraction and the analysis of high magnification micrographs. This value is one of the highest electrical conductivities registered so far for polymer-based foamed systems containing carbon-based conductive nanofillers. The experimental results fitted well to a percolative conduction model, with a percolation value threshold as low as 0.06 vol % (0.1 wt %) CNT, indicating that the combination of GnP and CNT formed an effective network for electrical conduction.

Due to the combination of high electrical conductivity and reduced density, these foams can target sectors such as telecommunications and aerospace for applications related to ESD and EMI shielding.

Acknowledgments: The authors would like to acknowledge the Spanish Ministry of Economy and Competitiveness for the financial support of project MAT2014-56213-P and MAT2017-89787-P.

Author Contributions: Hooman Abbasi performed the experiments; Hooman Abbasi, Marcelo Antunes and José Ignacio Velasco analyzed the data; Hooman Abbasi and Marcelo Antunes contributed reagents/materials/analysis tools; Hooman Abbasi, Marcelo Antunes and José Ignacio Velasco wrote the paper.

Conflicts of Interest: The authors declare no conflict of interest.

References

1. Antunes, M.; Gedler, G.; Abbasi, H.; Velasco, J.I. Graphene Nanoplatelets as a Multifunctional Filler for Polymer Foams. *Mater. Today Proc.* **2016**, *3*, S233–S239. [CrossRef]
2. Ling, J.; Zhai, W.; Feng, W.; Shen, B.; Zhang, J.; Zheng, W. Ge Facile Preparation of Lightweight Microcellular Polyetherimide/Graphene Composite Foams for Electromagnetic Interference Shielding. *ACS Appl. Mater. Interfaces* **2013**, *5*, 2677–2684. [CrossRef] [PubMed]
3. Shen, B.; Zhai, W.; Tao, M.; Ling, J.; Zheng, W. Lightweight, Multifunctional Polyetherimide/Graphene@Fe$_3$O$_4$ Composite Foams for Shielding of Electromagnetic Pollution. *ACS Appl. Mater. Interfaces* **2013**, *5*, 11383–11391. [CrossRef] [PubMed]
4. Zhai, W.; Chen, Y.; Ling, J.; Wen, B.; Kim, Y.-W. Fabrication of lightweight, flexible polyetherimide/nickel composite foam with electromagnetic interference shielding effectiveness reaching 103 dB. *J. Cell. Plast.* **2014**, *50*, 537–550. [CrossRef]
5. Li, B.; Olson, E.; Perugini, A.; Zhong, W.-H. Simultaneous enhancements in damping and static dissipation capability of polyetherimide composites with organosilane surface modified graphene nanoplatelets. *Polymer* **2011**, *52*, 5606–5614. [CrossRef]
6. Wu, H.; Drzal, L.T. Graphene nanoplatelet-polyetherimide composites: Revealed morphology and relation to properties. *J. Appl. Polym. Sci.* **2013**, *130*, 4081–4089. [CrossRef]
7. Zhang, H.-B.; Yan, Q.; Zheng, W.-G.; He, Z.; Yu, Z.-Z. Tough Graphene−Polymer Microcellular Foams for Electromagnetic Interference Shielding. *ACS Appl. Mater. Interfaces* **2011**, *3*, 918–924. [CrossRef] [PubMed]
8. Gedler, G.; Antunes, M.; Velasco, J.I.; Ozisik, R. Enhanced electromagnetic interference shielding effectiveness of polycarbonate/graphene nanocomposites foamed via 1-step supercritical carbon dioxide process. *Mater. Des.* **2016**, *90*, 906–914. [CrossRef]
9. Gedler, G.; Antunes, M.; Velasco, J.I.; Ozisik, R. Electromagnetic shielding effectiveness of polycarbonate/graphene nanocomposite foams processed in 2-steps with supercritical carbon dioxide. *Mater. Lett.* **2015**, *160*, 41–44. [CrossRef]
10. Lee, L.J.; Zeng, C.; Cao, X.; Han, X.; Shen, J.; Xu, G. Polymer nanocomposite foams. *Compos. Sci. Technol.* **2005**, *65*, 2344–2363. [CrossRef]
11. Kostopoulos, V.; Vavouliotis, A.; Karapappas, P.; Tsotra, P.; Paipetis, A. Damage Monitoring of Carbon Fiber Reinforced Laminates Using Resistance Measurements. Improving Sensitivity Using Carbon Nanotube Doped Epoxy Matrix System. *J. Intell. Mater. Syst. Struct.* **2009**, *20*, 1025–1034. [CrossRef]
12. Badamshina, E.; Estrin, Y.; Gafurova, M. Nanocomposites based on polyurethanes and carbon nanoparticles: Preparation, properties and application. *J. Mater. Chem. A* **2013**, *1*, 6509–6529. [CrossRef]
13. Potts, J.R.; Dreyer, D.R.; Bielawski, C.W.; Ruoff, R.S. Graphene-based polymer nanocomposites. *Polymer* **2011**, *52*, 5–25. [CrossRef]
14. Tung, V.C.; Chen, L.-M.; Allen, M.J.; Wassei, J.K.; Nelson, K.; Kaner, R.B.; Yang, Y. Low-Temperature Solution Processing of Graphene−Carbon Nanotube Hybrid Materials for High-Performance Transparent Conductors. *Nano Lett.* **2009**, *9*, 1949–1955. [CrossRef] [PubMed]
15. Nguyen, D.D.; Tiwari, R.N.; Matsuoka, Y.; Hashimoto, G.; Rokuta, E.; Chen, Y.-Z.; Chueh, Y.-L.; Yoshimura, M. Low Vacuum Annealing of Cellulose Acetate on Nickel Towards Transparent Conductive CNT−Graphene Hybrid Films. *ACS Appl. Mater. Interfaces* **2014**, *6*, 9071–9077. [CrossRef] [PubMed]

16. Nguyen, D.D.; Tai, N.-H.; Chen, S.-Y.; Chueh, Y.-L. Controlled growth of carbon nanotube-graphene hybrid materials for flexible and transparent conductors and electron field emitters. *Nanoscale* **2012**, *4*, 632–638. [CrossRef] [PubMed]

17. Kim, S.H.; Song, W.; Jung, M.W.; Kang, M.-A.; Kim, K.; Chang, S.-J.; Lee, S.S.; Lim, J.; Hwang, J.; Myung, S.; et al. Carbon Nanotube and Graphene Hybrid Thin Film for Transparent Electrodes and Field Effect Transistors. *Adv. Mater.* **2014**, *26*, 4247–4252. [CrossRef] [PubMed]

18. Fan, Z.; Yan, J.; Zhi, L.; Zhang, Q.; Wei, T.; Feng, J.; Zhang, M.; Qian, W.; Wei, F. A Three-Dimensional Carbon Nanotube/Graphene Sandwich and Its Application as Electrode in Supercapacitors. *Adv. Mater.* **2010**, *22*, 3723–3728. [CrossRef] [PubMed]

19. Deng, J.; Zheng, R.; Zhao, Y.; Cheng, G. Vapor–Solid Growth of Few-Layer Graphene Using Radio Frequency Sputtering Deposition and Its Application on Field Emission. *ACS Nano* **2012**, *6*, 3727–3733. [CrossRef] [PubMed]

20. Kim, Y.-S.; Kumar, K.; Fisher, F.T.; Yang, E.-H. Out-of-plane growth of CNTs on graphene for supercapacitor applications. *Nanotechnology* **2012**, *23*, 15301. [CrossRef] [PubMed]

21. Hu, Y.; Li, X.; Wang, J.; Li, R.; Sun, X. Free-standing graphene-carbon nanotube hybrid papers used as current collector and binder free anodes for lithium ion batteries. *J. Power Sources* **2013**, *237*, 41–46. [CrossRef]

22. Chen, S.; Yeoh, W.; Liu, Q.; Wang, G. Chemical-free synthesis of graphene-carbon nanotube hybrid materials for reversible lithium storage in lithium-ion batteries. *Carbon* **2012**, *50*, 4557–4565. [CrossRef]

23. Maxian, O.; Pedrazzoli, D.; Manas-Zloczower, I. Conductive polymer foams with carbon nanofillers-Modeling percolation behavior. *Express Polym. Lett.* **2017**, *11*, 406–418. [CrossRef]

24. Gedler, G.; Antunes, M.; Velasco, J.I. Enhanced electrical conductivity in graphene-filled polycarbonate nanocomposites by microcellular foaming with sc-CO$_2$. *J. Adhes. Sci. Technol.* **2016**, *30*, 1017–1029. [CrossRef]

25. Abbasi, H.; Antunes, M.; Velasco, J.I. Graphene nanoplatelets-reinforced polyetherimide foams prepared by water vapor-induced phase separation. *eXPRESS Polym. Lett.* **2015**, *9*, 412–423. [CrossRef]

26. Abbasi, H.; Antunes, M.; Velasco, J.I. Influence of polyamide-imide concentration on the cellular structure and thermo-mechanical properties of polyetherimide/polyamide-imide blend foams. *Eur. Polym. J.* **2015**, *69*, 273–283. [CrossRef]

27. Song, P.A.; Liu, L.; Fu, S.; Yu, Y.; Jin, C.; Wu, Q.; Li, Q. Striking multiple synergies created by combining reduced graphene oxides and carbon nanotubes for polymer nanocomposites. *Nanotechnology* **2013**, *24*, 125704. [CrossRef] [PubMed]

28. Safdari, M.; Al-Haik, M. Electrical conductivity of synergistically hybridized nanocomposites based on graphite nanoplatelets and carbon nanotubes. *Nanotechnology* **2012**, *23*, 405202. [CrossRef] [PubMed]

29. Yue, L.; Pircheraghi, G.; Monemian, S.A.; Manas-Zloczower, I. Epoxy composites with carbon nanotubes and graphene nanoplatelets-Dispersion and synergy effects. *Carbon* **2014**, *78*, 268–278. [CrossRef]

30. Li, W.; Xu, Z.; Chen, L.; Shan, M.; Tian, X.; Yang, C.; Lv, H.; Qian, X. A facile method to produce graphene oxide-g-poly (L-lactic acid) as a promising reinforcement for PLLA nanocomposites. *Chem. Eng. J.* **2014**, *237*, 291–299. [CrossRef]

31. Sims, G.L.A.; Khunniteekool, C. Cell-size measurement of polymeric foams. *Cell. Polym.* **1994**, *13*, 137–146.

32. Abbasi, H.; Antunes, M.; Velasco, J.I. Enhancing the electrical conductivity of polyetherimide-based foams by simultaneously increasing the porosity and graphene nanoplatelets dispersion. *Polym. Compos.* **2018**. Accepted.

33. Huxtable, S.T.; Cahill, D.G.; Shenogin, S.; Xue, L.; Ozisik, R.; Barone, P.; Usrey, M.; Strano, M.S.; Siddons, G.; Shim, M. Interfacial heat flow in carbon nanotube suspensions. *Nat. Mater.* **2003**, *2*, 731–734. [CrossRef] [PubMed]

34. Gupta, A.; Choudhary, V. Effect of multiwall carbon nanotubes on thermomechanical and electrical properties of poly(trimethylene terephthalate). *J. Appl. Polym. Sci.* **2012**, *123*, 1548–1556. [CrossRef]

35. Gedler, G.; Antunes, M.; Realinho, V.; Velasco, J.I. Thermal stability of polycarbonate-graphene nanocomposite foams. *Polym. Degrad. Stab.* **2012**, *97*, 1297–1304. [CrossRef]

36. Wang, X.; Yang, H.; Song, L.; Hu, Y.; Xing, W.; Lu, H. Morphology, mechanical and thermal properties of graphene-reinforced poly(butylene succinate) nanocomposites. *Compos. Sci. Technol.* **2011**, *72*, 1–6. [CrossRef]

37. Liu, L.; Gu, A.; Fang, Z.; Tong, L.; Xu, Z. The effects of the variations of carbon nanotubes on the micro-tribological behavior of carbon nanotubes/bismaleimide nanocomposite. *Compos. Part A* **2007**, *38*, 1957–1964. [CrossRef]

38. Velasco-Santos, C.; Martínez-Hernández, A.L.; Fisher, F.; Ruoff, R.; Castano, V.M. Dynamical-mechanical and thermal analysis of carbon nanotube-methyl-ethyl methacrylate nanocomposites. *J. Phys. D Appl. Phys.* **2003**, *36*, 1423–1428. [CrossRef]

39. Sung, Y.T.; Kum, C.K.; Lee, H.S.; Byon, N.S.; Yoon, H.G.; Kim, W.N. Dynamic mechanical and morphological properties of polycarbonate/multi-walled carbon nanotube composites. *Polymer* **2005**, *46*, 5656–5661. [CrossRef]

40. Agarwal, G.; Patnaik, A.; Sharma, R.K. Mechanical and Thermo–Mechanical Properties of Bi-Directional and Short Carbon Fiber Reinforced Epoxy Composites. *J. Eng. Sci. Technol.* **2014**, *9*, 590–604.

41. Ornaghi, H.L.; Bolner, A.S.; Fiorio, R.; Zattera, A.J.; Amico, S.C. Mechanical and dynamic mechanical analysis of hybrid composites molded by resin transfer molding. *J. Appl. Polym. Sci.* **2010**, *118*, 887–896. [CrossRef]

42. Idicula, M.; Malhotra, S.K.; Joseph, K.; Thomas, S. Dynamic mechanical analysis of randomly oriented intimately mixed short banana/sisal hybrid fibre reinforced polyester composites. *Compos. Sci. Technol.* **2005**, *65*, 1077–1087. [CrossRef]

43. Wu, D.; Lv, Q.; Feng, S.; Chen, J.; Chen, Y.; Qiu, Y.; Yao, X. Polylactide composite foams containing carbon nanotubes and carbon black: Synergistic effect of filler on electrical conductivity. *Carbon* **2015**, *95*, 380–387. [CrossRef]

44. Antunes, M.; Velasco, J.I. Multifunctional polymer foams with carbon nanoparticles. *Prog. Polym. Sci.* **2014**, *39*, 486–509. [CrossRef]

45. Antunes, M.; Mudarra, M.; Velasco, J.I. Broad-band electrical conductivity of carbon nanofibre-reinforced polypropylene foams. *Carbon* **2011**, *49*, 708–717. [CrossRef]

46. Ameli, A.; Kazemi, Y.; Wang, S.; Park, C.B.; Pötschke, P. Process-microstructure-electrical conductivity relationships in injection-molded polypropylene/carbon nanotube nanocomposite foams. *Compos. Part A Appl. Sci. Manuf.* **2017**, *96*, 28–36. [CrossRef]

47. Ameli, A.; Nofar, M.; Park, C.B.; Pötschke, P.; Rizvi, G. Polypropylene/carbon nanotube nano/microcellular structures with high dielectric permittivity, low dielectric loss, and low percolation threshold. *Carbon* **2014**, *71*, 206–217. [CrossRef]

48. Bauhofer, W.; Kovacs, J.Z. A review and analysis of electrical percolation in carbon nanotube polymer composites. *Compos. Sci. Technol.* **2009**, *69*, 1486–1498. [CrossRef]

49. Stankovich, S.; Dikin, D.A.; Dommett, G.H.B.; Kohlhaas, K.M.; Zimney, E.J.; Stach, E.A.; Piner, R.D.; Nguyen, S.T.; Ruoff, R.S. Graphene-based composite materials. *Nature* **2006**, *442*, 282–286. [CrossRef] [PubMed]

50. Stauffer, D.; Aharony, A. *Introduction to Percolation Theory*; CRC Press: Boca Raton, FL, USA, 1994; ISBN 1420074792.

51. Sahimi, M. *Applications of Percolation Theory*; CRC Press: Boca Raton, FL, USA, 1994; ISBN 0748400761.

52. Kilbride, B.E.; Coleman, J.N.; Fraysse, J.; Fournet, P.; Cadek, M.; Drury, A.; Hutzler, S.; Roth, S.; Blau, W.J. Experimental observation of scaling laws for alternating current and direct current conductivity in polymer-carbon nanotube composite thin films. *J. Appl. Phys.* **2002**, *92*, 4024–4030. [CrossRef]

53. Balberg, I.; Anderson, C.H.; Alexander, S.; Wagner, N. Excluded volume and its relation to the onset of percolation. *Phys. Rev. B* **1984**, *30*, 3933. [CrossRef]

54. Fraysse, J.; Plane, J. Interplay of Hopping and Percolation in Organic Conducting Blends. *Phys. Status Solidi* **2000**, *273*, 273–278. [CrossRef]

55. Liu, Z.; Bai, G.; Huang, Y.; Ma, Y.; Du, F.; Li, F.; Guo, T.; Chen, Y. Reflection and absorption contributions to the electromagnetic interference shielding of single-walled carbon nanotube/polyurethane composites. *Carbon* **2007**, *45*, 821–827. [CrossRef]

56. Yang, Y.; Gupta, M.C.; Dudley, K.L.; Lawrence, R.W. Novel Carbon Nanotube-Polystyrene Foam Composites for Electromagnetic Interference Shielding. *Nano Lett.* **2005**, *5*, 2131–2134. [CrossRef] [PubMed]

57. Yang, Y.; Gupta, M.C.; Dudley, K.L.; Lawrence, R.W. A comparative study of EMI shielding properties of carbon nanofiber and multi-walled carbon nanotube filled polymer composites. *J. Nanosci. Nanotechnol.* **2005**, *5*, 927–931. [CrossRef] [PubMed]

58. Verdejo, R.; Bernal, M.M.; Romasanta, L.J.; Lopez-Manchado, M.A. Graphene filled polymer nanocomposites. *J. Mater. Chem.* **2011**, *21*, 3301–3310. [CrossRef]

MDPI

St. Alban-Anlage 66

4052 Basel

Switzerland

Tel. +41 61 683 77 34

Fax +41 61 302 89 18

www.mdpi.com

Polymers Editorial Office

E-mail: polymers@mdpi.com

www.mdpi.com/journal/polymers